Theory of Molecular Collisions

RSC Theoretical and Computational Chemistry Series

Titles in the Series:

1: Knowledge-based Expert Systems in Chemistry: Not Counting on Computers
2: Non-Covalent Interactions: Theory and Experiment
3: Single-Ion Solvation: Experimental and Theoretical Approaches to Elusive Thermodynamic Quantities
4: Computational Nanoscience
5: Computational Quantum Chemistry: Molecular Structure and Properties in Silico
6: Reaction Rate Constant Computations: Theories and Applications
7: Theory of Molecular Collisions

How to obtain future titles on publication:
A standing order plan is available for this series. A standing order will bring delivery of each new volume immediately on publication.

For further information please contact:
Book Sales Department, Royal Society of Chemistry, Thomas Graham House, Science Park, Milton Road, Cambridge, CB4 0WF, UK
Telephone: +44 (0)1223 420066, Fax: +44 (0)1223 420247
Email: booksales@rsc.org
Visit our website at www.rsc.org/books

Theory of Molecular Collisions

Gabriel G. Balint-Kurti
University of Bristol, UK
Email: gabriel.balint-kurti@bristol.ac.uk

and

Alexander P. Palov
Moscow State University, Russia
Email: a.palov@mics.msu.su

ROYAL SOCIETY OF CHEMISTRY

THE QUEEN'S AWARDS
FOR ENTERPRISE:
INTERNATIONAL TRADE
2013

RSC Theoretical and Computational Chemistry Series No. 7

Print ISBN: 978-1-84973-830-9
PDF eISBN: 978-1-78262-019-8
ISSN: 2041-3181

A catalogue record for this book is available from the British Library

Published by The Royal Society of Chemistry,
Thomas Graham House, Science Park, Milton Road,
Cambridge CB4 0WF, UK

Registered Charity Number 207890

For further information see our web site at www.rsc.org

Printed and bound by CPI Group (UK) Ltd, Croydon, CR0 4YY

Preface

The book provides an introduction to the theory of molecular collisions and reactive scattering and relates it to current experiments. The emphasis is on the quantum theory of scattering, although classical mechanics for atom–atom scattering is also briefly described. While there is inevitably a considerable amount of mathematics in the book, it is kept to the minimum required, and the more difficult topics are relegated to appendices. The motivation for studying the topic is to provide the tools needed for investigating some of the most basic processes of nature: how chemical reactions occur and how molecular bonds are formed and broken.

The first two chapters discuss the classical and quantum theory of atom–atom scattering. These discussions introduce important concepts used later in the book. Chapters 3 and 4 deal with the quantum theory of non-reactive atom–diatom collisions. The use of different coordinate reference frames, the calculation of cross sections and the role of the scattering **S** matrix are described. The following chapters deal mainly with reactive scattering. Both historic and modern experimental methods for measuring detailed reactive cross sections and reaction probabilities are reviewed. The current theoretical methods for the calculation of reaction cross sections, using both time-independent and time-dependent quantum treatments, are described. The use of lasers in modern experiments is reviewed and the underlying theory of polarization and alignment experiments, which measure the correlation of, say, the vector directions of the rotational angular momentum of the products of a reaction with their angle of scattering, is explained. The final chapter of the book discusses the difficulties encountered in applying exact quantum mechanics to systems of many atoms. The various methods available for the approximate quantum treatment of reactions involving many

RSC Theoretical and Computational Chemistry Series No. 7
Theory of Molecular Collisions
By Gabriel G. Balint-Kurti and Alexander P. Palov

Published by the Royal Society of Chemistry, www.rsc.org

atoms are discussed. Also discussed are methods for computing thermal rate constants without the need to compute the scattering **S** matrix.

In addition to the main body of the book, a glossary of terms is included and extensive references to literature sources are provided.

The authors would like to thank and acknowledge their debt to past supervisors, colleagues and students who have helped them in the field of molecular scattering theory. On behalf of GGBK they include: M. Karplus, R. D. Levine, M. Shapiro, R. N. Dixon, C. C. Marston, F. Gögtas, A. R. Offer, S. K. Gray, M. Hankel, E. M. Goldfield, G. C. Schatz, O. S. Vasyutinskii, L. Eno, B. M. D. D. Jansen op de Haar, M. M. Novak, J. van Lenthe, S. Bosanac and R. Saktreger. On behalf of APP they comprise: V. A. Mitrofanov, V. V. Pletnev, B. I. Klain, T. V. Rakhimova, K. S. Klopovsky, A. A. Chukalovsky and A. A. Sycheva.

G. Balint-Kurti and A. Palov

Contents

RSC Theoretical and Computational Chemistry Series No. 7
Theory of Molecular Collisions
By Gabriel G. Balint-Kurti and Alexander P. Palov

Published by the Royal Society of Chemistry, www.rsc.org

CHAPTER 1

Scattering Experiments and Classical Theory of Atom–Atom Scattering

1.1 Crossed Atomic and Molecular Beams

We wish to investigate how atoms and molecules interact with each other in the most fundamental manner. What happens when they closely approach each other or collide? As molecules are far too small to see and the time-scale of their collisions is very short, we resort to making them collide with each other under strictly controlled conditions and examining the results in as much detail as we can. Figure 1.1 shows a schematic of a crossed atomic beam apparatus.[1]

To the left hand side of the diagram there is a sophisticated analyzer which detects the products of the collision. The "quadrapole mass filter" selects products of a defined mass and the chopper permits the determination as to when the products arrive at the detector. From this it will be possible to determine the speed and kinetic energy of the products. The detector assembly remains fixed as it is the most complex component of the apparatus. To the lower right hand side of the diagram are the two atomic or molecular beam sources. These are at some fixed angle, normally 90°, relative to each other and can be rotated with respect to the detector assembly. This permits the measurement of the angular variation of the reaction probability (*i.e.* the differential cross section).

The heart of the apparatus is the collision region where the two beams cross and the collisions take place. This region is maintained under the greatest attainable vacuum so as to exclude any unwanted molecules.

RSC Theoretical and Computational Chemistry Series No. 7
Theory of Molecular Collisions
By Gabriel G. Balint-Kurti and Alexander P. Palov

Published by the Royal Society of Chemistry, www.rsc.org

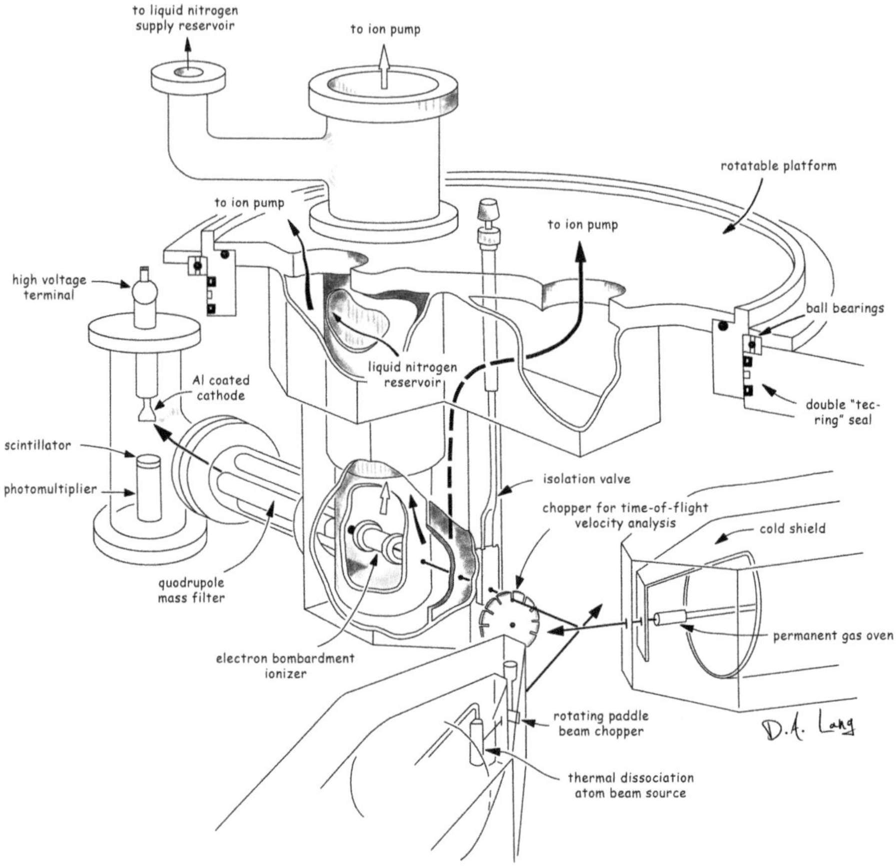

Figure 1.1 Diagram of a crossed molecular beam machine. Redrawn figure reproduced with permission from J. D. McDonald, P. R. LeBreton, Y. T. Lee and D. R. Herschbach, *J. Chem. Phys.*, 1972, **56**, 769. Copyright 1972, AIP Publishing LLC.[1] DOI: 10.1063/1.1677230.

1.2 Classical Theory of Atom–Atom Elastic Scattering

1.2.1 Hard Sphere Collisions

Let us now consider what happens when two atoms collide. Figure 1.2 shows two atoms traveling towards each other. This is what a collision would look like in the center-of-mass reference frame. We will discuss the difference between the "laboratory frame" and the "center-of-mass" frame a little later. In the center-of-mass frame the collision partners travel directly towards each other. The distance "b" in the figure is called the impact parameter and is the hypothetical distance of what would be the closest approach of the centers of the two atoms if there were no interaction between them. If the atoms were hard spheres then they would either collide or miss each other entirely.

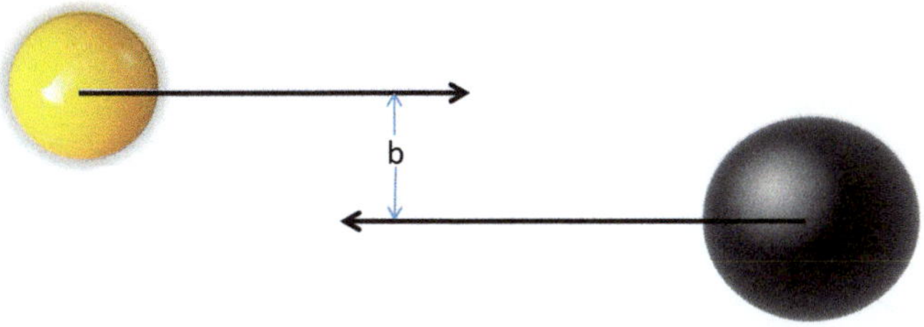

Figure 1.2 Two atoms colliding; "b" is the impact parameter.

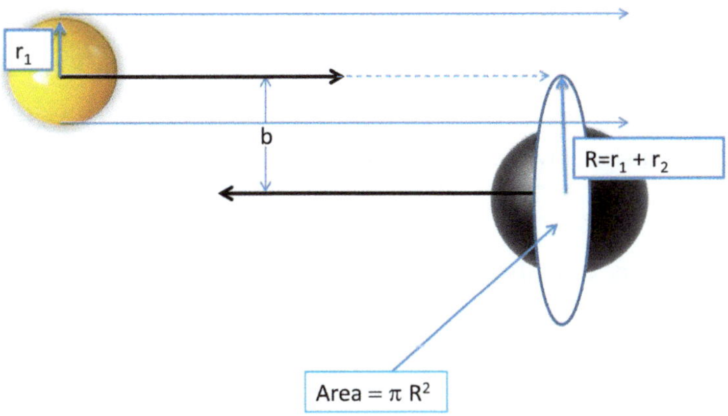

Figure 1.3 Grazing collision between two hard sphere atoms illustrating the concept of a collision cross section.

They would collide if the impact parameter was smaller than the sum of the radii of the two spheres.

Figure 1.3 shows a collision in which the two atoms just graze each other. If the impact parameter were any larger, then the atoms would miss each other entirely and there would be no collision. The maximum value of the impact parameter for a collision to occur is the sum of the radii of the two atoms, $b = r_1 + r_2$. If the projected path of atom 1 lies anywhere within the area of the circle shown in Figure 1.3 then a collision will take place. The area of this circle is called the *collision cross section*, $\sigma = \pi R^2$, where $R = r_1 + r_2$.

Let us now consider hard sphere collisions in greater detail. What is the angle of scattering of the atoms after the collision? Figure 1.4 illustrates what happens when two "hard sphere" atoms collide. When their surfaces touch they bounce off each other like two billiard balls. The angle that the direction of impact makes with the normal to the surface (α) is the same as that made by the recoil velocity direction to this normal. This is termed specular reflection. The "scattering angle" is shown as θ in Figure 1.4. From the figure we see that

$$b = R \sin \alpha, \tag{1.1}$$

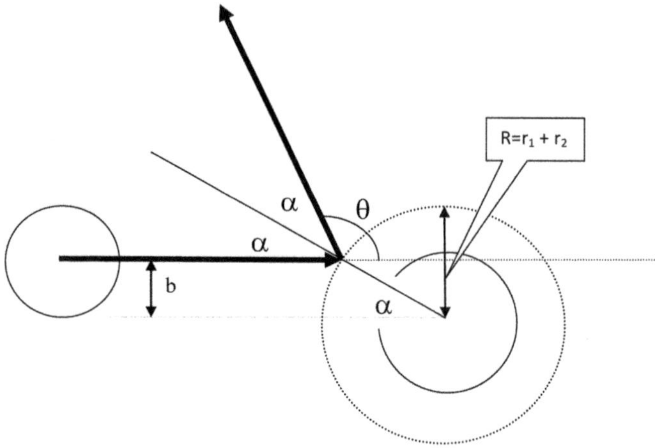

Figure 1.4 Collision between two hard sphere atoms showing the scattering angle θ.

where $R = r_1 + r_2$, and also

$$\theta + 2\alpha = \pi, \tag{1.2}$$

and therefore

$$b = R\sin\left(\frac{\pi - \theta}{2}\right) = R\cos\frac{\theta}{2}. \tag{1.3}$$

All the atoms colliding with an impact parameter b to $b + \mathrm{d}b$ are scattered into scattering angles θ to $\theta + \mathrm{d}\theta$. The cross section for scattering into polar angles θ to $\theta + \mathrm{d}\theta$ is therefore

$$\begin{aligned} 2\pi b|\mathrm{d}b| &= 2\pi R\left(\cos\frac{\theta}{2}\right)\left|R\left(\sin\frac{\theta}{2}\right)\frac{\mathrm{d}\theta}{2}\right| \\ &= \frac{1}{2}\pi R^2 \sin\theta\,\mathrm{d}\theta. \end{aligned} \tag{1.4}$$

The absolute value signs have been introduced as $\frac{\mathrm{d}\theta}{\mathrm{d}b}$ is negative and the cross sectional area is a positive quantity.

The "differential cross section" is defined as the flux of atoms scattered into a given solid angle ($\mathrm{d}\Omega = \sin\theta\,\mathrm{d}\theta\,\mathrm{d}\phi$) divided by the incident flux of atoms. In the present case we have not considered the azimuthal angle, ϕ. We have in effect integrated over this angle. The solid angle, integrated over all ϕ, is $2\pi\sin\theta\,\mathrm{d}\theta$. The differential cross section for scattering into a solid angle $\mathrm{d}\Omega$ is therefore

$$\frac{\mathrm{d}\sigma}{\mathrm{d}\Omega} = \frac{R^2}{4}. \tag{1.5}$$

The integral of the differential cross section over all angles must be equal to thc total or integral cross section

$$\int \mathrm{d}\Omega \frac{\mathrm{d}\sigma}{\mathrm{d}\Omega} = \int_0^{\pi} \sin\theta \, \mathrm{d}\theta \int_0^{2\pi} \mathrm{d}\phi \, \frac{\mathrm{d}\sigma}{\mathrm{d}\Omega}$$
$$= \frac{R^2}{4} \int_0^{\pi} \sin\theta \, \mathrm{d}\theta \int_0^{2\pi} \mathrm{d}\phi$$
$$= \frac{R^2}{4} 4\pi = \pi R^2. \tag{1.6}$$

1.2.2 Scattering Under the Influence of a Potential

Figure 1.2 shows a schematic of two atoms colliding. The position of each atom is described by a vector with three spatial components $\boldsymbol{r}_1$ and $\boldsymbol{r}_2$. Thus the system overall requires six position coordinates to fully describe it. As the forces determining the scattering of the atoms depend on the magnitude of the distance between the two atoms, $r = |\boldsymbol{r}_2 - \boldsymbol{r}_1|$, it seems reasonable to change coordinates to $\boldsymbol{r}$, the position of atom 2 relative to atom 1, and $\boldsymbol{R}$, the position of the center-of-mass

$$\boldsymbol{r} = \boldsymbol{r}_2 - \boldsymbol{r}_1,$$
$$\boldsymbol{R} = \frac{m_1\boldsymbol{r}_2 + m_2\boldsymbol{r}_1}{m_1 + m_2}. \tag{1.7}$$

Classical mechanics is based on Newton's laws of motion and these are generally expressed in terms of the Cartesian (*i.e.* x, y, z) coordinates of each of the particles involved. When other types of coordinates are used or when there are constraints on the motion then it is advantageous to use alternative formulations of classical mechanics. In the present case we will use Lagrange's equations of motion.[2,3] These equations can be written as

$$\frac{\mathrm{d}}{\mathrm{d}t}\left(\frac{\partial L}{\partial \dot{q}_i}\right) - \frac{\partial L}{\partial q_i} = 0 \tag{1.8}$$

where

$$L = \text{Kinetic energy} - \text{Potential energy}$$
$$= T - V, \tag{1.9}$$

q_i is a generalized coordinate and $\dot{q}_i = \dfrac{\mathrm{d}q_i}{\mathrm{d}t}$ is its time derivative.

Returning now to the specific case of the scattering of two atoms, we can write the kinetic energy in the form

$$T = \frac{1}{2}m_1 \mathrm{v}_1^2 + \frac{1}{2}m_2 \mathrm{v}_2^2$$
$$= \frac{1}{2}m_1 \dot{\boldsymbol{r}}_1^2 + \frac{1}{2}m_2 \dot{\boldsymbol{r}}_2^2$$
$$= \frac{1}{2}M\dot{\boldsymbol{R}}^2 + \frac{1}{2}\mu\dot{\boldsymbol{r}}^2 \tag{1.10}$$

where

$$M = m_1 + m_2,$$
$$\mu = \frac{m_1 m_2}{m_1 + m_2} \tag{1.11}$$

and the potential energy V is a function of only $r = |\boldsymbol{r}|$, so

$$L = \frac{1}{2}M\dot{\boldsymbol{R}}^2 + \frac{1}{2}\mu\dot{\boldsymbol{r}}^2 - V(r). \tag{1.12}$$

Let us first examine the equations of motion for the motion of the center-of-mass of the system, $\boldsymbol{R} = \boldsymbol{R}(X, Y, Z)$. The derivatives of the Lagrangian with respect to the X coordinate and its time derivative are

$$\frac{\partial L}{\partial \dot{X}} = M\dot{X},$$
$$\frac{\partial L}{\partial X} = 0. \tag{1.13}$$

The equation of motion for X is therefore (see eqn (1.8))

$$\frac{\mathrm{d}}{\mathrm{d}t}(M\dot{X}) = 0 \tag{1.14}$$

or

$$\frac{\mathrm{d}}{\mathrm{d}t}X = \dot{X} = \text{constant}. \tag{1.15}$$

Similar equations apply to the Y and Z components of the center-of-mass vector. We see therefore that the velocity of the center-of-mass is constant. The center-of-mass of the colliding pair moves through space with a constant velocity which is not changed by the collision. The variables X, Y and Z are called cyclic coordinates. The potential function does not depend on them and this leads to the properties just mentioned.

There remain the three coordinates which describe the vector $\boldsymbol{r}$. As the potential depends only on $r = |\boldsymbol{r}|$ it is natural to use the spherical polar coordinates r, θ and ϕ to define the vector $\boldsymbol{r}$. In terms of these coordinates the expression for the kinetic energy is[2,3]

$$\begin{aligned} T &= \frac{1}{2}M\dot{\boldsymbol{R}}^2 + \frac{1}{2}\mu\dot{\boldsymbol{r}}^2 \\ &= \left\{\frac{1}{2}M\left(\dot{X}^2 + \dot{Y}^2 + \dot{Z}^2\right)\right\} + \left\{\frac{1}{2}\mu\left(\dot{r}^2 + \left(r\dot{\theta}\right)^2\right)\right\}. \end{aligned} \tag{1.16}$$

We notice immediately that the azimuthal angle, ϕ, does not appear anywhere. It is therefore constant and is not changed by the collision. The polar angle, θ, appears only as a time derivative in the kinetic energy. The Lagrangian equation of motion for θ is therefore

$$\frac{\mathrm{d}}{\mathrm{d}t}\left(\frac{\partial L}{\partial \dot{\theta}}\right) - \frac{\partial L}{\partial \theta} = 0. \tag{1.17}$$

As $\frac{\partial L}{\partial \theta} = 0$ we see that

$$\frac{\partial L}{\partial \dot{\theta}} = \text{constant}. \tag{1.18}$$

The quantity $\frac{\partial L}{\partial \dot{\theta}}$ is known as p_θ, the generalized momentum conjugate to θ

$$\begin{aligned} p_\theta &= \frac{\partial L}{\partial \dot{\theta}} \\ &= \mu r^2 \dot{\theta}. \end{aligned} \tag{1.19}$$

This quantity p_θ is the orbital angular momentum and in the present case of a collision between two partners without any internal structure it is a conserved quantity. We will, for the moment, denote it by the symbol ℓ. If the interaction potential was zero, then the closest distance two atoms would approach each other would be the impact parameter, b, and this would be equal to r at this point. The speed of the one atom relative to the other at this point is $r\dot{\theta} = b\dot{\theta}$. As there is no potential in this hypothetical case the relative speed of the two atoms is just the original velocity v. So we can write

$$\begin{aligned} \ell &= p_\theta \\ &= \mu r^2 \dot{\theta} \\ &= b\mu \mathrm{v}. \end{aligned} \tag{1.20}$$

Now let us consider the Lagrangian equations of motion for r. From eqn (1.8), (1.9) and (1.16) we obtain

$$\begin{aligned} &\frac{\mathrm{d}}{\mathrm{d}t}\left(\frac{\partial L}{\partial \dot{r}}\right) - \frac{\partial L}{\partial r} = 0, \\ &\frac{\mathrm{d}}{\mathrm{d}t}(\mu \dot{r}) - \mu r \dot{\theta}^2 + \frac{\partial V}{\partial r} = 0 \end{aligned} \tag{1.21}$$

or, using eqn (1.20), we obtain

$$\begin{aligned} &\frac{\mathrm{d}}{\mathrm{d}t}(\mu \dot{r}) - \mu r \left\{\frac{\ell}{\mu r^2}\right\}^2 + \frac{\partial V}{\partial r} = 0, \\ &\frac{\mathrm{d}}{\mathrm{d}t}(\mu \dot{r}) = -\frac{\partial \left\{V + \dfrac{\ell^2}{2\mu r^2}\right\}}{\partial r}. \end{aligned} \tag{1.22}$$

This is the equation of motion for a particle of mass μ in one dimension moving in an effective potential of $V + \frac{\ell^2}{2\mu r^2}$. The total energy of the system is also conserved. As the energy associated with the motion of the center-of-mass is constant (see eqn (1.15)), the energy associated with the relative motion must also be constant. The kinetic energy for the relative motion of the two atoms is given by the second set of curly brackets in eqn (1.16). We see therefore that the energy of relative motion is

$$\begin{aligned} E &= \frac{1}{2}\mu\left(\dot{r}^2 + \left(r\dot{\theta}\right)^2\right) + V(r) \\ &= \frac{1}{2}\mu\dot{r}^2 + \frac{1}{2}\mu r^2\dot{\theta}^2 + V(r) \\ &= \frac{1}{2}\mu\dot{r}^2 + \frac{1}{2}\mu r^2\left\{\frac{\ell}{\mu r^2}\right\}^2 + V(r) \\ &= \frac{1}{2}\mu\dot{r}^2 + \left\{V(r) + \frac{\ell^2}{2\mu r^2}\right\}. \end{aligned} \tag{1.23}$$

From eqn (1.22) and (1.23) we see that the one-dimensional equation of motion for r corresponds to a particle of mass μ moving in an effective potential of $\left\{V(r) + \frac{\ell^2}{2\mu r^2}\right\}$. The *extra* term, $\frac{\ell^2}{2\mu r^2}$, is referred to as the centrifugal potential.

Typically, two atoms attract each other at large distances and repel each other when they approach each other very closely or collide. This leads to a form of potential such as the analytic Lennard-Jones[4] model potential in eqn (1.24) and shown in Figure 1.5

$$V(r) = 4\,\varepsilon\left(\left(\frac{\sigma}{r}\right)^{12} - \left(\frac{\sigma}{r}\right)^{6}\right). \tag{1.24}$$

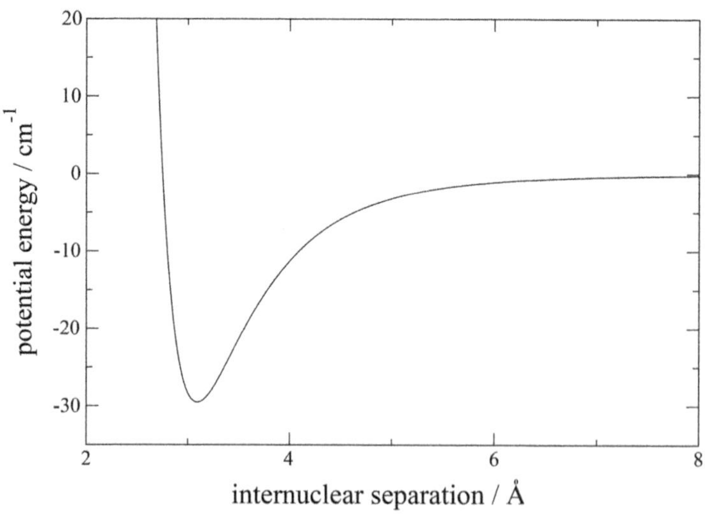

Figure 1.5 Qualitative form of the inter-atomic potential between two atoms. The analytic form used here is the so-called Lennard-Jones potential and the parameters used are those applicable to the Ne + Ne collision pair of atoms:[5] $\varepsilon = 29.52\ \text{cm}^{-1}$, $\sigma = 2.756\,\text{Å}$.

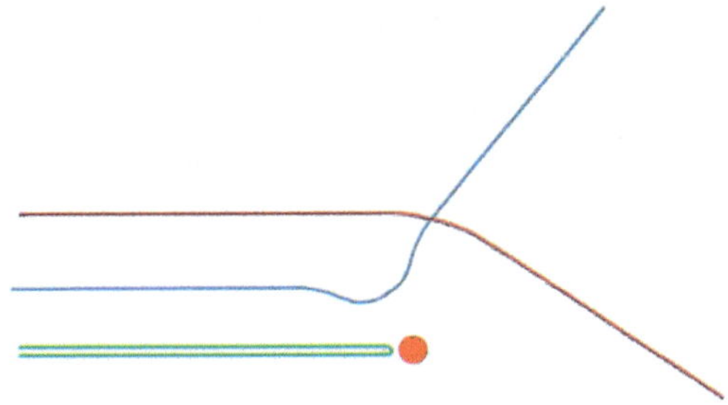

Figure 1.6 Schematic trajectories of two colliding atoms for various values of the impact parameter. The qualitative form of the interaction energy between the atoms is shown in Figure 1.5.

Let us now look at the scattering trajectories which arise for different impact parameters. In Figure 1.6 we show three characteristic trajectories for different values of the impact parameter. One of the atoms is fixed, and the trajectory of the other atom relative to it is shown. Trajectory (A) shows a *head-on* collision with an impact parameter of $b=0$. We see that in this case the incident atom bounces straight backwards and the relative motion of the scattering partners has therefore been deflected by an angle of 180°. Trajectory (B) shows a collision at an intermediate impact parameter. The two atoms initially attract each other and the trajectory is deflected from its original direction towards the other atom. They approach to a distance such that the initial energy of the relative motion $\left(\frac{1}{2}m\text{v}^2\right)$ equals the effective potential $\left\{V(r)+\frac{\ell^2}{2\mu r^2}\right\}=\left\{V(r)+\frac{\mu b^2\text{v}^2}{2r^2}\right\}$. At this point the radial velocity of the colliding atom is reversed and the atom ends up being deflected from its original direction by some positive angle, which we denote here by θ. The point of closest approach of the two atoms is called the "classical turning point". At larger impact parameters (shown in trajectory (C)) the colliding atom never approaches the other atom sufficiently closely to sample their mutual repulsion, but only feels the long distance attraction. This leads to a negative deflection.

The angle of deflection as a function of the impact parameter is called the deflection function $\chi(E,b)$. For very large impact parameters the angle of deflection is zero (see Figure 1.6). The polar angle θ on the other hand starts at 180° or π, as this corresponds to the angle of the negative Z axis. For this large impact parameter, straight line trajectory the final polar angle is 0, the angle of the positive Z axis. The change in the angle θ is therefore π when the deflection function $\chi(E,b)=0$. From Figure 1.6 we see that the change of deflection angle during the course of the trajectory is made up of two equal parts. If we start looking at the trajectory from the point at which the collision partners are closest to each other, *i.e.* the classical turning point r_0, then the total change of angle during the trajectory is twice that which occurs during the final half of the trajectory. Let us call this total change of angle $\Delta\theta$. The deflection function is therefore $\chi(E,b) = \pi - \Delta\theta$. The most straightforward way to compute the deflection function is to recast eqn (1.20) and (1.23) so as to provide equations for $\dot{\theta}$ and $\dot{r}$ and to integrate them numerically from the classical turning point outwards.

From eqn (1.20) we obtain

$$\frac{\mathrm{d}\theta}{\mathrm{d}t} = \dot{\theta} = \frac{b\mathrm{v}}{r^2}. \tag{1.25}$$

The relative velocity is related to the relative collision energy by

$$E = \frac{1}{2}\mu\mathrm{v}^2. \tag{1.26}$$

and combining eqn (1.25) and (1.26) we obtain

$$\frac{\mathrm{d}\theta}{\mathrm{d}t} = \frac{b}{r^2}\left(\frac{2E}{\mu}\right)^{\frac{1}{2}}. \tag{1.27}$$

From eqn (1.23) we can obtain an equation for $\dot{r}$

$$\frac{\mathrm{d}r}{\mathrm{d}t} = \dot{r} = \pm\left(\frac{2}{\mu}\right)^{\frac{1}{2}}\left[E - \left(V(r) + \frac{\ell^2}{2\mu r^2}\right)\right]^{\frac{1}{2}}. \tag{1.28}$$

From eqn (1.20) we can write

$$\ell = b\mu\mathrm{v} = b\mu\left(\frac{2E}{\mu}\right)^{\frac{1}{2}} = b\,(2\mu E)^{\frac{1}{2}}. \tag{1.29}$$

and substituting for ℓ in eqn (1.28) we obtain

$$\frac{\mathrm{d}r}{\mathrm{d}t} = \pm\left(\frac{2}{\mu}\right)^{\frac{1}{2}}\left[E - \left(V(r) + \frac{b^2E}{r^2}\right)\right]^{\frac{1}{2}}. \tag{1.30}$$

Dividing eqn (1.27) by eqn (1.30) eliminates time from the equations and we can then integrate from the classical turning point outwards to obtain an equation for the deflection function[2,3]

$$\chi(E,b) = \pi - 2b\int_{r_0}^{\infty}\frac{\mathrm{d}r}{r^2\sqrt{1 - b^2/r^2 - V(r)/E}} \tag{1.31}$$

where r_0 is the classical turning point.

Three deflection functions for a Lennard-Jones[4] model potential calculated using different collision energies are shown in Figure 1.7. At lower energies the deflection function shows deflection angles more negative than $-180°$ or even $-360°$. This indicates the presence of orbiting type collisions in which the atoms either circle around each other and finally spin off in the direction opposite to their original direction of motion ($\chi = -180°$) or completely orbit each other so as to exit the collision moving in their original directions ($\chi = -360°$).

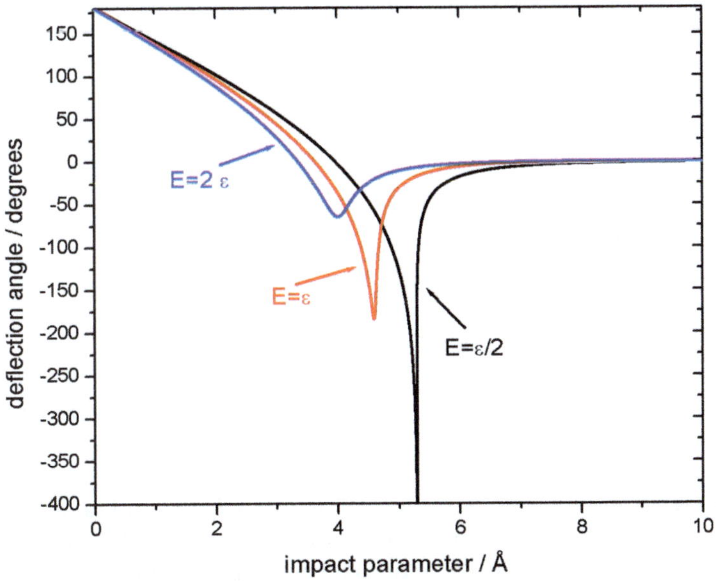

Figure 1.7 The deflection function for the collision of two Ne atoms at various different relative collision energies, where ε is the well depth (29.52 cm^{-1}). A Lennard-Jones potential is used to model the Ne + Ne interaction, see Figure 1.5.

1.2.3 Singularities of a Differential Scattering Cross Section

As in the case of the collision of two hard spheres (see eqn (1.5)) we may now define a *differential scattering cross section.* This is defined as the number of atoms scattered into an increment of solid angle ($d\Omega$), per unit time and per unit incident flux of the collision partners. For an atom–atom collision, in which the interaction potential is independent of angles, the azimuthal angle ϕ (*i.e.* the angle around the initial relative collision velocity) plays no role and we in effect integrate the cross section over this angle. Indeed this is the case in all situations where the target atom or molecule is not in some way oriented in space (see however Chapter 10).

There is one, or maybe more than one, way that an incident atom can end up being scattered through an angle θ. This is illustrated in Figure 1.8 which shows the incident atom being scattered into the same final direction by the target atom while following two different scattering trajectories. In the upper trajectory the collision has a large impact parameter and the deflection is negative. The lower trajectory passes the target atom on the other side. It suffers a positive deflection but ends up being scattered into the same scattering angle as the upper trajectory. The scattering experiment cannot distinguish between a positive or a negative deflection. It just measures the absolute value of the angle of deflection, $|\theta|$. For example, for an absolute value of the deflection of 40°, we see from Figure 1.7 that there will be three different values of the impact parameter which can lead to this scattering angle.

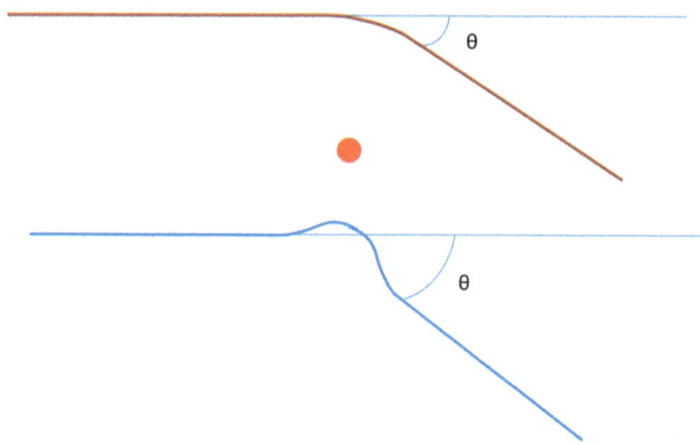

Figure 1.8 Schematic illustration of how two different trajectories can end up being scattered into the same final scattering angle.

If we consider a small annulus of width db around a given impact parameter b, all the incident projectiles passing through this annulus will be scattered into a direction θ to $\theta + \mathrm{d}\theta$. The contribution of this small annulus to the scattering cross section will therefore be

$$2\pi \frac{\mathrm{d}\sigma}{\mathrm{d}\Omega} \sin\theta \, \mathrm{d}\theta = 2\pi b \, \mathrm{d}b \tag{1.32}$$

or

$$\frac{\mathrm{d}\sigma}{\mathrm{d}\Omega} = \frac{b}{\sin\theta \frac{\mathrm{d}\theta}{\mathrm{d}b}}. \tag{1.33}$$

We see that the differential cross section will posses a singularity, *i.e.* blows up, whenever $\sin\theta = 0$ or $\frac{\mathrm{d}\theta}{\mathrm{d}b} = 0$. The first condition is known as the glory effect which can be either forward for $|\theta| = 0,\ 2\pi,\ 4\pi, \ldots$ or backward for $|\theta| = \pi,\ 3\pi,\ 5\pi, \ldots$. The second condition $\left(\frac{\mathrm{d}\theta}{\mathrm{d}b} = 0\right)$ is called the rainbow effect and corresponds to the slope of the deflection function (see Figure 1.7) being zero. Clearly when the impact parameter, b, is very large there will be no deflection; consequently, $\theta = 0$ and the differential cross section will become infinite. From Figure 1.7 we see that θ is also zero for chosen energies when the deflection function passes from positive to negative in the region of an impact parameter between 3 and 4 Å. The range of impact parameters around this value also contributes to the glory singularity in the differential cross section in the vicinity of $\theta = 0$. We will see later that the quantum mechanical description of scattering leads to an interference between the scattering corresponding to these two ranges of impact parameters (*i.e.* $b = \infty$ and $b \approx$ 3–4 Å).

Figure 1.7 demonstrates that for scattering at a collision energy of $2\varepsilon = 59.04\ \mathrm{cm}^{-1}$ at impact parameters a little displaced to either side of the minimum of the deflection function (at around 4 Å) there are two slightly different values of b which lead to scattering into the same scattering angle, close to but a little smaller than the rainbow angle (at which the slope of the deflection function is zero). When we discuss this phenomenon using quantum mechanics, we will see that the contributions from these two different impact parameters lead to a characteristic interference pattern.

As we have seen from Figure 1.7, scattering into an angle of 40° can receive contributions from three different impact parameters: two corresponding to negative deflections and one to a positive deflection. We therefore need to modify the expression for the differential cross section (eqn (1.33)) to take account of these multiple contributions

$$\frac{\mathrm{d}\sigma}{\mathrm{d}\Omega} = \sum_i \frac{b_i}{\sin\theta\left(\dfrac{\mathrm{d}\theta}{\mathrm{d}b}\right)_i} \qquad (1.34)$$

where the summation is over all values of b which contribute to scattering into the angle θ.

Figure 1.9 shows the classical differential cross section for Ne + Ne scattering modelled using a Lennard-Jones interaction potential at the same three collision energies as shown in Figure 1.7. For the high energy cross section (collision energy $= 2\varepsilon \equiv 59.04\ \mathrm{cm}^{-1}$) the main features of the cross section are the two regions, at 0° and 62°, where the cross section becomes infinite. The first of these arises because $\sin\theta$ in eqn (1.34) is zero and is referred to as glory scattering, while the second comes about because $\frac{\mathrm{d}\theta}{\mathrm{d}b}$ is zero and is

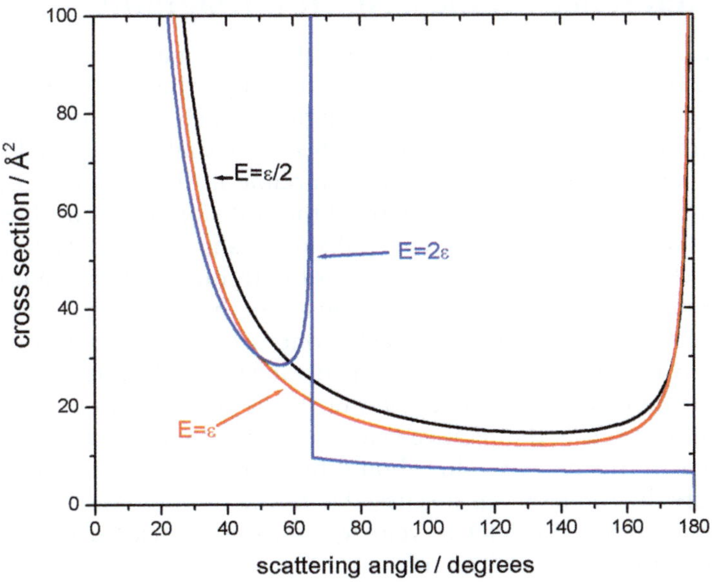

Figure 1.9 Differential cross section for Ne + Ne scattering as modelled using a Lennard-Jones potential at the same three collision energies as in Figure 1.7.

referred to as rainbow scattering. We see that the rainbow scattering angle corresponds to the "bottom" of the deflection function plot (see Figure 1.7). At lower energies the forward and backward glories dominate.

Another phenomenon that might occur at lower collision energies is an orbiting collision. In such a collision the incident atom would circle completely around the target atom. This would show up in the deflection function plot (Figure 1.7) as a much deeper "well" with the deflection function ($\chi(E,b)$) reaching negative values of $-180°$ or more. This would give rise to another singularity or infinity in the cross section at $\theta = |\chi(E,b)| = 180°$ termed a backward glory. When treated quantum mechanically the glory scattering gives rise to maxima rather than infinities in the differential cross section and interference terms are present from the different contributions to the scattering at 0° or at 180°.[6]

The integral cross section is obtained as the integral of the differential cross section over all solid angles. When the potential extends to infinite atom–atom separations, as do all realistic model potentials, and if we use classical mechanics, then the integral cross section will always be infinite. This is because no matter how large the impact parameter, the incident atom will always experience a force and will be scattered to at least a small extent. We will see that this problem is easily resolved with the use of quantum mechanics.

1.3 The Laboratory and the Center-of-Mass Reference Frames: Newton Diagrams

Figure 1.1 shows a typical crossed molecular beam machine. The two atomic or molecular beams are shown at the bottom right hand side of the apparatus. They are shown as being at right angles to each other. While this is the most common and simplest configuration to visualize, the angle between the two beams may also be used as a variable in the design of the experiment. Figure 1.10 shows a schematic of the collision between a beam of "A" and "B" atoms, or molecules. The collision is shown in the "laboratory" or "space-fixed" frame of reference. This is the frame in which we must make our measurements, but it is not the most suitable frame for analyzing and discussing the collision process. Each atom "sees" the other only in terms of its motion relative to itself. Thus, for instance, the absolute velocity of the atoms in the laboratory is not relevant to the mechanism or the outcome of the collision. It is the relative velocity of the atoms to each other that is important in this respect. This should be clear from our preceding discussion where the collision was discussed in the center-of-mass reference frame and it was always the relative velocity and the associated relative kinetic energy that played the central role.

The coordinate of the center-of-mass of the system and the position of atom "B" relative to atom "A" is defined in the same way as in eqn (1.7). The relationship of the velocity of the center-of-mass and of the relative

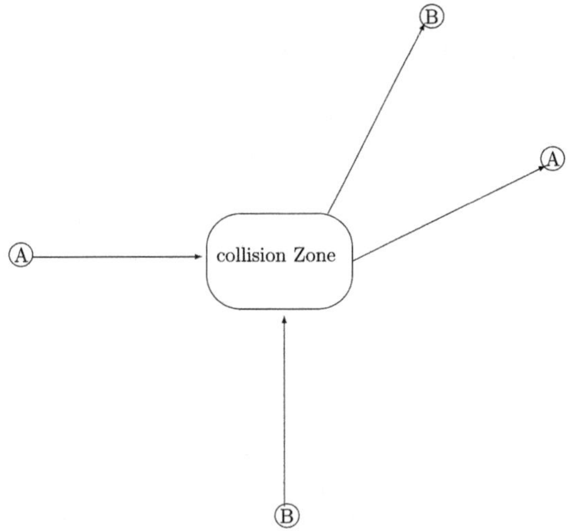

Figure 1.10 Schematic diagram showing collision between atomic beams "A" and "B" in laboratory frame.

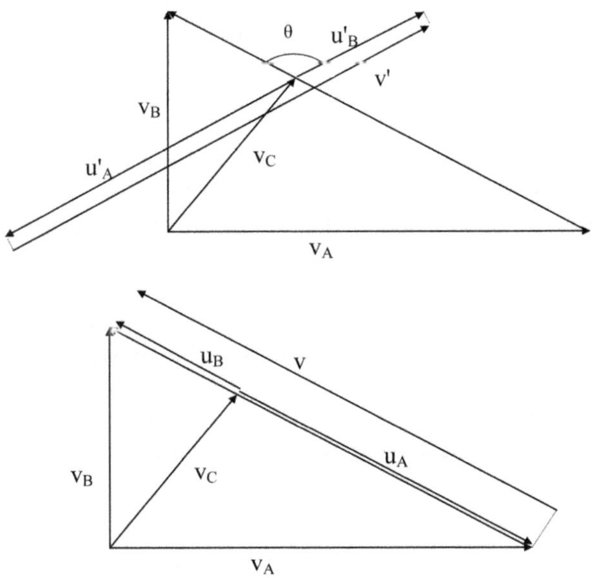

Figure 1.11 Newton diagram showing relationship between center-of-mass and laboratory velocities.

velocity to the initial velocities of the two atoms can be represented in a very useful diagram known as a Newton diagram.[7,8] The diagram depicts the relationship between the laboratory and center-of-mass reference frames and the limitations placed on the scattering process by the conservation rules of energy and linear momentum. Figure 1.11 presents such a diagram.

$\mathbf{v}_A$ and $\mathbf{v}_B$ are the velocities of atoms "A" and "B" in the laboratory reference frame. $\mathbf{v}_C$ is the velocity of the center-of-mass (see eqn (1.7)). $\mathbf{v}_C$ remains constant throughout the collision even when inelastic and reactive collisions are considered. $\mathbf{v}$ is the relative velocity (velocity of "B" relative to "A"). $\boldsymbol{u}_A$ and $\boldsymbol{u}_B$ are the velocities of "A" and "B" in the center-of-mass reference frame (*i.e.* relative to the center-of-mass velocity). The magnitudes of the vectors $\boldsymbol{u}_A$, $\boldsymbol{u}_B$ and $\mathbf{v}$ are related by

$$|\boldsymbol{u}_A| = \frac{m_B}{m_A + m_B}|\mathbf{v}|$$
$$|\boldsymbol{u}_B| = \frac{m_A}{m_A + m_B}|\mathbf{v}|. \tag{1.35}$$

After the collision we denote the velocities of "A" and "B" in the center-of-mass by $\boldsymbol{u}'_A$ and $\boldsymbol{u}'_B$ and the relative velocity by $\mathbf{v}'$. These post-collision velocities are shown in the upper panel of Figure 1.11. As we are currently considering atom–atom elastic collisions the magnitudes of the velocities of "A" and "B" in the center-of-mass reference frame remain the same after the collision, but the directions of these velocities are altered by the collision. So we can write

$$|\boldsymbol{u}'_A| = |\boldsymbol{u}_A|$$
$$|\boldsymbol{u}'_B| = |\boldsymbol{u}_B|$$
$$|\mathbf{v}'| = |\mathbf{v}|. \tag{1.36}$$

The center-of-mass scattering angle θ for the scattering of atom "B" is shown in Figure 1.11. Figure 1.12 shows how the Newton diagram permits us to relate the center-of mass and the laboratory scattering angle. We see from the figure that there may be two different center-of-mass scattering angles which lead to the same laboratory scattering angle. Figure 1.13 shows two different final center-of-mass velocities of atom "B" which lead to the same measured laboratory scattering angle. The Newton diagrams we have presented here are somewhat simplified in that we have shown only in-plane scattering. This is valid for atom–atom scattering, but in a more general case there may also be some scattering out of the plane defined by the two initial velocity vectors.

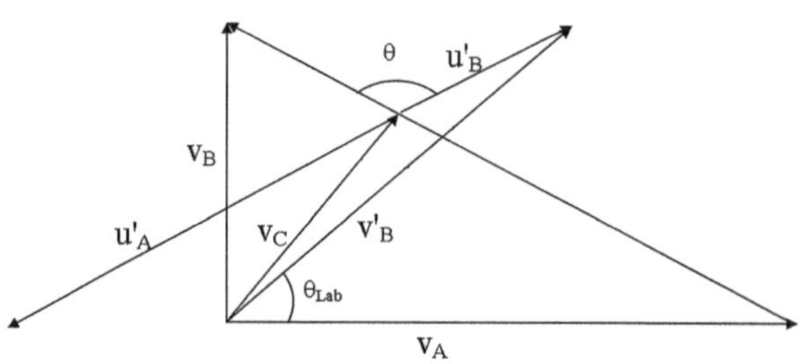

Figure 1.12 Newton diagram showing relationship between center-of-mass and laboratory scattering angles.

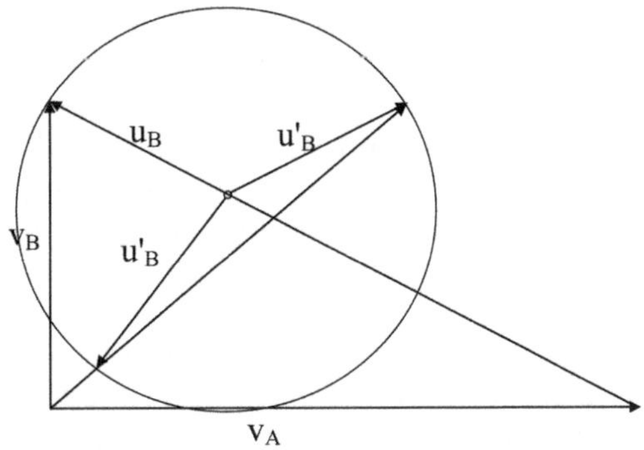

Figure 1.13 Newton diagram showing the two different center-of-mass velocities of atom "B" which lead to the same laboratory scattering angle.

The aim of a crossed molecular beam experiment is to measure the scattering cross section. The theoretical discussion of this cross section and its interpretation in terms of molecular interactions is most conveniently, or maybe necessarily, undertaken in the center-of-mass reference frame. The transformation of the measured differential laboratory cross section to the center-of-mass reference frame is quite a technical subject and we refer the reader to the literature[9–12] for further discussion.

References

1. J. D. McDonald, P. R. LeBreton, Y. T. Lee and D. R. Herschbach, Molecular beam kinetics: reactions of deuterium atoms with halogen molecules, *J. Chem. Phys.*, 1972, **56**, 769.
2. H. Goldstein, *Classical Mechanics*, Addison-Wesley, Reading, Massachusetts, USA, 1950.
3. H. Goldstein, C. Poole Jr and J. Safko, *Classical Mechanics*, Addison-Wesley, San Francisco, 2002.
4. D. M. Hirst, *Potential Energy Surfaces, Molecular Structure and Dynamics*, Taylor and Francis, 1985.
5. F. Pirani, S. Brizi, L. F. Roncaratti, P. Casavecchia, D. Cappelletti and F. Vecchiocattivi, Beyond the Lennard-Jones model: a simple and accurate potential function probed by high resolution scattering data useful for molecular dynamics simulations, *Phys. Chem. Chem. Phys.*, 2008, **10**, 5489.
6. K. W. Ford and J. A. Wheeler, Semiclassical description of scattering, *Ann. Phys.*, 1959, **7**, 259.
7. D. R. Herschbach, Reactive collisions in crossed molecular beams, *Discuss. Faraday Soc.*, 1962, **33**, 149.
8. D. R. Herschbach, Reactive scattering in molecular beams, *Adv. Chem. Phys.*, 1966, **10**, 319.

9. F. A. Morse and R. B. Bernstein, Velocity dependence of the differential cross sections for the scattering of atomic beams of K and Cs by Hg, *J. Chem. Phys.*, 1962, **37**, 2019.
10. S. Datz, D. R. Herschbach and E. H. Taylor, Collision mechanics in crossed Maxwellian molecular beams, *J. Chem. Phys.*, 1961, **35**, 1549.
11. R. K. B. Helbing, Transformations and averaging processes for differential cross-section measurements at thermal energies, *J. Chem. Phys.*, 1968, **48**, 472.
12. T. T. Warnock and R. B. Bernstein, Transformation relationships from center-of-mass cross section and excitation functions to observable angular and velocity distributions of scattered flux, *J. Chem. Phys.*, 1968, **49**, 1878.

CHAPTER 2

Quantum Theory of Atom–Atom Elastic Scattering

2.1 The Wave Picture of the Scattering Process

In the previous chapter we discussed how the motion of the two atoms in an atom–atom scattering process could be represented in terms of the motion of the center-of-mass of the combined system and the motion of one of the atoms relative to the other. The same analysis applies in quantum mechanics (see the treatment of the hydrogen atom problem in any quantum mechanics text book).[1–3] The wavefunction of the two atoms is then represented as a product of the wavefunction of the center-of-mass and the wavefunction for the motion of one atom relative to the other. Just as in Chapter 1, we may ignore the motion of the center-of-mass of the system for all purposes other than transforming from the center-of-mass to laboratory reference frames.

The relative motion of two colliding atoms approaching each other is represented as a plane wave in quantum mechanics. Such a plane wave has a well defined linear momentum and, as a result of the uncertainty principle, is therefore totally delocalized in space. Its wavefunction may be written in the form

$$\psi_0(\boldsymbol{r},E)=\left(\frac{k\mu}{2\pi h^2}\right)^{\frac{1}{2}}\mathrm{e}^{i\mathbf{k}\cdot\boldsymbol{r}} \tag{2.1}$$

where the subscript on ψ_0 indicates that this is the solution of the problem in the absence of the interaction potential, k is the "wavenumber" and μ is the reduced mass of the colliding atoms.

RSC Theoretical and Computational Chemistry Series No. 7
Theory of Molecular Collisions
By Gabriel G. Balint-Kurti and Alexander P. Palov

Published by the Royal Society of Chemistry, www.rsc.org

Unlike the wavefunction of a bound state, a plane wave or a scattering wavefunction cannot be normalized in the conventional sense. The normalization factor in eqn (2.1) is somewhat arbitrary. We discuss the problem in Appendix A. In this book we will consistently normalize scattering wavefunctions on the energy scale.[4]

Following the normal convention, we consider the incident particle to be a plane wave approaching from the negative Z direction. The target atom or scattering center is located at the origin. The interaction potential is taken to be spherically symmetric, and the azimuthal angle ϕ therefore plays no role in the collision. After the collision the particle is scattered into different polar angles with a scattering amplitude of $f(\theta)$. At large distances from the scattering center the wavefunction therefore takes the form

$$\psi(\boldsymbol{r}, E) \widetilde{_{r\to\infty}} \left(\frac{k\mu}{2\pi\hbar^2}\right)^{\frac{1}{2}} \left\{ \mathrm{e}^{ikz} + f(\theta)\,\frac{\mathrm{e}^{ikr}}{r} \right\}. \tag{2.2}$$

The term e^{ikz} is the initial incident wave, while $f(\theta)\,\mathrm{e}^{ikr}$ represents the scattered wave going radially outwards from the scattering center. The momentum operator is $-i\hbar\frac{\partial}{\partial z}$. If we operate with this on the incident wave we see that it has a momentum of $\hbar k$ and consequently the incident flux of atoms (*i.e.* the number of atoms crossing a unit area per unit time) corresponding to the plane wave e^{ikz} is $\frac{k\hbar}{\mu}|\mathrm{e}^{ikz}|^2 = \frac{k\hbar}{\mu}$. The solid angle $\sin\theta\,\mathrm{d}\theta\,\mathrm{d}\phi$ subtends an area of $r^2\sin\theta\,\mathrm{d}\theta\,\mathrm{d}\phi$ on the surface of a sphere of radius r centered on the scattering center. As the collision is elastic, the radial momentum remains unchanged by the collision and is equal to $k\hbar$. The flux of atoms passing through the small area $r^2\sin\theta\,\mathrm{d}\theta\,\mathrm{d}\phi$ on the surface of the sphere arising from the scattered wave $f(\theta)\,\frac{\mathrm{e}^{ikr}}{r}$ is therefore $\frac{k\hbar}{\mu}\left|f(\theta)\,\frac{\mathrm{e}^{ikr}}{r}\right|^2 r^2\sin\theta\,\mathrm{d}\theta\,\mathrm{d}\phi = \frac{k\hbar}{\mu}|f(\theta)|^2\sin\theta\,\mathrm{d}\theta\,\mathrm{d}\phi$.

The "differential cross section" is defined as the flux of atoms scattered into a given solid angle divided by the incident flux of atoms (see Chapter 1). Note that any normalization factor cancels out in this definition of the cross section. From the above definition we may write the differential cross section in the form

$$\sigma(\Omega)\,\mathrm{d}\Omega = \frac{\dfrac{k\hbar}{\mu}|f(\theta)|^2\sin\theta\,\mathrm{d}\theta\,\mathrm{d}\phi}{\dfrac{k\hbar}{\mu}}. \tag{2.3}$$

We conclude that the differential cross section is given by

$$\sigma(\Omega) = |f(\theta)|^2. \tag{2.4}$$

2.2 The Phase Shift

Close to the scattering center the wavefunction can be found by solving the time-independent Schrödinger equation. As with all second order differential equations there are two linearly independent solutions to this equation. We will mainly be interested in the physically reasonable or "regular" solution. This is the one which tends to zero near the origin, where the interaction potential becomes infinite. The full time-independent Schrödinger equation is

$$\left\{-\frac{\hbar^2}{2\mu}\nabla^2 + V(r)\right\}\psi(\boldsymbol{r}) = E\psi(\boldsymbol{r}). \tag{2.5}$$

Expressing the Laplacian operator (∇^2) in spherical polar coordinates[1–3,5,6] yields

$$-\frac{\hbar^2}{2\mu}\left\{\frac{1}{r^2}\frac{\partial}{\partial r}\left(r^2\frac{\partial\psi}{\partial r}\right) + \frac{1}{r^2}\left[\frac{1}{\sin\theta}\frac{\partial}{\partial\theta}\left(\sin\theta\frac{\partial\psi}{\partial\theta}\right) + \frac{1}{\sin^2\theta}\frac{\partial^2\psi}{\partial\phi}\right]\right\} + V(r)\psi = E\psi. \tag{2.6}$$

If we multiply this equation by r^2 we see that the Hamiltonian operator becomes a sum of radial and angular parts. The Hamiltonian is said to be "separable" in radial and angular coordinates and the wavefunction may therefore be written as a product of radial and angular parts. The angular wavefunctions are the eigenfunctions of the orbital angular momentum operator

$$\begin{aligned}&-\hbar^2\left[\frac{1}{\sin\theta}\frac{\partial}{\partial\theta}\left(\sin\theta\frac{\partial}{\partial\theta}\right) + \frac{1}{\sin^2\theta}\frac{\partial^2}{\partial\phi}\right]Y_{\ell m}(\theta,\phi)\\ &= \ell(\ell+1)\hbar^2 Y_{\ell m}(\theta,\phi)\end{aligned} \tag{2.7}$$

where $Y_{\ell m}(\theta,\phi)$ are the spherical harmonics.[7,8] Utilizing eqn (2.7) we now write the total wavefunction in the form

$$\psi_{\ell,m}(\boldsymbol{r}) = \frac{1}{r}\chi_\ell(r)Y_{\ell m}(\theta,\phi) \tag{2.8}$$

where ℓ and m are both "good" quantum numbers, as in the case of the motion of an electron in the hydrogen atom.[2,3]

The extra factor of $\frac{1}{r}$ is introduced so as to simplify the ensuing differential equation in the radial coordinate, r. Substituting eqn (2.8) into eqn (2.6) we obtain

$$\frac{\hbar^2}{2\mu}\left[-\frac{\partial^2}{\partial r^2} + \frac{\ell(\ell+1)}{r^2}\right]\chi_\ell(r) + V(r)\chi_\ell(r) = E\chi_\ell(r) \tag{2.9}$$

or alternatively

$$-\frac{\hbar^2}{2\mu}\frac{\partial^2}{\partial r^2}\chi_\ell(r) + \left\{V(r) + \frac{\hbar^2}{2\mu}\frac{\ell(\ell+1)}{r^2}\right\}\chi_\ell(r) = E\chi_\ell(r). \tag{2.10}$$

The term in curly brackets in eqn (2.10) is called the "effective potential". At large r both the centrifugal potential $\left(\frac{\hbar^2}{2\mu}\frac{\ell(\ell+1)}{r^2}\right)$ and the interaction potential $(V(r))$ tend to zero. The general solution of eqn (2.9) is then a combination of sines and cosines and can be written in the form

$$\begin{aligned}\chi(r) \underset{r\to\infty}{\longrightarrow} & A \sin\left(kr - \frac{\ell\pi}{2} + \eta_\ell\right) \\ &= A\frac{1}{2i}\left\{e^{i\left(kr - \frac{\ell\pi}{2} + \eta_\ell\right)} - e^{-i\left(kr - \frac{\ell\pi}{2} + \eta_\ell\right)}\right\} \\ &= A\frac{i^{\ell+1}e^{-i\eta_\ell}}{2}\left\{e^{-ikr} + (-1)^{-(\ell+1)}e^{i(kr+2\eta_\ell)}\right\}\end{aligned} \tag{2.11}$$

where k is the wavenumber $\left(k = \frac{\sqrt{2\mu E}}{\hbar}\right)$ and η_ℓ is the phase shift which represents the difference in phase compared to the solution in the absence of a potential.

The radial Schrödinger equation (eqn (2.9)) for free motion when $V(r) = 0$ has two linearly independent solutions. These are called the spherical Riccati–Bessel functions[9] $\hat{j}_\ell(kr) = kr\, j_\ell(kr)$ and $\hat{n}_\ell(kr) = kr\, n_\ell(kr)$, where $j_\ell(kr)$ and $n_\ell(kr)$ are the closely related spherical Bessel functions.[6,10] The extra "r" in front of the spherical Bessel functions of the first (j_ℓ) and second (n_ℓ) kinds arises from the fact that we introduced an extra factor of $\frac{1}{r}$ when deriving eqn (2.9). The regular solution $\hat{j}_\ell(kr)$ goes to zero at the origin where the centrifugal potential $\frac{\hbar^2}{2\mu}\frac{\ell(\ell+1)}{r^2}$ becomes infinite. The irregular solution $\hat{n}_\ell(kr)$ becomes infinite as $r \to 0$. At large r values the solutions behave as below

$$\begin{aligned}\hat{j}_\ell(kr) &= kr\, j_\ell(kr) \underset{r\to\infty}{\longrightarrow} \sin\left(kr - \frac{\ell\pi}{2}\right) \\ \hat{n}_\ell(kr) &= kr\, n_\ell(kr) \underset{r\to\infty}{\longrightarrow} -\cos\left(kr - \frac{\ell\pi}{2}\right).\end{aligned} \tag{2.12}$$

2.3 Partial Wave Expansion

A plane wave traveling in the positive Z direction can be expanded in terms of the regular spherical Bessel functions

$$e^{ikz} = \sum_{\ell=0}^{\infty}(2\ell+1)i^\ell j_\ell(kr)P_\ell(\cos\theta) \tag{2.13}$$

where $P_\ell(\cos\theta)$ are the Legendre polynomials.[7,8]

Using eqn (2.12) and (2.13) in eqn (2.2) we obtain

$$\psi(\boldsymbol{r},E) \widetilde{_{r\to\infty}} \left(\frac{k\mu}{2\pi h^2}\right)^{\frac{1}{2}} \left\{\sum_{\ell=0}^{\infty}(2\ell+1)i^{\ell}\left(\frac{1}{kr}\right) \times \sin\left(kr-\frac{\ell\pi}{2}\right)P_\ell(\cos\theta)+f(\theta)\frac{\mathrm{e}^{ikr}}{r}\right\}$$

$$= \left(\frac{k\mu}{2\pi h^2}\right)^{\frac{1}{2}} \left\{\sum_{\ell=0}^{\infty}(2\ell+1)i^{\ell}\left(\frac{1}{kr}\right)\frac{1}{2i}\left(\mathrm{e}^{i\left(kr-\frac{\ell\pi}{2}\right)}-\mathrm{e}^{-i\left(kr-\frac{\ell\pi}{2}\right)}\right) \times P_\ell(\cos\theta)+f(\theta)\frac{\mathrm{e}^{ikr}}{r}\right\}$$

$$= \left(\frac{k\mu}{2\pi h^2}\right)^{\frac{1}{2}} \left\{\sum_{\ell=0}^{\infty}(2\ell+1)\left(\frac{1}{kr}\right)\frac{(-1)^{\ell+1}}{2i}\left\{\mathrm{e}^{-ikr}+(-1)^{-(\ell+1)}\mathrm{e}^{ikr}\right\} \times P_\ell(\cos\theta)+f(\theta)\frac{\mathrm{e}^{ikr}}{r}\right\}. \tag{2.14}$$

In order to relate the phase shift of eqn (2.11) to the scattering amplitude $f(\theta)$ we need to expand the scattering amplitude in terms of the Legendre polynomials

$$f(\theta)=\sum_{\ell=0}^{\infty}f_\ell P_\ell(\cos\theta). \tag{2.15}$$

It is this expansion which is called the "partial wave expansion" and the f_ℓ constitute the partial waves or amplitudes.

Substituting eqn (2.15) into eqn (2.14) we obtain

$$\psi(\boldsymbol{r},E) \widetilde{_{r\to\infty}} \left(\frac{k\mu}{2\pi h^2}\right)^{\frac{1}{2}} \left\{\sum_{\ell=0}^{\infty}(2\ell+1)\left(\frac{1}{kr}\right)\frac{(-1)^{\ell+1}}{2i} \times \left\{\mathrm{e}^{-ikr}+(-1)^{-(\ell+1)}\mathrm{e}^{ikr}\right\}P_\ell(\cos\theta)+\sum_{\ell=0}^{\infty}f_\ell P_\ell(\cos\theta)\frac{\mathrm{e}^{ikr}}{r}\right\}$$

$$= \left(\frac{k\mu}{2\pi h^2}\right)^{\frac{1}{2}}\frac{1}{r}\left\{\sum_{\ell=0}^{\infty}\left[(2\ell+1)\frac{(-1)^{\ell+1}}{2ik}\mathrm{e}^{-ikr} + \left((2\ell+1)\frac{1}{2ik}+f_\ell\right)\mathrm{e}^{ikr}\right]P_\ell(\cos\theta)\right\}$$

$$= \left(\frac{k\mu}{2\pi h^2}\right)^{\frac{1}{2}}\frac{1}{2ikr}\left\{\sum_{\ell=0}^{\infty}\left[(2\ell+1)(-1)^{\ell+1}\mathrm{e}^{-ikr} + \left((2\ell+1)+2ikf_\ell\right)\mathrm{e}^{ikr}\right]P_\ell(\cos\theta)\right\}. \tag{2.16}$$

Incoming and outgoing radial waves must have equal amplitudes because of the particle numbers conservation law. Consequently, comparing eqn (2.16) to eqn (2.11) we can equate the relative coefficients of these waves

$$(-1)^{-(\ell+1)}e^{i(2\eta_\ell)} = \frac{((2\ell + 1) + 2ikf_\ell)}{(2\ell + 1)(-1)^{\ell+1}} \tag{2.17}$$

and therefore

$$f_\ell = \frac{(2\ell + 1)}{k}e^{i\eta_\ell} \sin \eta_\ell. \tag{2.18}$$

Combining eqn (2.18) with eqn (2.4) and (2.15) provides an expression for the differential cross section in terms of the phase shift

$$\sigma(\Omega) = \left| \sum_{\ell=0}^{\infty} \frac{(2\ell + 1)}{k} e^{i\eta_\ell} \sin \eta_\ell P_\ell(\cos\theta) \right|^2. \tag{2.19}$$

The integral cross section is obtained by integrating the differential cross section over all solid angles.

The normalization condition for the Legendre polynomials is

$$\int_0^\pi P_\ell(\cos\theta)P_{\ell'}(\cos\theta)\sin\theta\, d\theta = \frac{2\delta_{\ell\ell'}}{2\ell + 1} \tag{2.20}$$

where $\delta_{\ell\ell'}$ is a Kronecker delta function (*i.e.* $\delta_{\ell\ell'} = 1$ if $\ell = \ell'$ but is zero otherwise).

The integral cross section is therefore given by

$$\begin{aligned}
\sigma &= \int \sigma(\Omega)\, d\Omega \\
&= \int_0^{2\pi} d\phi \int_0^\pi \left| \sum_{\ell=0}^{\infty} \frac{(2\ell + 1)}{k} e^{i\eta_\ell} \sin \eta_\ell P_\ell(\cos\theta) \right|^2 \sin\theta\, d\theta \\
&= 2\pi \int_0^\pi \left| \sum_{\ell=0}^{\infty} \frac{(2\ell + 1)}{k} e^{i\eta_\ell} \sin \eta_\ell P_\ell(\cos\theta) \right|^2 \sin\theta\, d\theta \\
&= 2\pi \sum_{\ell=0,\ell'=0}^{\infty} \frac{(2\ell + 1)}{k}\frac{(2\ell' + 1)}{k} e^{i(\eta_\ell - \eta'_\ell)} \sin \eta_\ell \sin \eta_{\ell'} \\
&\quad \times \int_0^\pi P_\ell(\cos\theta)P_{\ell'}(\cos\theta)\sin\theta\, d\theta \\
&= \frac{4\pi}{k^2}\sum_{\ell=0}^{\infty}(2\ell + 1)\sin^2 \eta_\ell.
\end{aligned} \tag{2.21}$$

The integral cross section can be also presented as a sum of partial integral cross sections

$$\sigma = \sum_{\ell=0}^{\infty} \sigma_\ell \tag{2.22}$$

where σ_ℓ is given by

$$\sigma_\ell = \frac{4\pi}{k^2}(2\ell + 1)\sin^2\eta_\ell. \tag{2.23}$$

2.4 Scattering of Identical Particles

When two identical atoms or molecules are scattered off each other, their symmetry with respect to interchange must be considered. The situation is exactly the same as that leading to the formation of *ortho*- and *para*-hydrogen.[2,3,5,11,12] If the particles have integer spin they are bosons and the corresponding wavefunction must be symmetric with respect to their interchange. Fermions on the other hand have half integer spins and their wavefunctions must be odd with respect to their interchange. We have chosen to illustrate atom–atom scattering using the scattering of two ^{20}Ne atoms. As these are bosons we will discuss the effect of interchange symmetry on the scattering for this case.

When the two atoms are interchanged the main effect is that their relative position vector is changed in sign, *i.e.* $\boldsymbol{r} \rightarrow -\boldsymbol{r}$. In addition to changing the sign of the Z component of the relative position vector the interchange of the two atoms affects the scattering angle θ in eqn (2.2). Figure 2.1 shows a schematic illustration of the classical trajectories of two identical particles colliding with each other in the center-of-mass reference frame. Only the initial and final velocities are illustrated. In Figure 2.1 (top) the solid red line shows the trajectory of the incident particle and the solid blue line shows that of the target. The incident particle is shown as being deflected through an angle of θ. In Figure 2.1 (bottom) the two particles have been interchanged and we see that the effect is that the incident particle is now scattered into an angle $(\pi - \theta)$ as compared with the original direction of motion of the particles before interchange. Our conclusion is that the process of interchanging the two indistinguishable particles is to change the angle of scattering in the manner $\theta \rightarrow (\pi - \theta)$. The figure also illustrates the fact that the differential cross section for the scattering of identical atoms will be symmetric about $\theta = \pi/2$, *i.e.* $\sigma(\theta) = \sigma(\pi - \theta)$.

If we denote the particle exchange or permutation operator by the symbol $\hat{\boldsymbol{P}}$ we can write the symmetrized wavefunction as

$$\begin{aligned}\psi_{sym}(\boldsymbol{r}, E) &= \frac{1}{\sqrt{2}}\left[\psi(\boldsymbol{r}, E) + \hat{\boldsymbol{P}}\psi(\boldsymbol{r}, E)\right] \\ &= \frac{1}{\sqrt{2}}\left[\psi(\boldsymbol{r}, E) + \psi(-\boldsymbol{r}, E)\right].\end{aligned} \tag{2.24}$$

The factor of $\frac{1}{\sqrt{2}}$ is required to renormalize the wavefunction. Referring back to eqn (2.2) we can now write the asymptotic form of the symmetrized wavefunction as

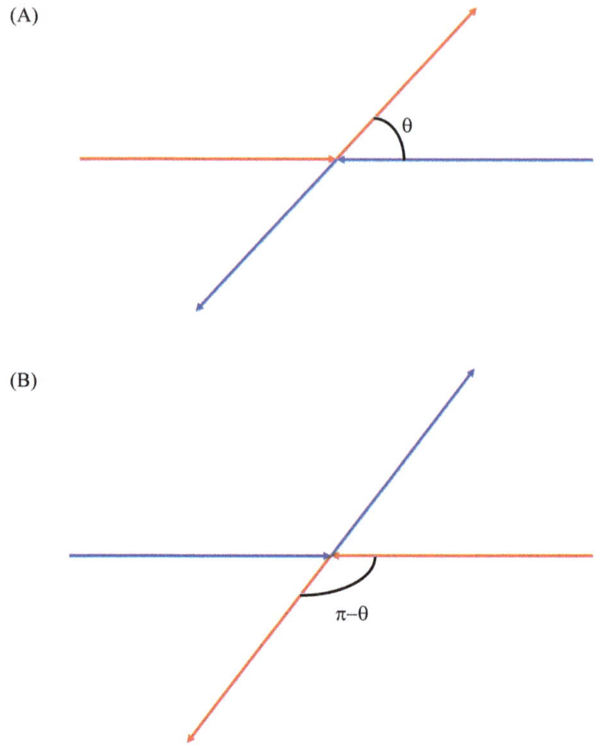

Figure 2.1 Schematic illustration of the trajectories of two colliding atoms in the center-of-mass frame. (A) Before interchange of identical atoms. (B) After interchange.

$$\psi(\boldsymbol{r}, E) \underset{r\to\infty}{\widetilde{}} \frac{1}{\sqrt{2}}\left(\frac{k\mu}{2\pi h^2}\right)^{\frac{1}{2}}\left\{\left[e^{ikz} + e^{-ikz}\right] + \left[f(\theta) + f(\pi - \theta)\right]\frac{e^{ikr}}{r}\right\}. \tag{2.25}$$

The differential cross section is then found in the same way as before (see the discussion leading to eqn (2.4)) and is found to be given by

$$\sigma(\Omega) = \frac{1}{2}\left|f(\theta) + f(\pi - \theta)\right|^2. \tag{2.26}$$

Note the new wavefunction in eqn (2.25) contains flux incident from both the negative and positive Z directions.[11] As the incident particle and the target particle are indistinguishable, both the incident particle wave, approaching from negative z, and the target wave, approaching from positive z, now contribute to the scattering into the solid angle Ω. If we had treated these two particle beams as independent beams, without insisting that the wavefunction be symmetric with respect to exchange of the two atoms, then we would have obtained the differential cross section expression $\sigma(\Omega) = \frac{1}{2}\{|f(\theta)|^2 + |f(\pi - \theta)|^2\}$, which as we will see leads to very different results.

The scattering amplitude for the angle $(\pi - \theta)$ can be expressed as (see eqn (2.15))

$$f(\pi - \theta) = \sum_{\ell=0}^{\infty} f_\ell P_\ell(\cos(\pi - \theta)) = \sum_{\ell=0}^{\infty} f_\ell P_\ell(-\cos\theta). \tag{2.27}$$

The Legendre polynomials are symmetric or antisymmetric with respect to the change of sign of their argument depending on their order

$$P_\ell(-\cos\theta) = (-1)^\ell P_\ell(\cos\theta). \tag{2.28}$$

Using eqn (2.18) along with eqn (2.27) and (2.28) we may write the total scattering amplitude of eqn (2.26) as

$$\begin{aligned} f(\theta) + f(\pi - \theta) &= \sum_{\ell=0}^{\infty} \frac{(2\ell + 1)}{k} e^{i\eta_\ell} \sin\eta_\ell \left(P_\ell(\cos\theta) + (-1)^\ell P_\ell \cos\theta)\right) \\ &= \sum_{\ell=0,2,4,\ldots}^{\infty} \frac{(2\ell + 1)}{k} e^{i\eta_\ell} \sin\eta_\ell \, 2P_\ell(\cos\theta) \end{aligned} \tag{2.29}$$

where the odd terms have cancelled due to the parity property of the Legendre polynomials.

Combining eqn (2.29) with (2.26) now yields an expression for the differential cross section in terms of the phase shift

$$\sigma(\Omega) = 2\left| \sum_{\ell=0,2,4,\ldots}^{\infty} \frac{(2\ell + 1)}{k} e^{i\eta_\ell} \sin\eta_\ell P_\ell(\cos\theta)\right|^2. \tag{2.30}$$

If we consider scattering into an angle of $\theta = \pi/2$ we can immediately see a clear difference between the scattering predicted by the above equation and that which would have arisen if we had not used a symmetric wavefunction. As $P_\ell(\cos\pi/2) = 0$ for odd values of ℓ, the cross section predicted by eqn (2.30) will be a factor of 2 greater than that which we would have obtained without invoking the use of symmetry.

We obtain the integral cross section by integrating over all solid angles. Integration over ϕ yields a factor of 2π. Using the orthogonality relationship of the Legendre polynomials (see eqn (2.20)) the integral cross section is

$$
\begin{aligned}
\sigma &= \int \sigma(\Omega)\,\mathrm{d}\Omega \\
&= \int_0^{2\pi} \mathrm{d}\phi \int_0^{\pi} 2 \left| \sum_{\ell=0,2,4,\dots}^{\infty} \frac{(2\ell+1)}{k} \mathrm{e}^{i\eta_\ell} \sin\eta_\ell P_\ell(\cos\theta) \right|^2 \sin\theta\,\mathrm{d}\theta \\
&= 4\pi \sum_{\ell=0,2,4\dots}^{\infty} \sum_{\ell'=0,2,4,\dots}^{\infty} \frac{(2\ell+1)}{k}\frac{(2\ell'+1)}{k} \mathrm{e}^{i(\eta_\ell-\eta'_\ell)} \sin\eta_\ell \sin\eta_{\ell'} \\
&\quad \times \int_0^{\pi} P_\ell(\cos\theta) P_{\ell'}(\cos\theta) \sin\theta\,\mathrm{d}\theta \\
&= \frac{8\pi}{k^2} \sum_{\ell=0,2,4,\dots}^{\infty} (2\ell+1)\sin^2\eta_\ell. \qquad (2.31)
\end{aligned}
$$

2.5 The S and T Matrices

In this section we introduce the **S** and **T** matrices. These matrices are most useful in the case of inelastic and reactive scattering. The discussion here serves simply as an introduction.

The large distance, asymptotic form, of the scattering wavefunction (see eqn (2.11)) may be written as

$$
\begin{aligned}
\chi(r) \underset{r\to\infty}{\widehat{\ \ }} A \sin\left(kr - \frac{\ell\pi}{2} + \eta_\ell\right) \\
&= -A\frac{1}{2i}\mathrm{e}^{-i\eta_\ell}\left\{\mathrm{e}^{-i\left(kr-\frac{\ell\pi}{2}\right)} - \mathrm{e}^{2i\eta_\ell}\mathrm{e}^{i\left(kr-\frac{\ell\pi}{2}\right)}\right\} \\
&= -A\frac{1}{2i}\mathrm{e}^{-i\eta_\ell}\left\{\mathrm{e}^{-i\left(kr-\frac{\ell\pi}{2}\right)} - \mathbf{S}_\ell\,\mathrm{e}^{i\left(kr-\frac{\ell\pi}{2}\right)}\right\}. \qquad (2.32)
\end{aligned}
$$

The **S** matrix is the ratio of the radial outgoing wave to that of the radial incoming wave. The radial incoming wave is $\mathrm{e}^{-i\left(kr-\frac{\ell\pi}{2}\right)}$ and the corresponding outgoing wave is $\mathrm{e}^{i\left(kr-\frac{\ell\pi}{2}\right)}$. These radial waves come from the asymptotic form of the spherical Bessel functions which are the solutions of the scattering problem in the absence of a potential. From eqn (2.32) we can identify the **S** matrix as

$$\mathbf{S}_\ell = \mathrm{e}^{2i\eta_\ell}. \qquad (2.33)$$

An important property of the **S** matrix is its unitarity, *i.e.* $\mathbf{S}_\ell\mathbf{S}_\ell^\dagger = \mathbf{S}_\ell^\dagger\mathbf{S}_\ell = 1$. This property is a reflection of the conservation of particle flux.

We will define the **T** matrix by[†]

$$\mathbf{S} = 1 - \mathbf{T}. \qquad (2.34)$$

[†]The **T** matrix is sometimes defined by the alternative relationship $\mathbf{S} = 1 - i\mathbf{T}$ or $\mathbf{S} = 1 - i2\pi\mathbf{T}$.

Substituting this into eqn (2.32) we see that

$$\begin{aligned}\chi(r) \underset{r\to\infty}{\widetilde{\quad}} A \sin\left(kr - \frac{\ell\pi}{2} + \eta_\ell\right) \\ = A\,\frac{1}{2i}\mathrm{e}^{-i\eta_\ell}\left\{(1 - T_\ell)\mathrm{e}^{i\left(kr-\frac{\ell\pi}{2}\right)} - \mathrm{e}^{-i\left(kr-\frac{\ell\pi}{2}\right)}\right\} \\ = A\,\mathrm{e}^{-i\eta_\ell}\left\{\sin\left(kr - \frac{\ell\pi}{2}\right) - \frac{1}{2i}\mathbf{T}_\ell\,\mathrm{e}^{i\left(kr-\frac{\ell\pi}{2}\right)}\right\}.\end{aligned} \tag{2.35}$$

The first term in the curly bracket is just the asymptotic form of the spherical Bessel function. This would be the entire wavefunction if there were no potential. The second term is pre-multiplied by the **T** matrix and represents the extra outgoing spherical wave resulting from the scattering caused by the presence of the potential.

In the present elastic scattering case discussed in this chapter the **S** and **T** are just complex numbers. We will see later that for inelastic and reactive scattering they become full multi-dimensional matrices.

There is another matrix which needs to be included here for completeness. This is the **K** matrix. This matrix relates $\cos\left(kr-\frac{\ell\pi}{2}\right)$ to $\sin\left(kr-\frac{\ell\pi}{2}\right)$ in the asymptotic form of the wavefunction (see eqn (2.35) and (2.32)). (Note Mott and Massey[11] refer to this matrix by the letter **K** and call it the reactance matrix.) It is related to the **S** matrix by the equation (see ref. 13)

$$\mathbf{S} = (\mathbf{I} + i\mathbf{K})\,(\mathbf{I} - i\mathbf{K})^{-1}. \tag{2.36}$$

2.6 Quasi-Bound States and the Lifetime Matrix

On occasions we will wish to discuss scattering resonances. They show up as sharp peaks in cross sections and generally correspond to quasi-bound states. These are states of the collision system which became trapped for a relatively long period of time compared with the normal duration of a collision. For atom–atom collisions such states come about through the coincidence of the scattering energy with a state of the two particle system that forms a nearly bound quantum level within an inner well created by the centrifugal potential (see the $\ell = 15$ curve in Figure 2.3). In such a case the resonance is termed a "shape" resonance as it results entirely from the shape of a single potential. In more complex situations, such as arise in atom–molecule scattering, the resonances may have their origin in an exchange of energy from relative translational motion to the internal excitation of the molecule resulting in the system temporarily not possessing sufficient energy of relative motion to break up. Such resonances are termed "Feschbach" resonances. Yet another type of resonance which we will encounter in Chapter 6 is a reactive scattering resonance in which the motion of the whole system is temporarily trapped in a cooperative motion in such a way that no one atom can escape from the scattering complex.

In all cases the key feature of a resonance is the fact that the scattering partners stay close to each other for a considerably longer period than would be anticipated. The problem has been elegantly analyzed by Smith[14] who discusses the time the collision partners spend within the collision region as compared with the time they would spend there in the absence of any interaction between them. This leads him to define a "lifetime matrix"

$$\mathbf{Q} = -i\hbar\mathbf{S}^* \frac{\mathrm{d}\mathbf{S}}{\mathrm{d}E}. \tag{2.37}$$

The eigenvalues of this matrix give the lifetimes of the states involved. As $\mathbf{Q}$ is a Hermitian matrix all its eigenvalues will be real. Some may be negative as they measure the lifetime compared to that expected without any interaction. If we denote the eigenvalues of $\mathbf{Q}$ by q_i, then a resonance is present when $2Eq_i/\hbar \gg 1$.

For the case of elastic atom–atom scattering when, there is just a single "channel" and there are no inelastic or reactive processes, the **S** matrix reduces to a single complex number as in eqn (2.33). Substituting this into eqn (2.37) we obtain‡

$$\mathbf{Q} = \tau = 2\hbar\frac{\mathrm{d}\eta}{\mathrm{d}E} \tag{2.38}$$

where τ is the lifetime of the resonance at energy E. Some examples of shape resonances are given in the section below.

2.7 Illustrative Examples: Ne + Ne and Ar^+ + Ar Scattering

A Lennard-Jones potential energy function[15,16] is used as a model for the Ne + Ne interaction. This is the same model as was used in Chapter 1. We have written two different computer codes to evaluate the phase shifts. One of these is based directly on the numerical solution of eqn (2.10) followed by an analysis of the resulting wavefunction at large internuclear separations.[17] This code uses the Numerov[18] method to perform the integration. This approach does not yield the absolute value of the phase shift, but rather results in phase shifts between $-\pi/2$ and $\pi/2$. Careful examination of the formulae shows that this uncertainty in the absolute values of the phase shifts is not reflected in any ambiguity in the value of the cross section. The second program we have written enables us to evaluate the absolute value of the phase shifts. This code is based on the Variable Phase Approach[19–21] and is described in Appendix B.

The calculated phase shifts as a function of collision energy for a model Ne + Ne collision system are shown in Figure 2.2 while Figure 2.3 shows the corresponding effective potentials (see eqn (2.10)).

‡Eqn (2.38) has an extra factor of 2 as compared with Smith's paper. This arises from the fact that Smith uses a nonstandard definition of phase shift.

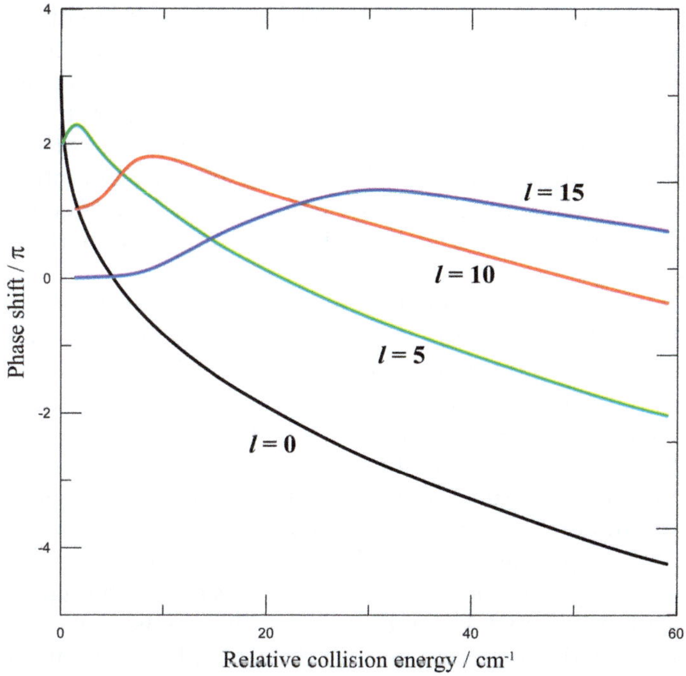

Figure 2.2 Phase shift as a function of the relative collision energy of two Ne atoms. The phase shift, divided by π, is shown for four different values of the orbital angular momentum ℓ. The well depth of the potential is $\varepsilon = 29.52\ \mathrm{cm}^{-1}$.

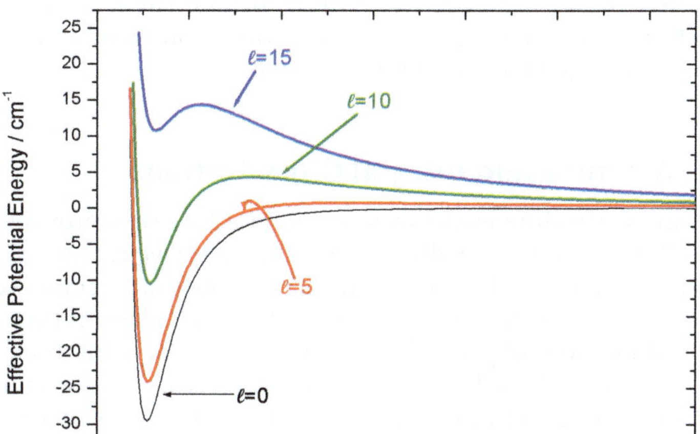

Figure 2.3 Effective potential energy as a function of the inter-nuclear separation of two Ne atoms. The effective potential energy is shown for four different values of the orbital angular momentum ℓ. The well depth of the potential is $\varepsilon = 29.52\ \mathrm{cm}^{-1}$.

The first thing to note is that the low energy values of the phase shifts on the left hand side of Figure 2.2 are all different. They are respectively equal to 3π, 2π, π and 0. According to Levinson's theorem[11,22] this indicates that the corresponding effective potentials (see Figure 2.3) support three, two, one and zero bound states[§]. Figure 2.3 illustrates the fact that as the orbital angular momentum, ℓ, increases the depth of the effective potential decreases correspondingly. Thus for sufficiently large values of ℓ the effective potential will not support any bound states and the low energy limit of the phase shift will be zero.

At large collision energies the the repulsive part of the potential dominates the behavior of the phase shift and it decreases steadily with increasing energy[¶].[19,20] At low collision energies and for $\ell > 0$, as the energy increases from zero, the wavefunction progressively samples more of the attractive part of the potential and this results in an increase of the phase shift.

Figure 2.4 shows the computed phase shift as a function of the orbital angular momentum quantum number for three different energies. Although the orbital angular momentum quantum number takes on only discrete integer values, for the purposes of visualization we have connected up the values on the graph using a smooth curve. At low values of the angular momentum the phase shift is negative. This indicates that the repulsive part of the potential is dominating the scattering. As the orbital angular momentum increases the wavefunction is prevented from penetrating to small values of r and samples the repulsive part of the potential to a progressively smaller extent. So at larger values of the orbital angular momentum the scattering is determined to a smaller extent by the repulsive part of the potential and to a larger degree by the attractive part of the potential. This leads to more positive values of the phase shift. At very large values of the orbital angular momentum the centrifugal potential prevents the wavefunction from sampling any part of the potential and the phase shift tends to zero.

2.7.1 Differential and Integral Cross Sections

Figure 2.5 shows the differential cross section for Ne + Ne scattering where we have treated the Ne atoms as distinguishable and have calculated the cross sections by the direct application of eqn (2.19). Note that because we have not taken account of the exchange symmetry in these calculations they should be regarded as illustrative of general theory and not suitable for direct comparison with experimental results. The figure displays cross sections for the same three collision energies as were shown in the figures of the classical deflection functions and cross sections in Chapter 1 (see Figures (1.7) and (1.9)).

---

§The proof of Levinson's theorem is complicated but in Appendix B a simple argument is set out to support its reasonableness.

¶For a Lennard-Jones 12-6 potential whose inner repulsive wall varies as A/r^{12} the phase shift varies as $\eta \sim -C\,k^{5/6}$.

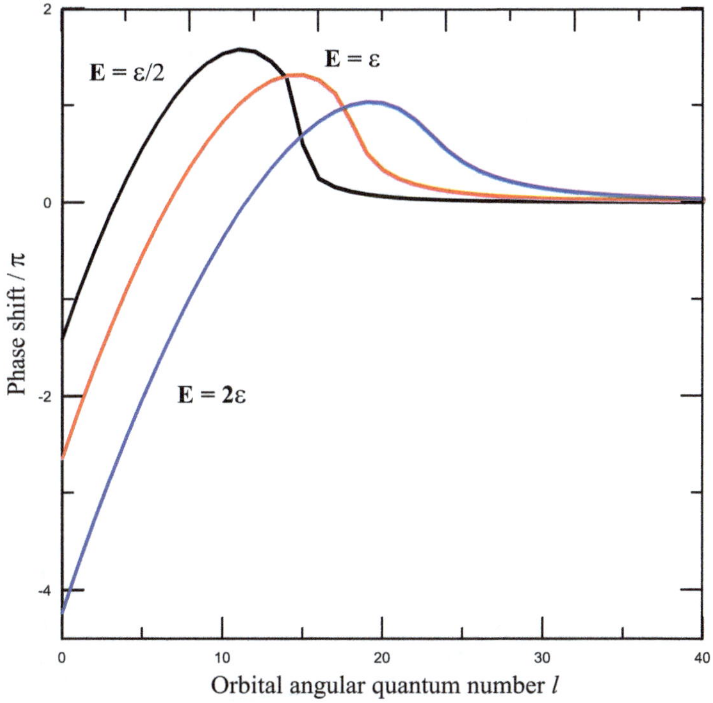

Figure 2.4 The phase shifts divided by π as a function of the orbital angular momentum quantum number ℓ. The calculated phase shift is shown for three separate energies.

Figure 1.7 shows that orbiting collisions are present for the two lower energies examined ($E = \varepsilon/2$ and $E = \varepsilon$), but not for the highest energy ($E = 2\varepsilon$). Both the classical and quantum cross sections display backward glories (at $\theta = 180°$) for the two lower energies but only a forward glory (at $\theta = 0°$) for the highest energy. The cross sections are plotted on a logarithmic scale to permit the display of the large glory maxima. As discussed in Chapter 1, the differential cross section becomes infinite for the classical glories, but we see from Figure 2.5 that the quantum cross sections yield large but finite values at the glory maxima.

The most noticeable differences between the classical and quantum cross sections in Figure 2.5 are the oscillations present in the quantum cross sections. In an attempt to clarify the origin of these oscillations we have performed a calculation using only the repulsive part of the Lennard-Jones potential. These calculations are shown in Figure 2.6. We see that while the differential cross section computed using only the repulsive part of the potential does show a small amount of oscillatory structure, this is far less than that for the calculation using the full potential. The backward glory scattering is also totally absent from the cross section computed with only

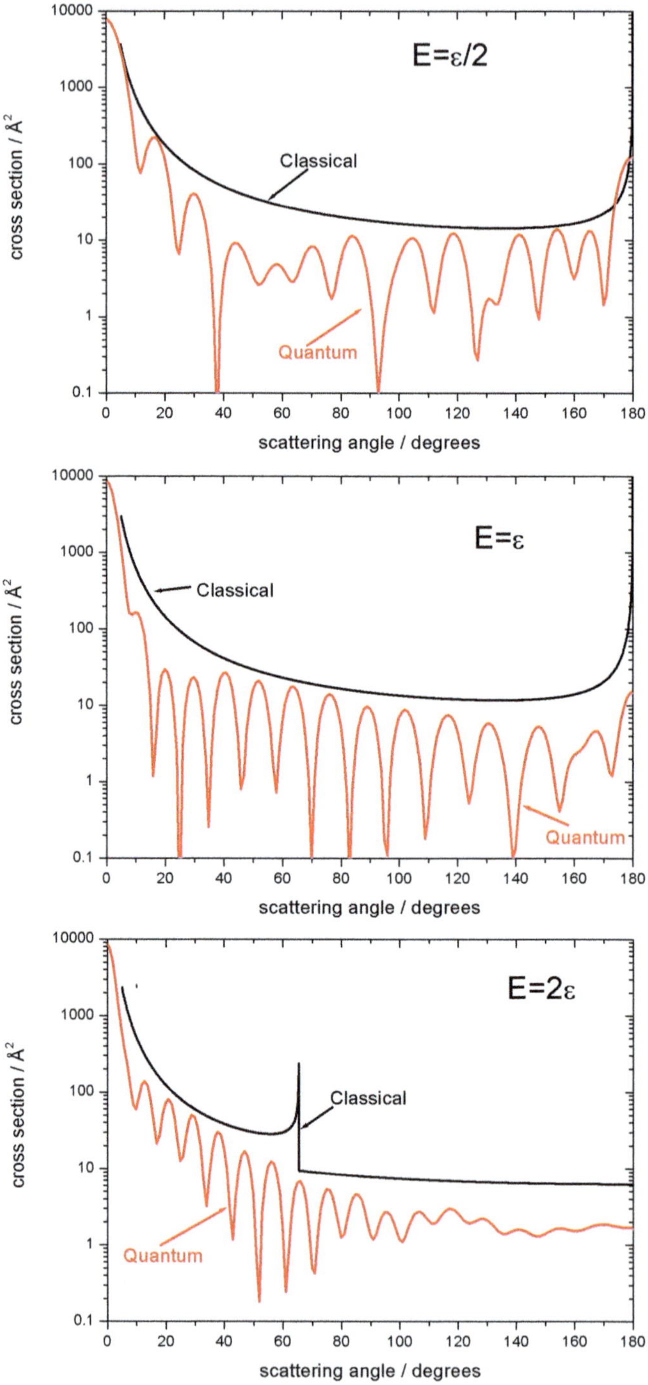

Figure 2.5 Differential cross sections for model of Ne + Ne scattering ignoring symmetry and indistinguishability of Ne atoms. Also shown are the classical cross sections.

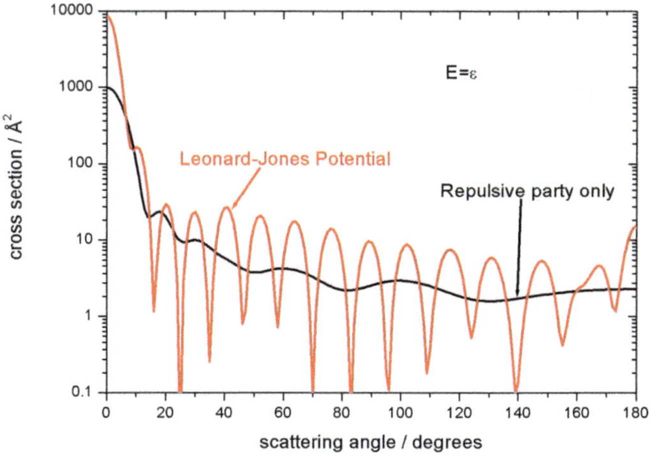

Figure 2.6 Differential cross sections for model of Ne + Ne scattering ignoring symmetry and the indistinguishability of Ne atoms. The figure shows cross sections computed with the full Lennard-Jones potential and also using just its repulsive part. See the Figure 1.5 caption for the Lennard-Jones parameters.

the repulsive part of the potential. The small amount of oscillatory structure in the cross section calculated using only the repulsive part of the potential may be thought of as arising from diffraction effects at the edges of a hard sphere potential of radius approximately equal to $R = \sqrt{\sigma/\pi}$ (see Figure 1.5 and eqn (1.24)[11,12,23]). From the classical deflection functions (Figure 1.7) we see that for the lower two energies shown there are multiple classical trajectories that can lead to scattering at all angles. For the highest energy ($E = 2\varepsilon$) there is a rainbow angle at about 62° and we see from Figure 2.5 that for this energy the oscillations for angles beyond 62° are damped out. This confirms that the origin of the quantum oscillations comes from interference effects between classical trajectories leading to scattering into the same angle.

The classical cross section for $E = 2\varepsilon$ shows a rainbow maximum, but this is not really noticeable in the quantum cross sections. The reason for this arises from the weak interactions in the Ne + Ne case and consequently in the very low collision energies we are examining. This leads to very marked quantum oscillations which mask the typical signatures of rainbow scattering in this case. Figure 2.7 shows differential cross sections computed using a Lennard-Jones potential with parameters derived for $Ar^+ + Ar$ scattering.[24] Here we see the main rainbow maximum at around 83°. It is accompanied by many so-called supernumary rainbows at smaller angles. There are many rapid quantum oscillations underlying the rainbow maxima. Note that the potential used (see caption to Figure 2.7) has a far deeper well and the masses of the atoms are far greater.

The integral cross section for Ne + Ne, calculated without taking account of the indistinguishability of the two Ne atoms, is shown in Figure 2.8. The

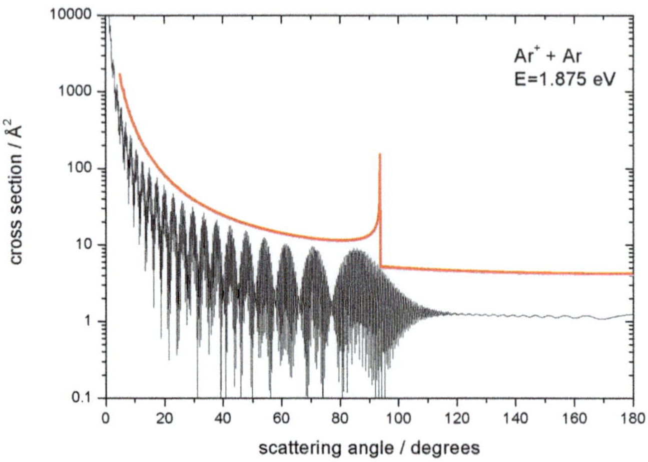

Figure 2.7 Calculated differential cross section using potential parameters derived for $Ar^+ + Ar$ scattering.[24] The cross sections shows rainbow and supernumary rainbow maxima. A Lennard-Jones potential with parameters $\varepsilon = 1.25$ eV and $\sigma = 2.17$ Å was used in the calculations.

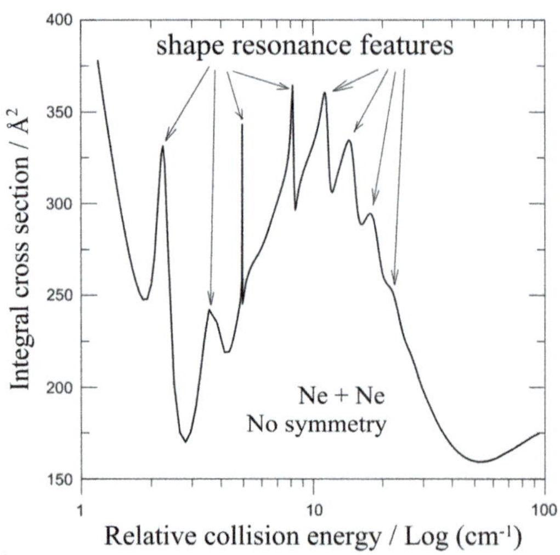

Figure 2.8 Calculated integral cross section for a model of Ne + Ne scattering ignoring the indistinguishability of the two Ne atoms. See caption to Figure 1.5 for the Lennard-Jones parameters.

cross section displays considerable structure. In all there are eight sharp features superimposed on a broader peak centered at around 17 cm^{-1}. Seven of the features give rise to sharp peaks and one to a small shoulder. All eight marked features arise from "shape resonances", which are associated with quasi-bound vibrational states trapped within the effective potential (see Figure 2.3). The peaks are located at 2.25, 3.54, 4.96, 8.16, 11.21, 14.36, 17.71 and 21.35 cm^{-1} and arise from shape resonances associated with orbital angular momenta $\ell = 8$, 9, 10, (11 & 13), (12 & 14), 15, (16 & 14) and 17 respectively. Note that the shape resonance associated with $\ell = 14$ contributes to two different sharp peaks. We will discuss the $\ell = 12$ shape resonance and its contribution to the integral cross section in detail below.

From Figure 2.8 the "shape resonances" are seen to be located on a broad peak in what appears to be part of an oscillatory pattern. There is a minimum at around 2.8 cm^{-1}, followed by the broad peak at 17 cm^{-1} and subsequently by another minimum at 47 cm^{-1}. To understand the nature of these oscillations in the integral cross section we need to look at Figure 2.9 which illustrates the dependence of the phase shifts on ℓ at three different scattering energies. It is seen that for a collision energy of 17 cm^{-1}, when the cross section is at the maximum of an oscillatory peak, the phase shift reaches its maximum value of around $\frac{3\pi}{2}$ for $\ell = 12$. The classical deflection function written as a function of ℓ $\left(\left(\ell + \frac{1}{2}\right)\hbar \approx b \times k\hbar\right)$ passes through zero at the same value of ℓ. We commented in Chapter 1 (see Figures 1.7 and 1.8 and the ensuing discussion) that the classical cross section displayed a glory maximum when the deflection function passed through zero. From a quantum mechanical perspective when the phase shift, plotted as a function of ℓ, passes through a maximum and that maximum value is equal to $\frac{(2n+1)\pi}{2}$ so that $\sin^2(\eta_\ell)$ is also maximal, then several ℓ values will make

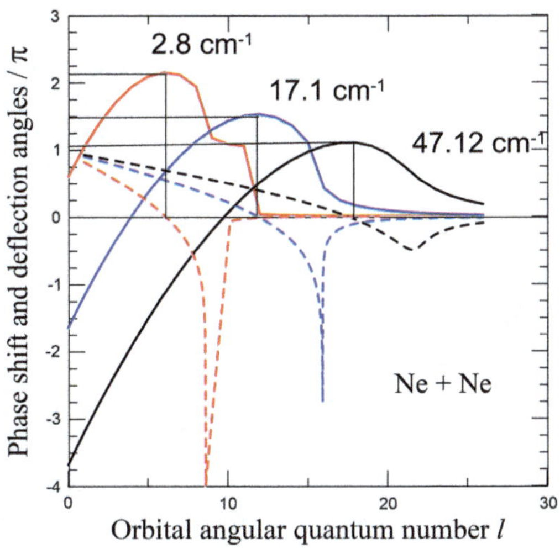

Figure 2.9 The phase shifts (solid lines) and the classic deflection functions (broken lines) versus angular orbital number calculated for a model of Ne + Ne scattering.

the largest possible contribution to the integral cross section at that energy, leading to a glory maximum (see eqn (2.31)). In contrast, at energies of 2.8 and 47.12 cm^{-1} the cross sections have glory minima because the maximum phase shift in the phase shift $\sim\ell$ plots occurs at a value close to $\eta_\ell \approx \frac{2n\pi}{2}$. Thus a large number of ℓ values with a minimal, near zero, value of $\sin^2(\eta_\ell)$ contribute to the integral cross section and a glory minimum results. By the same logic we expect that another two broad maxima in integral cross section should exist, one for an energy higher than 47 cm^{-1} and another for an energy lower than 2.8 cm^{-1}, when the phase shift maxima will be located around $\frac{\pi}{2}$ and $\frac{5\pi}{2}$. We recall that the maximum possible value of the phase shift for Ne + Ne scattering is equal to 3π (see Figure 2.2) and occurs for $k = 0$ and $\ell = 0$. This means that we can expect exactly three glory maxima for this system.[25]

Figure 2.10 shows the integral cross section for Ar^+ + Ar collisions. In this case we see a much larger number of glory oscillations. The number of these oscillations has been shown by Bernstein[26] to be equal to the number of bound states supported by the potential. This is just three for the Ne + Ne case, but much larger for Ar^+ + Ar scattering. "Shape resonance" features are superimposed on some of the glory oscillations. One of these is shown in the inset in Figure 2.10.

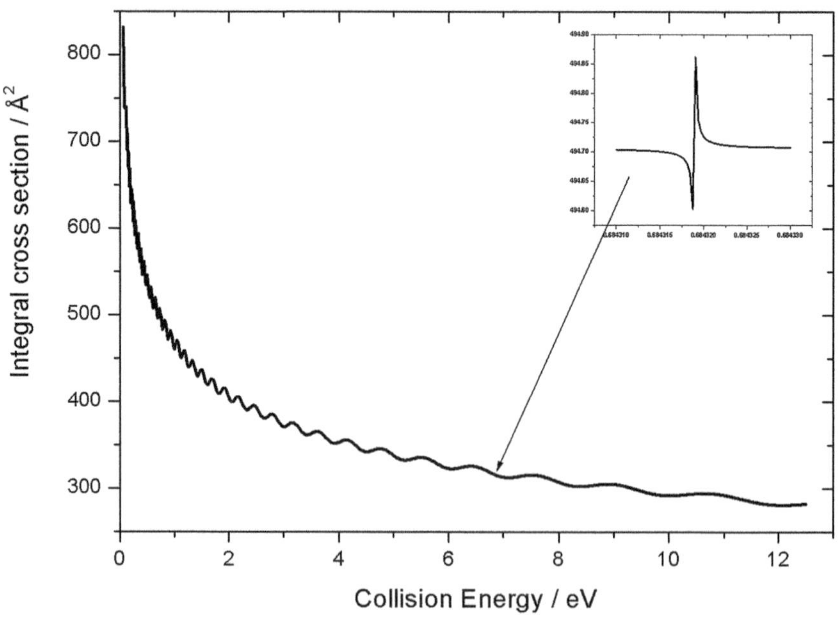

Figure 2.10 Calculated integral cross section for model of Ar^+ + Ar scattering. See caption to Figure 2.7 for the Lennard-Jones parameters.

2.7.2 Shape Resonances

The very sharp resonance at 4.96 cm^{-1} in Figure 2.8 is found to arise from a quasi-bound state for the effective potential with $\ell = 12$. Figure 2.11 shows the wavefunction for this resonance superimposed on the effective potential. The plot of the wavefunction has been shifted upwards so that its zero value lies at the energy of the resonance. We see clearly that the amplitude of the wavefunction is very large in the region of the potential well but that it tunnels through the centrifugal barrier and "leaks" out. As the wavefunction has no nodes within the well region, it is clearly the first and only quasi-bound state supported by the effective potential. The phase shift for $\ell = 12$ is shown by a red line in Figure 2.12 as a function of energy. We see clearly that as the energy increases through the resonance at $E = 4.96$ cm^{-1} the phase shift increases sharply by π. The energy interval over which the phase shift changes sharply is designated by Γ in the figure. This is the "width" of the resonance. The blue line in Figure 2.12 shows the contribution to the integral cross section (see Figure 2.8 and eqn (2.31)) of the partial cross section for $\ell = 12$. We see from the figure that the resonance contribution is sharply peaked just below the energy of the quasi-bound energy level. The width of the resonance, Γ, coincides with the energy region over which the resonance contributes significantly to the cross section.

The **S** matrix is defined as $\mathbf{S} = \exp 2i\eta_\ell$ (see eqn (2.33)), consequently if we draw an Argand diagram, *i.e.* Re**S** versus Im**S**, of the **S** matrix then as the

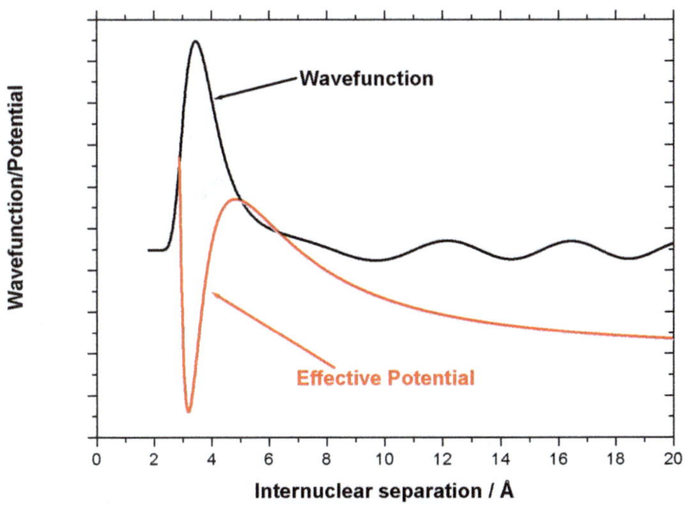

Figure 2.11 Wavefunction for Ne + Ne shape resonance at 4.96 cm^{-1}. The shape resonance occurs for $\ell = 12$. The wavefunction is superimposed on a plot of the effective potential. See caption to Figure 1.5 for the Lennard-Jones potential parameters.

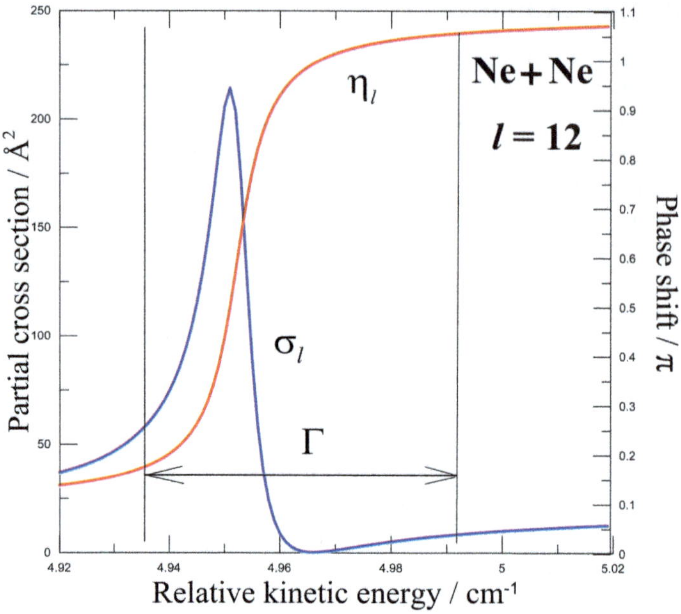

Figure 2.12 The phase shift for $\ell=12$ divided by π, for Ne + Ne scattering as a function of collision energy in the vicinity of the shape resonance at 4.96 cm^{-1} is shown by a red line. The contribution of orbital angular momentum quantum number $\ell = 12$ to the integral cross section is shown as a blue line.

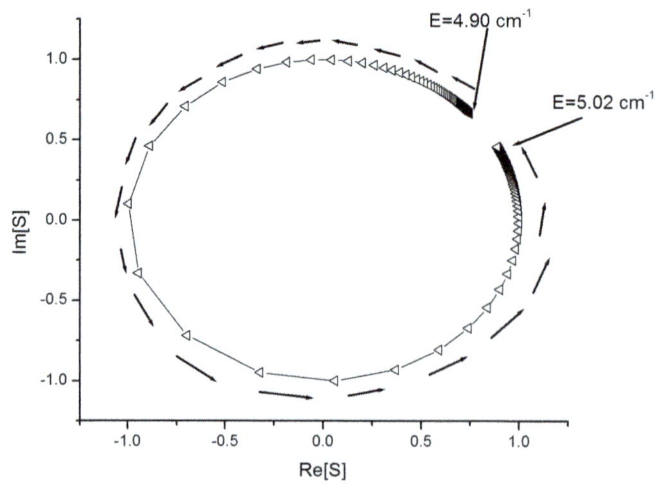

Figure 2.13 An Argand diagram of the **S** matrix for relative collision energies around the shape resonance at 4.96 cm^{-1}.

energy increases through the resonance the Argand diagram will map out a circle going counter-clockwise with energy. This is illustrated in Figure 2.13. This behavior may be generalized to more complex situations including collisions involving more than two atoms.[27,28]

2.7.3 Exchange Symmetry

Section 2.4 discussed the formulae applicable in the case of the collision of two indistinguishable atoms, such as Ne + Ne. In the figures displaced above we have treated the two Ne atoms as distinguishable. This provided a good illustration of the general formalism and of many aspects of atom–atom scattering. In Figure 2.14 we see a comparison of the differential cross sections computed when taking full account of the exchange symmetry and when not taking it into account. The cross sections seem quite similar, but at larger collision energies when a far greater number of partial waves (ℓ values) contribute, they become very noticeably different.[16]

Figure 2.15 compares the integral cross sections computed with and without taking account of the exchange symmetry. In this case, when proper account is taken of the exchange symmetry, there are fewer sharp shape resonance peaks because only even ℓ values contribute and consequently the shape resonances arising from odd ℓ values are absent. The remaining peaks are doubled in amplitude in accordance with eqn (2.31).

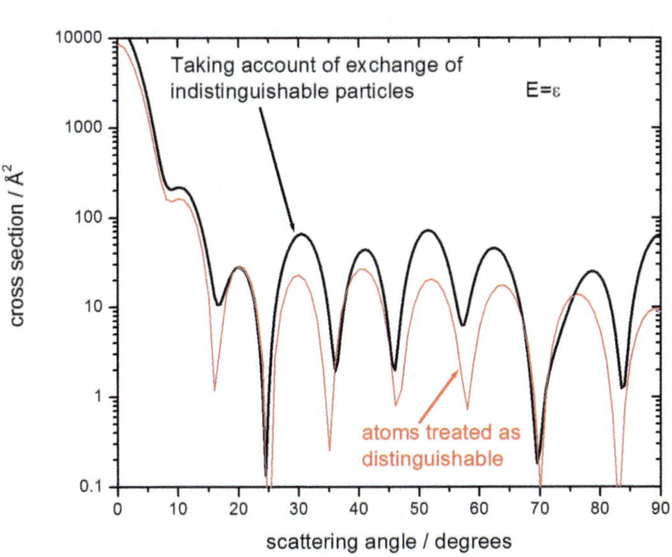

Figure 2.14 Comparison of differential cross sections for Ne + Ne scattering computed both taking account of exchange symmetry of indistinguishable atoms and ignoring this symmetry.

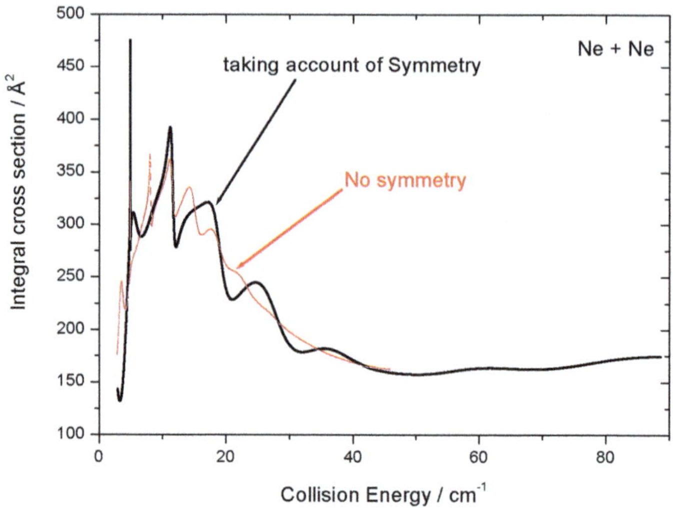

Figure 2.15 Comparison of integral cross sections for Ne + Ne scattering computed both taking account of exchange symmetry of indistinguishable atoms and ignoring this symmetry.

2.8 Extraction of Interaction Potentials from Crossed Beam Experiments

In the sections above we have illustrated the integral and differential cross sections which may arise in the case of atom–atom scattering. The details of the cross sections are very sensitive to the potential energy function governing the interaction of the collision partners and consequently measurement and subsequent modeling of the cross sections provides a reliable method of determining the interaction potential. Several reviews have been written on this subject and the reader is referred to them.[16,29–34]

2.9 Low Energy Scattering and the Scattering Length

At very low energies the centrifugal potential creates an impenetrable barrier for all orbital angular momentum quantum numbers $\ell > 0$ (see eqn (2.10) and Figure 2.3). Because of this only orbital angular momentum $\ell = 0$ contributes to the scattering. Let us consider a hard sphere of radius a. The wavefunction has an asymptotic form of $\sin(kr + \eta_0)$ (see eqn (2.11)). To find the phase shift η_0 we need to compare the wavefunction with that for free motion ($\sin(kr)$). The presence of the hard sphere moves the effective origin outwards by a distance a leading to a phase shift of $\eta_0 = -ka$. So the equation for the integral cross section becomes (see eqn (2.23))

$$\begin{aligned}\sigma &= \frac{4\pi}{k^2}\sin^2\eta_0 \\ &= \frac{4\pi}{k^2}\sin^2(ka).\end{aligned} \tag{2.39}$$

In the limit $k \to 0$ we may therefore write

$$\lim_{k\to 0}\{\sigma(k)\} = 4\pi a^2. \tag{2.40}$$

Note that the quantum mechanical cross section is four times as large as the classical cross section.

Here we can identify a as the scattering length or the range of the potential. More generally, the scattering length, a, is defined as (see ref. 35)

$$-\frac{1}{a} = \lim_{k\to 0}\{k\cot(\eta_0(k))\}. \tag{2.41}$$

The concept of scattering length is widely used especially in low temperature physics of Bose–Einstein condensates.

References

1. L. I. Schiff, *Quantum Mechanics*, McGraw-Hill Book Co. Inc., New York, 1955.
2. L. Pauling and E. B. Wilson, *Introduction to Quantum Mechanics*, McGraw-Hill Book Co., 1935.
3. H. Eyring, J. Walter and G. E. Kimball, *Quantum Chemistry*, John Wiley and Sons Inc., 1st edn, 1944.
4. R. D. Levine, *Quantum Mechanics of Molecular Rate Processes*, Dover Publications, Oxford, 2nd edn, 1999.
5. L. D. Landau and E. M. Lifshitz, *Quantum Mechanics: Non-Relativistic Theory*, Butterworth-Heinemann, Oxford, 3rd edn, 1977 (Reprinted with corrections 1991).
6. A. Messiah, *Quantum Mechanics*, North Holland Publishing Co., Amsterdam, 1965.
7. A. R. Edmonds, *Angular Momentum in Quantum Mechanics*, Princeton University Press, Princeton, New Jersey, 1960.
8. R. N. Zare, *Angular Momentum*, John Wiley and Sons, New York, 1988. See also list of corrections at web address: http://web.stanford.edu/group/Zarelab/Errata.pdf.
9. M. Abromowitz and I. A. Stegun, *Handbook on Mathematical Functions, Applied Mathematics Series 55*, National Bureau of Standards, 10th edn, 1972.
10. P. M. Morse and H. Feschbach, *Methods of Theoretical Physics*, McGraw-Hill Book Co. Inc., New York, 1st edn, 1953. See Section 9.3.
11. N. F. Mott and H. S. W. Massey, *The Theory of Atomic Collisions*, Clarendon Press, Oxford, 2nd edn, 1949.

12. H. S. W. Massey and C. B. O. Mohr, Free paths and transport phenomena in gases and the quantum theory of collisions. I. The rigid sphere model, *Proc. R. Soc. A*, 1933, **141**, 434.
13. R. G. Newton, *Scattering Theory of Waves and Particle*, David and Charles, 2nd edn, 1982.
14. F. T. Smith, Lifetime matrix in collision theory, *Phys. Rev.*, 1960, **118**, 349.
15. D. M. Hirst, *Potential Energy Surfaces, Molecular structure and dynamics*, Taylor and Francis, 1985.
16. F. Pirani, S. Brizi, L. F. Roncaratti, P. Casavecchia, D. Cappelletti and F. Vecchiocattivi, Beyond the Lennard-Jones model: a simple and accurate potential function probed by high resolution scattering data useful for molecular dynamics simulations, *Phys. Chem. Chem. Phys.*, 2008, **10**, 5489.
17. R. B. Bernstein, Quantum mechanical (phase shift) analysis of differential elastic scattering of molecular beams, *J. Chem. Phys.*, 1960, **33**, 795.
18. B. Numerov, Note on the numerical integration of $\mathrm{d}^2x/\mathrm{d}t^2 = f(x,t)$, *Astron. Nachr.*, 1927, **230**, 359.
19. F. Calogero, *Variable Phase Approach to Potential Scattering*, Academic Press, New York, 1967.
20. F. Calogero, Approximation for the phase shifts produced by repulsive potentials strongly singular in the origin, *Phys. Rev.*, 1964, **135**, B693.
21. V. Babikov, *Variable Phase Approach in Quantum Mechanics*, Nauka, Moscow, 1968 [in Russian].
22. N. Levinson, On the uniqueness of the potential in a Schrodinger equation for a given asymptotic phase, *K. Dan. Vid. Selsk. Mat-Fys. Medd.*, 1949, **25**, 1.
23. M. S. Child, *Molecular Collision Theory*, Academic Press, 1974. Corrected Dover Publication, 1996.
24. D. C. Lorents, R. E. Olson and G. M. Conklin, Rainbow scattering for Ar^+ + Ar and Xe^+ + Xe, *Chem. Phys. Lett.*, 1973, **20**, 589.
25. R. B. Bernstein, Semiclassical analysis of the extrema in the velocity dependence of total elastic-scattering cross sections: relation to the bound states, *J. Chem. Phys.*, 1963, **38**, 2599.
26. F. A. Morse and R. B. Bernstein, Velocity dependence of the differential cross sections for the scattering of atomic beams of K and Cs by Hg, *J. Chem. Phys.*, 1962, **37**, 2019.
27. R. D. Levine, M. Shapiro and B. R. Johnson, Transition probabilities in molecular collisions: computational studies of rotational excitation, *J. Chem. Phys.*, 1970, **52**, 1755.
28. B. C. Garrett, D. G. Truhlar, R. S. Grev, G. C. Schatz and R. B. Walker, Reaction probabilities, resonances, and thermal rate constants for the collinear reactions H + FH and D + FD on a low-barrier surface. Close-coupling and tunneling calculations, variational transition-state theory, and the unified statistical model, *J. Phys. Chem.*, 1981, **85**, 3806.
29. U. Buck, Elastic scattering, *Adv. Chem. Phys.*, 1975, **30**, 313.

30. U. Buck, Inversion of molecular scattering data, *Rev. Mod. Phys.*, 1974, **46**, 369.
31. J. P. Toennies, W. Welz and G. Wolf, The determination of the H–He potential well depth from low energy elestic scattering, *Chem. Phys. Lett.*, 1976, **44**, 5.
32. P. E. Siska, J. M. Parson, T. P. Schafer and Y. T. Lee, Intermolecular potentials from crossed beam differential elastic scattering measurements. III. He + He and Ne + Ne, *J. Chem. Phys.*, 1971, **55**, 5762.
33. M. Shapiro and R. B. Gerber, Extraction of interaction potentials from the elastic scattering amplitudes: an accurate quantum mechanical procedure, *Chem. Phys.*, 1976, **13**, 235.
34. M. S. Child and R. B. Gerber, Inversion of inelastic atom–atom scattering data: recovery of the interaction function, *Mol. Phys.*, 1979, **38**, 421.
35. T. y. Wu and T. Omura, *Quantum Theory of Scattering*, Prentice-Hall, 1962.

CHAPTER 3

Inelastic Scattering: Basic Theory

3.1 Atom–Diatom Collisions

In order to develop the underlying theory of inelastic and reactive scattering processes we will use the example of an atom–diatom collision. Figure 3.1 shows the four possible outcomes, or channels, that can arise from this process. When an atom (A) collides with a diatomic molecule (BC) one of the atoms of the diatomic may be exchanged with atom A leading to a reactive scattering process or if there is no such exchange then the scattering is non-reactive and the process is said to be either elastic or inelastic (shown in red in Figure 3.1). At higher energies, there may be sufficient energy for a three-body break-up process.

Chapters 3 and 4 are devoted to the theory of inelastic scattering, but the discussions contained within these chapters serve to introduce much theory and methodology also needed in the theory of reactive scattering. The subsequent chapters are mainly concerned with reactive scattering theory.

We will use Jacobi coordinates in the current chapter. These coordinates are natural ones to use for inelastic scattering processes as one of the coordinates is the separation of the two atoms of the diatomic, and consequently the energy eigenfunctions of the isolated diatomic form a natural basis set for the expansion of the full wavefunction. Another major advantage of Jacobi coordinates is that the Hamiltonian has no cross second derivative terms[1–3] when expressed in terms of these coordinates. Figure 3.2 shows a schematic of an atom colliding with a diatom in Jacobi coordinates. A detailed description of Jacobi coordinates and their use is given in Appendix C.

RSC Theoretical and Computational Chemistry Series No. 7
Theory of Molecular Collisions
By Gabriel G. Balint-Kurti and Alexander P. Palov

Published by the Royal Society of Chemistry, www.rsc.org

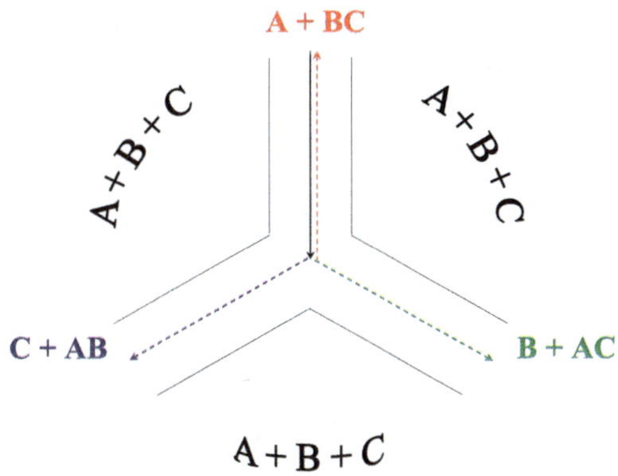

Figure 3.1 Diagram of an atom–diatom collision. The red line shows the inelastic scattering channel, the blue and green ones correspond to reactive channels. The regions corresponding to the three-body break-up process are labeled "A + B + C".

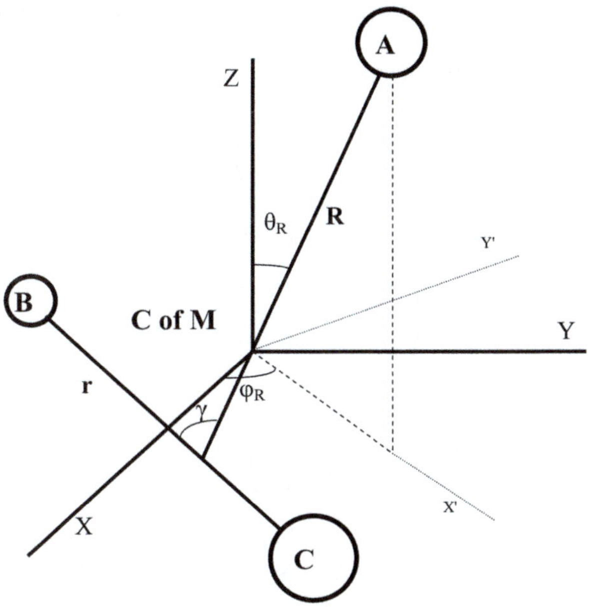

Figure 3.2 Schematic picture of an atom–diatom collision. The coordinates shown are called Jacobi coordinates. R is the distance from the incident atom (A) to the center-of-mass of the diatomic (B–C). The internuclear separation in the diatomic is denoted by r and the angle between the two coordinates is γ.

We will initially develop atom–diatom collision theory using a space-fixed reference frame. This corresponds to the manner in which the theory was historically developed. Most applications are now made in a body-fixed frame, where the direction of the coordinate axes is fixed relative to the atoms, rather than relative to the laboratory. Because the coordinate frame is itself rotating in the body-fixed reference frame, the algebra involved in this case is more complex and it will be described in detail only in Appendix D.

3.2 Space-Fixed Formulation

The Schrödinger equation for a atom–diatom system in the center-of-mass reference frame is (see eqn (2.5))

$$\begin{aligned}\hat{H}\psi(\boldsymbol{R},\boldsymbol{r},E) &= \left\{-\frac{\hbar^2}{2\mu_R}\nabla^2_{\boldsymbol{R}} - \frac{\hbar^2}{2\mu_r}\nabla^2_{\boldsymbol{r}} + V(R,r,\gamma)\right\}\psi(\boldsymbol{R},\boldsymbol{r},E)\\ &= E\,\psi(\boldsymbol{R},\boldsymbol{r},E). \end{aligned} \tag{3.1}$$

The angular parts of the two Laplacian operators are angular momentum operators (see eqn (2.7)). We will denote the orbital angular momentum of the atom relative to the diatomic by ℓ and the angular momentum of the rotation of the diatomic by $\boldsymbol{j}$. The two parts of the kinetic energy operator in eqn (3.1) may therefore be written as

$$\begin{aligned}-\frac{\hbar^2}{2\mu_R}\nabla^2_{\boldsymbol{R}} = &-\frac{\hbar^2}{2\mu_R}\left\{\frac{1}{R^2}\frac{\partial}{\partial R}\left(R^2\frac{\partial}{\partial R}\right)\right.\\ &\left.+\frac{1}{R^2}\left[\frac{1}{\sin\theta_R}\frac{\partial}{\partial\theta_R}\left(\sin\theta_R\frac{\partial}{\partial\theta_R}\right)+\frac{1}{\sin^2\theta_R}\frac{\partial^2}{\partial\phi_R}\right]\right\}\\ = &-\frac{\hbar^2}{2\mu_R}\left\{\frac{1}{R^2}\frac{\partial}{\partial R}\left(R^2\frac{\partial}{\partial R}\right)\right\}+\frac{\hat{\ell}^2}{2\mu_R R^2}\end{aligned} \tag{3.2}$$

and

$$\begin{aligned}-\frac{\hbar^2}{2\mu_r}\nabla^2_{\boldsymbol{r}} = &-\frac{\hbar^2}{2\mu_r}\left\{\frac{1}{r^2}\frac{\partial}{\partial r}\left(r^2\frac{\partial}{\partial r}\right)\right.\\ &\left.+\frac{1}{r^2}\left[\frac{1}{\sin\theta_r}\frac{\partial}{\partial\theta_r}\left(\sin\theta_r\frac{\partial}{\partial\theta_r}\right)+\frac{1}{\sin^2\theta_r}\frac{\partial^2}{\partial\phi_r}\right]\right\}\\ = &-\frac{\hbar^2}{2\mu_r}\left\{\frac{1}{r^2}\frac{\partial}{\partial r}\left(r^2\frac{\partial}{\partial r}\right)\right\}+\frac{\hat{\boldsymbol{j}}^2}{2\mu_r r^2}\end{aligned} \tag{3.3}$$

where μ_R and μ_r are respectively the reduced mass of the atom relative to the diatom and the reduced mass of the diatom, θ_R and ϕ_R are the polar and azimuthal angles of the vector $\boldsymbol{R}$ in a space-fixed or laboratory coordinate system and θ_r and ϕ_r are those of the diatomic internuclear axis.

Let us now consider the situation when the collision partners are far away from each other (*i.e.* when $R\to\infty$). In this situation the interaction potential $V(R\to\infty, r, \gamma)$ no longer depends on the distance between the atom and the diatomic, R, and becomes just the potential for the vibrational motion of the diatomic, $V_{\text{diatom}}(r)$. At these large separations we may therefore write the Schrödinger equation for the system as

$$\begin{aligned}
&\lim_{R\to\infty}\left\{-\frac{\hbar^2}{2\mu_R}\nabla_R^2-\frac{\hbar^2}{2\mu_r}\nabla_r^2+V(R,r,\gamma)\right\}\psi_o(\boldsymbol{R},\boldsymbol{r},E)\\
&=\left\{-\frac{\hbar^2}{2\mu_R}\nabla_R^2-\frac{\hbar^2}{2\mu_r}\nabla_r^2+V_{\text{diatom}}(r)\right\}\psi_o(\boldsymbol{R},\boldsymbol{r},E)\\
&=E\,\psi_o(\boldsymbol{R},\boldsymbol{r},E)
\end{aligned}\tag{3.4}$$

where the subscript "o" on the wavefunction ψ_o indicates the solution to the problem in the absence of the atom–diatom interaction potential.

In this large R limit the Hamiltonian is separable and the wavefunction can be written as a product of a wavefunction in the two vector coordinates $\boldsymbol{R}$ and $\boldsymbol{r}$. The wavefunction for the diatom is $\chi_{vj}(r)\,Y_{j,m_j}(\theta_r,\phi_r)$, where $\chi_{vj}(r)$ is the vibrational and $Y_{j,m_j}(\theta_r,\phi_r)$ the rotational wavefunction. $Y_{j,m_j}(\theta_r,\phi_r)$ is a spherical harmonic.[4,5] It is an eigenfunction of the angular momentum operator $\hat{\boldsymbol{j}}^2$ and obeys the equation

$$\hat{\boldsymbol{j}}^2 Y_{j,m_j}(\theta_r,\phi_r)=j(j+1)\hbar^2 Y_{j,m_j}(\theta_r,\phi_r)\tag{3.5}$$

while the vibrational wavefunction, $\chi_{vj}(r)$, obeys the equation

$$\left[-\frac{\hbar^2}{2\mu_r}\left\{\frac{1}{r^2}\frac{\partial}{\partial r}\left(r^2\frac{\partial}{\partial r}\right)\right\}+\frac{j(j+1)\hbar^2}{2\mu_r r^2}+V_{\text{diatom}}(r)\right]\chi_{vj}(r)=\varepsilon_{vj}\chi_{vj}(r).\tag{3.6}$$

The overall wavefunction, in the absence of the interaction potential, may therefore be written in the form

$$\psi_o(\boldsymbol{R},\boldsymbol{r},E)=\left(\frac{k_{vj}\mu}{2\pi h^2}\right)^{\frac{1}{2}}\mathrm{e}^{i\boldsymbol{k}\cdot\boldsymbol{R}}\chi_{vj}(r)\,Y_{j,m_j}(\theta_r,\phi_r),\tag{3.7}$$

i.e. a plane wave for the relative motion of the atom and diatomic multiplied by the rotational–vibrational wavefunction of the diatomic.

This wavefunction is comparable to eqn (2.1) of Chapter 2, but the big difference is that the molecule has an "internal" wavefunction and also possesses an angular momentum. The consequences of this is that we will not be able to perform the "partial wave expansion" (see Section 2.3) in the same way as in Chapter 2.

We now consider the form of the wavefunction at large separations of the scattering partners. If we assume that the collision partners approach each other along the space-fixed Z axis, then the plane wave $\mathrm{e}^{i\boldsymbol{k}\cdot\boldsymbol{R}}$ becomes e^{ikZ} and in analogy with eqn (2.2) we can write the form of the wavefunction at large separations as

$$\psi(\boldsymbol{R},\boldsymbol{r},E)\underset{R\to\infty}{\widetilde{\quad}}\left(\frac{k_{vj}\mu}{2\pi h^2}\right)^{\frac{1}{2}}\left[e^{ik_{vj}Z}\chi_{vj}(r)\,Y_{j,m_j}(\theta_r,\phi_r)\right.$$
$$\left.+\sum_{v'j'm'_j}f_{v'j'm'_j\leftarrow vjm_j}(\theta_R,\phi_R)\frac{e^{ik_{v'j'}R}}{R}\chi_{v'j'}(r)\,Y_{j',m'_j}(\theta_r,\phi_r)\right] \quad (3.8)$$

where E is the total energy, k_{vj} is the wavenumber associated with the vibrational rotational level vj whose energy is ε_{vj}, $k_{vj}=\sqrt{2\mu(E-\varepsilon_{vj})}/\hbar$, and $f_{v'j'm'_j\leftarrow vjm_j}(\theta_R,\phi_R)$ is the scattering amplitude.

The first term in eqn (3.8) corresponds to an incident plane wave representing the two collision partners in their initial quantum states moving in the positive Z direction. The second term describes the scattered wave moving radially outwards with the diatom now in all the possible final quantum states.[6–8]

The "inelastic differential cross section" is defined as the flux of collision partners in state $v',j',m_{j'}$ scattered into a solid angle element $\sin\theta_R\,d\theta_R\,d\phi_R$ about the direction $\boldsymbol{R}$ divided by the incident flux of scattering partners in initial state v,j,m_j (see Chapter 1). The incident flux of particles represented by the term $e^{ik_{vj}Z}$ is $k_{vj}\hbar/\mu$, while the flux of scattering partners in state $v',j',m_{j'}$ scattered into a unit solid angle element $\sin\theta_R\,d\theta_R\,d\phi_R$ arising from the second term in eqn (3.8) is $k_{v'j'}\hbar\left|f_{v'j'm_{j'}\leftarrow vjm_j}(\theta_R,\phi_R)\right|^2\sin\theta_R\,d\theta_R\,d\phi_R/\mu$ (see ref. 9).

The inelastic differential cross section is therefore given by

$$\sigma_{v'j'm_{j'}\leftarrow vjm_j}(\theta_R,\phi_R,E)=\frac{k_{v'j'}}{k_{vj}}\left|f_{v'j'm_{j'}\leftarrow vjm_j}(\theta_R,\phi_R)\right|^2. \quad (3.9)$$

3.3 The Total Angular Momentum

In the case of atom–atom scattering, treated in the previous chapter, there was just one angular momentum to take into account. In the present case we have two different angular momenta to consider, the orbital angular momentum, ℓ, and the rotational angular momentum of the diatomic molecule, j. They are not independent of each other as they are coupled by the interaction potential. If we wish to break the overall problem down into more manageable parts we must utilize the fact that the the total angular momentum J is a conserved quantity and is termed a "good" quantum number. Many values of J are needed to solve the scattering problem. This results from the large range of impact parameters which contribute and the correspondingly large number of orbital angular momentum quantum numbers, ℓ, that are involved. As J is a good quantum number, we can break the problem down into many smaller problems, one for each value of J.

If the potential were not present, the first term of eqn (3.8) would represent the solution to the scattering problem. The plane wave in this term can be expanded in the manner (see eqn (2.13))

$$e^{ikZ} = \sum_{\ell=0}^{\infty} \{4\pi(2\ell + 1)\}^{\frac{1}{2}} i^{\ell} j_{\ell}(kR) Y_{\ell,0}(\theta_R, \phi_R) \tag{3.10}$$

where we have used the relationship $Y_{\ell,0}(\theta,\phi) = \left(\frac{2\ell + 1}{4\pi}\right)^{\frac{1}{2}} P_{\ell}(\cos\theta)$ (see ref. 4 and 5).

Note that our choice of the direction of the relative momentum, $\boldsymbol{k}$, as the Z axis has led to an immediate simplification in that only $Y_{\ell,0}(\theta,\phi)$ appears in eqn (3.10) and there is no summation over m_ℓ, the Z components of the orbital angular momentum ℓ.

Combining eqn (3.7) and (3.10) we obtain for the plane wave wavefunction, which would be the solution of the problem in the absence of any potential

$$\begin{aligned}\psi_0(\boldsymbol{R},\boldsymbol{r},E) = \left(\frac{k_{vj}\mu}{2\pi h^2}\right)^{\frac{1}{2}} &\sum_{\ell=0}^{\infty} \{4\pi(2\ell + 1)\}^{\frac{1}{2}} i^{\ell} j_{\ell}(k_{vj}R)\, \chi_{vj}(r) \\ &\times Y_{\ell,0}(\theta_R, \phi_R)\, Y_{j,m_j}(\theta_r, \phi_r).\end{aligned} \tag{3.11}$$

The angular functions $Y_{\ell,0}(\theta_R,\phi_R)\, Y_{j,m_j}(\theta_r,\phi_r) \equiv Y_{\ell,0}(\hat{\boldsymbol{R}})\, Y_{j,m_j}(\hat{\boldsymbol{r}})$ therefore form a natural angular basis set in which to expand the scattering problem. In order to utilize the conservation of the total angular momentum, we now form linear combinations of these functions so as to obtain angular functions which correspond to definite values of the total angular momentum[10]

$$\mathcal{Y}_{j\ell}^{JM}(\hat{\boldsymbol{R}},\hat{\boldsymbol{r}}) = \sum_{m_\ell m_j} Y_{\ell m_\ell}(\hat{\boldsymbol{R}}) Y_{jm_j}(\hat{\boldsymbol{r}})(jm_j\ell m_\ell|j\ell JM) \tag{3.12}$$

where $(jm_j\ell m_\ell|j\ell JM)$ are the Clebsch–Gordan coefficients.[4,5,11] We will use the notation of Edmonds[4] for these coefficients in this book. The angular functions $\mathcal{Y}_{j\ell}^{JM}(\hat{\boldsymbol{R}},\hat{\boldsymbol{r}})$ form an orthonormal set.

The inverse transformation is

$$Y_{\ell m_\ell}(\hat{\boldsymbol{R}}) Y_{jm_j}(\hat{\boldsymbol{r}}) = \sum_{JM} \mathcal{Y}_{j\ell}^{JM}(\hat{\boldsymbol{R}},\hat{\boldsymbol{r}})(j\ell JM|jm_j\ell m_\ell). \tag{3.13}$$

3.4 The Coupled Equations

Substituting eqn (3.13) into eqn (3.11) and recalling eqn (3.7) we can write

$$\begin{aligned}\psi_0(\boldsymbol{R},\boldsymbol{r},E) &= \left(\frac{k_{vj}\mu}{2\pi h^2}\right)^{\frac{1}{2}} e^{i\boldsymbol{k}\cdot\boldsymbol{R}}\, \chi_{vj}(r)\, Y_{j,m_j}(\theta_r,\phi_r) \\ &= \left(\frac{k_{vj}\mu}{2\pi h^2}\right)^{\frac{1}{2}} \sum_{JM\ell} \{4\pi(2\ell + 1)\}^{\frac{1}{2}} i^{\ell} j_{\ell}(k_{vj}R)\, \chi_{vj}(r) \\ &\quad \times \mathcal{Y}_{j\ell}^{JM}(\hat{\boldsymbol{R}},\hat{\boldsymbol{r}})(j\ell JM|jm_j\ell 0).\end{aligned} \tag{3.14}$$

Note that we have set $m_\ell = 0$ in the Clebsch–Gordan coefficient in eqn (3.14). This is because only $Y_{\ell,m_\ell=0}(\theta_R,\phi_R)$ appears in eqn (3.11). The wavefunction $\psi_0(\boldsymbol{R},\boldsymbol{r})$ is the solution of the Schrödinger equation in the absence of a potential. We can use the angular and vibrational functions which appear in eqn (3.14) as a basis in which to expand the solution of the full Schrödinger equation, eqn (3.1). We now consider the matrix elements of the Hamiltonian operator of eqn (3.1) in this basis. The matrix elements of the kinetic energy operator $-\dfrac{\hbar^2}{2\mu_R}\nabla_R^2$ are

$$\begin{aligned}
&\langle\chi_{vj}(r)\mathcal{Y}_{j\ell}^{JM}(\hat{\boldsymbol{R}},\hat{\boldsymbol{r}})|\left[-\frac{\hbar^2}{2\mu_R}\nabla_R^2\right]|\chi_{v'j'}(r)\mathcal{Y}_{j'\ell'}^{JM}(\hat{\boldsymbol{R}},\hat{\boldsymbol{r}})\rangle\\
&=\langle\chi_{vj}(r)\mathcal{Y}_{j\ell}^{JM}(\hat{\boldsymbol{R}},\hat{\boldsymbol{r}})|\left[-\frac{\hbar^2}{2\mu_R}\left\{\frac{1}{R^2}\frac{\partial}{\partial R}\left(R^2\frac{\partial}{\partial R}\right)\right\}-\frac{\hat{\ell}^2}{2\mu_R R^2}\right]\\
&\quad\times|\chi_{v'j'}(r)\mathcal{Y}_{j'\ell'}^{JM}(\hat{\boldsymbol{R}},\hat{\boldsymbol{r}})\rangle\\
&=\left[-\frac{\hbar^2}{2\mu_R}\left\{\frac{1}{R^2}\frac{\partial}{\partial R}\left(R^2\frac{\partial}{\partial R}\right)\right\}+\frac{\ell(\ell+1)\hbar^2}{2\mu_R R^2}\right]\delta_{v,v'}\delta_{j,j'}\delta_{\ell,\ell'}.
\end{aligned}\tag{3.15}$$

The matrix elements of the diatomic Hamiltonian are likewise given by

$$\begin{aligned}
&\langle\chi_{vj}(r)\mathcal{Y}_{j\ell}^{JM}(\hat{\boldsymbol{R}},\hat{\boldsymbol{r}})|\left(-\frac{\hbar^2}{2\mu_r}\nabla_r^2+V_{\text{diatom}}(r)\right)|\chi_{v'j'}(r)\mathcal{Y}_{j'\ell'}^{JM}(\hat{\boldsymbol{R}},\hat{\boldsymbol{r}})\rangle\\
&=\varepsilon_{vj}\,\delta_{v,v'}\delta_{j,j'}\delta_{\ell,\ell'}.
\end{aligned}\tag{3.16}$$

The total wavefunction in eqn (3.1) is now expanded in a form similar to the expansion of eqn (3.14) but with $j_\ell(k_{vj}R)$ replaced by $\dfrac{\phi_{v'j'\ell'}^{Jvj\ell}(R,E)}{k_{vj}R}$

$$\begin{aligned}
\psi(\boldsymbol{R},\boldsymbol{r},E)&=\left(\frac{k_{vj}\mu}{2\pi\hbar^2}\right)^{\frac{1}{2}}\sum_{JM\ell}\{4\pi(2\ell+1)\}^{\frac{1}{2}}i^\ell(j\ell JM|jm_j\ell 0)\\
&\quad\times\sum_{v'j'\ell'}\frac{\phi_{v'j'\ell'}^{Jvj\ell}(R,E)}{k_{vj}R}\chi_{v'j'}(r)\mathcal{Y}_{j'\ell'}^{JM}(\hat{\boldsymbol{R}},\hat{\boldsymbol{r}}).
\end{aligned}\tag{3.17}$$

$\phi_{v'j'\ell'}^{Jvj\ell}(R,E)$ are solutions of the set of coupled second order radial Schrödinger equations

$$\begin{aligned}
&\left[-\frac{\partial^2}{\partial R^2}+\frac{\ell'(\ell'+1)}{R^2}-k_{v'j'}^2+U_{v'j'\ell';v'j'\ell'}^J(R)\right]\phi_{v'j'\ell'}^{Jvj\ell}(R,E)\\
&+\sum_{v'',j'',\ell''\neq v',j',\ell'}U_{v'j'\ell';v''j''\ell''}^J(R)\phi_{v''j''\ell''}^{Jvj\ell}(R,E)=0
\end{aligned}\tag{3.18}$$

where $k_{v'j'}=\sqrt{\dfrac{2\mu(E-\varepsilon_{v'j'})}{\hbar^2}}$ and

$$\begin{aligned}
U_{v'j'\ell';v''j''\ell''}^J(R)&=\frac{2\mu}{\hbar^2}\langle\chi_{v'j'}(r)\mathcal{Y}_{j'\ell'}^{JM}(\hat{\boldsymbol{R}},\hat{\boldsymbol{r}})|\{V(R,r,\gamma)-V_{\text{diatom}}(r)\}\\
&\quad\times|\chi_{v''j''}(r)\mathcal{Y}_{j''\ell''}^{JM}(\hat{\boldsymbol{R}},\hat{\boldsymbol{r}})\rangle,
\end{aligned}\tag{3.19}$$

where the angular brackets $\langle . | . | . \rangle$ imply integration over all four angles and over the diatomic radial coordinate r.

The matrix elements $U^J_{vj\ell;v'j'\ell'}(R)$ are independent of M.[10,12,13] The superscript $vj\ell$ on the wavefunction $\phi^{Jvj\ell}_{v'j'\ell'}(R,E)$ labels the initial channel quantum numbers. In the absence of a potential the radial wavefunctions $(\phi^{Jvj\ell}_{v'j'\ell'}(R,E))$ are just the Riccati–Bessel functions (eqn (2.12)), and the expansion of eqn (3.17) reduces to a plane wave (see eqn (3.14)).

For a fixed value of J, there is a second order differential equation of the form of eqn (3.18) for every set of quantum numbers $vj\ell$. There is a separate set of coupled equations for every value of J and each of these sets is self-contained and independent. The orbital angular momentum is related to the impact parameter, as discussed in Chapter 1, and there is therefore a maximum value of the orbital angular momentum quantum number ℓ which needs to be included in the calculations. The magnitudes of the vibrational and rotational angular momentum quantum numbers v and j may also be limited either by the total energy or by a combination of the values in which we are interested in a particular case and by the degree of coupling through the potential. This coupling makes itself apparent through convergence tests using progressively larger basis sets of v, j.

Eqn (3.18) are a set of coupled second order differential equations commonly referred to as the "close-coupling equations". Each equation requires two boundary conditions in order to determine a particular solution from all the possible solutions. At small separations we require the solution to be "regular", that is $\phi^{Jvj\ell}_{v'j'\ell'}(R,E)$ in eqn (3.18) must tend to zero as $R \to 0$, so that $\psi(\boldsymbol{R}, \boldsymbol{r}, E)$ (see eqn (3.17)) should remain finite throughout, particularly at the origin. At large separations the interaction potential is zero and the scattering solutions take on the same form as in Chapter 1.

3.5 The S and T Matrices and Differential Cross Sections

The boundary conditions which must be obeyed by the wavefunctions of eqn (3.18) are

$$
\begin{aligned}
&\phi^{Jvj\ell}_{v'j'\ell'}(R,E) \underset{R\to 0}{\widetilde{\longrightarrow}} 0 \\
&\phi^{Jvj\ell}_{v'j'\ell'}(R,E) \underset{R\to\infty}{\widetilde{\longrightarrow}} \hat{j}_{\ell'}(k_{v'j'}R)\delta_{v,v'}\delta_{j,j'}\delta_{\ell,\ell'} - T^J_{v'j'\ell',vj\ell} \\
&\qquad \frac{1}{2i}\left(\frac{k_{vj}}{k_{v'j'}}\right)^{\frac{1}{2}} e^{i\left(k_{v'j'}R - \frac{\ell'\pi}{2}\right)}.
\end{aligned} \tag{3.20}
$$

The conditions have been expressed in this form so as to make contact with the form of the overall wavefunction in eqn (3.8).

Alternatively using the asymptotic form of the Riccati–Bessel functions, see eqn (2.12), we can rewrite the second boundary condition as

$$\phi^{J\nu j\ell}_{\nu' j'\ell'}(R,E) \underset{R\to\infty}{\widehat{\sim}} -\frac{1}{2i}\left\{ \mathrm{e}^{-i\left(k_{\nu' j'}R-\frac{\ell'\pi}{2}\right)}\delta_{\nu,\nu'}\delta_{j,j'}\delta_{\ell,\ell'} - S^{J}_{\nu' j'\ell',\nu j\ell}\left(\frac{k_{\nu j}}{k_{\nu' j'}}\right)^{\frac{1}{2}} \mathrm{e}^{i\left(k_{\nu' j'}R-\frac{\ell'\pi}{2}\right)} \right\} \tag{3.21}$$

where the **S** and **T** matrices are related by

$$S^{J}_{\nu' j'\ell',\nu j\ell} = \delta_{\nu,\nu'}\delta_{j,j'}\delta_{\ell,\ell'} - T^{J}_{\nu' j'\ell',\nu j\ell}. \tag{3.22}$$

We now need to relate eqn (3.20) and (3.21) to the scattering amplitude of eqn (3.8) and (3.9). The asymptotic form of the full wavefunction is obtained by substituting eqn (3.20) into eqn (3.17)

$$\psi(\boldsymbol{R},\boldsymbol{r},E) \underset{R\to\infty}{\widehat{\sim}} \left(\frac{k_{\nu j}\mu}{2\pi h^2}\right)^{\frac{1}{2}} \sum_{JM\ell}\{4\pi(2\ell+1)\}^{\frac{1}{2}} i^{\ell}(j\ell JM|jm_j\ell 0)\sum_{\nu' j'\ell'} \times \left\{\frac{1}{k_{\nu j}R}\hat{j}_{\ell'}(k_{\nu' j'}R)\delta_{\nu,\nu'}\delta_{j,j'}\delta_{\ell,\ell'} - T^{J}_{\nu' j'\ell',\nu j\ell}\frac{1}{2i}\frac{1}{\sqrt{k_{\nu j}k_{\nu' j'}}} \times \frac{1}{R}e^{i\left(k_{\nu' j'}R-\frac{\ell'\pi}{2}\right)}\right\}\chi_{\nu' j'}(r)\mathcal{Y}^{JM}_{j'\ell'}(\hat{\boldsymbol{R}},\hat{\boldsymbol{r}}). \tag{3.23}$$

If we now multiply out the two terms in the curly brackets, we see from eqn (3.14) that the first term just yields a plane wave. We should also recall the relationship between the Riccati–Bessel functions ($\hat{j}_{\ell}(kR)$) and the spherical Bessel function, namely, $j_{\ell}(kR) = \dfrac{\hat{j}_{\ell}(kR)}{kR}$. We further use eqn (3.12) so as to cast eqn (3.23) into a form similar to eqn (3.8) and obtain

$$\psi(\boldsymbol{R},\boldsymbol{r},E) \underset{R\to\infty}{\widehat{\sim}} \left(\frac{k_{\nu j}\mu}{2\pi h^2}\right)^{\frac{1}{2}} \left\{ \mathrm{e}^{ik_{\nu j}Z}\chi_{\nu j}(r)\,Y_{j,m_j}(\theta_r,\phi_r) + i\pi^{\frac{1}{2}} \times \sum_{\nu' j' m_{j'}}\sum_{JM\ell}(2\ell+1)^{\frac{1}{2}}\sum_{\ell'} i^{\ell-\ell'}T^{J}_{\nu' j'\ell',\nu j\ell}\left(\frac{1}{k_{\nu j}\,k_{\nu' j'}}\right)^{\frac{1}{2}} \times \sum_{m_{\ell'}} Y_{\ell' m_{\ell'}}(\hat{\boldsymbol{R}})(j' m_{j'}\ell' m_{\ell'}|j'\ell' JM)(j\ell JM|jm_j\ell 0) \times \frac{\mathrm{e}^{ik_{\nu' j'}R}}{R}\,\chi_{\nu' j'}(r)Y_{j' m_{j'}}(\hat{\boldsymbol{r}}) \right\}. \tag{3.24}$$

Comparing this equation with eqn (3.8) we see that the scattering amplitude is related to the **T** matrix through the equation

$$f_{v'j'm_{j'}\leftarrow vjm_j}(\theta_R,\phi_R) = i\pi^{\frac{1}{2}}\sum_{JM\ell}(2\ell+1)^{\frac{1}{2}}\sum_{\ell'} i^{\ell-\ell'}T^J_{v'j'\ell',vj\ell}\left(\frac{1}{k_{vj}k_{v'j'}}\right)^{\frac{1}{2}} \times \sum_{m_{\ell'}} Y_{\ell' m_{\ell'}}(\hat{\boldsymbol{R}})(j'm_{j'}\ell'm_{\ell'}|j'\ell'JM)(j\ell JM|jm_j\ell 0). \tag{3.25}$$

Note that the summation over M can be omitted as the Clebsch–Gordan coefficient $(j\ell JM|jm_j\ell 0)$ is zero unless $M = m_j$.

The inelastic differential cross section is then obtained by substituting eqn (3.25) into eqn (3.9)

$$\sigma_{v'j'm_{j'}\leftarrow vjm_j}(\theta_R,\phi_R,E) = \frac{\pi}{k^2_{vj}}\left|\sum_{J\ell}(2\ell+1)^{\frac{1}{2}}\sum_{\ell'} i^{\ell-\ell'}T^J_{v'j'\ell',vj\ell} \times \sum_{m_{\ell'}} Y_{\ell' m_{\ell'}}(\hat{\boldsymbol{R}})(j'm_{j'}\ell'm_{\ell'}|j'\ell'Jm_j)(j\ell Jm_j|jm_j\ell 0)\right|^2. \tag{3.26}$$

Note that on the left hand side of the equation, in keeping with general conventions, we have used the notation θ_R, ϕ_R to define the solid angle into which the collision partners are scattered in the center-of-mass coordinate system. On the right hand side of the equation we use the notation $\hat{\boldsymbol{R}}$ to represent the same angles.

3.6 Conservation of Particle Flux, Symmetry and Unitarity of S Matrix

Combining eqn (3.21) and (3.17) provides the large R asymptotic form of the wavefunction

$$\psi(\boldsymbol{R},\boldsymbol{r},E)\,\widetilde{R\rightarrow\infty}\, -\left(\frac{k_{vj}\mu}{2\pi h^2}\right)^{\frac{1}{2}}\sum_{JM\ell}\{4\pi(2\ell+1)\}^{\frac{1}{2}} i^{\ell}\frac{1}{2i}(j\ell JM|jm_j\ell 0) \times \sum_{v'j'\ell'}\left\{\frac{e^{-i\left(k_{v'j'}R-\frac{\ell'\pi}{2}\right)}}{k_{vj}R}\delta_{v,v'}\delta_{j,j'}\delta_{\ell,\ell'} - S^J_{v'j'\ell',vj\ell}\frac{e^{i\left(k_{v'j'}R-\frac{\ell'\pi}{2}\right)}}{\left(k_{vj}k_{v'j'}\right)^{\frac{1}{2}}R}\right\}\chi_{v'j'}(r)\mathcal{Y}^{JM}_{j'\ell'}(\hat{\boldsymbol{R}},\hat{\boldsymbol{r}}). \tag{3.27}$$

The first term in the curly brackets, $\dfrac{e^{-i\left(k_{v'j'}R-\frac{\ell'\pi}{2}\right)}}{k_{vj}R}$, represents an incoming radial wave with orbital angular momentum ℓ and the colliding partners in their initial v, j quantum states. The total incoming flux passing through the surface of a sphere of radius R arising from this term is $\frac{4\pi\hbar}{k_{vj}\mu}$. The second term in the curly brackets represents the radially outgoing flux. This flux is spread over many different $v'j'\ell'$ states. The total outgoing radial flux arising from this second term is $\frac{4\pi\hbar}{k_{vj}\mu}\sum_{v'j'\ell'}\left|S^J_{v'j'\ell',vj\ell}\right|^2$. The incoming and outgoing particle flux must be equal so as the preserve the number of particles or collision partners. This leads to the conclusion

$$\sum_{v'j'\ell'} S^{J*}_{v'j'\ell',vj\ell} S^J_{v'j'\ell',vj\ell} = 1. \tag{3.28}$$

In eqn (3.27) the wavefunction Ψ corresponds to an initial state v, j, m_j of the colliding atom–diatom system. It should therefore be labeled Ψ_{v,j,m_j}. The set of wavefunctions with different initial states forms an orthonormal set. This orthonormality condition leads to a unitary condition on the **S** matrix

$$\sum_{v''j''\ell''} S^{J*}_{v''j''\ell'',vj\ell} S^J_{v''j''\ell'',v'j'\ell'} = \delta_{v,v'}\delta_{j,j'}\delta_{\ell,\ell'} \tag{3.29}$$

or

$$\mathbf{S}^{\dagger}\mathbf{S} = \mathbf{I}. \tag{3.30}$$

The **S** matrix may further be shown to be symmetric using time reversal arguments.[14,15]

3.7 The Integral Cross Sections

The integral cross section is obtained by multiplying out the right hand side of eqn (3.26) and integrating over all solid angles. Making use of the orthogonality of the spherical harmonics we obtain

$$\begin{aligned}
\sigma_{v'j'm_{j'}\leftarrow vjm_j}(E) &= \int \sigma_{v'j'm_{j'}\leftarrow vjm_j}(\theta_R, \phi_R, E)\sin\theta_R \mathrm{d}\theta_R \mathrm{d}\phi_R \\
&= \frac{\pi}{k^2_{vj}} \sum_{J\ell} \sum_{\ell'=|J-j'|}^{|J+j'|} \sum_{m_{\ell'}} \sum_{\bar{J}} \sum_{\bar{\ell}=|\bar{J}-j'|}^{|\bar{J}+j'|} i^{\ell-\bar{\ell}} T^J_{v'j'\ell',vj\ell} T^{\bar{J}*}_{v'j'\ell',vj\bar{\ell}} \\
&\quad (j'm_{j'}\ell'm_{\ell'}|j'\ell'Jm_j)(j'm_{j'}\ell'm_{\ell'}|j'\ell'\bar{J}m_j) \\
&\quad (2\ell+1)^{\frac{1}{2}}(j\ell Jm_j|jm_j\ell 0)(2\bar{\ell}+1)^{\frac{1}{2}}(j\bar{\ell}\bar{J}m_j|jm_j\bar{\ell}0).
\end{aligned} \tag{3.31}$$

We can now use the following symmetry property of the Clebsch–Gordan coefficients (see ref. 4 eqn (3.5.16))

$$(2\bar{\ell}+1)^{\frac{1}{2}}(j\bar{\ell}\bar{J}m_j|jm_j\bar{\ell}0) = (-1)^{\bar{J}-m_j}(2\bar{J}+1)^{\frac{1}{2}}(\bar{J}m_j j-m_j|\bar{J}j\bar{\ell}0) \tag{3.32}$$

to obtain

$$\sigma_{v'j'm_{j'}\leftarrow vjm_j}(E) = \frac{\pi}{k_{vj}^2}\sum_{J\bar{J}\ell}\sum_{\ell'=|J-j'|}^{|J+j'|}\sum_{\bar{\ell}=|\bar{J}-j'|}^{|\bar{J}+j'|} i^{\ell-\bar{\ell}}(-1)^{J+\bar{J}}(2J+1)^{\frac{1}{2}}(2\bar{J}+1)^{\frac{1}{2}}$$
$$\times T^{J}_{v'j'\ell',vj\ell}T^{\bar{J}*}_{v'j'\ell',vj\bar{\ell}}(j'm_{j'}\ell'(m_j-m_{j'})|j'\ell'Jm_j)$$
$$\times (j'm_{j'}\ell'(m_j-m_{j'})|j'\ell'\bar{J}m_j)$$
$$\times (Jm_jj-m_j|Jj\ell 0)(\bar{J}m_jj-m_j|\bar{J}j\bar{\ell}0). \qquad (3.33)$$

This detailed integral inelastic cross section includes information on the probability that a collision results in a change in the orientation of the molecule's rotational angular momentum as specified by the m_j quantum number. A cross section which is generally of greater interest is one which is averaged over the initial m_j and summed over the final $m_{j'}$. From eqn (3.31) we have

$$\bar{\sigma}_{v'j'\leftarrow vj}(E) = \frac{1}{(2j+1)}\sum_{m_jm_{j'}}\sigma_{v'j'm_{j'}\leftarrow vjm_j}(E)$$
$$= \frac{1}{(2j+1)}\sum_{m_jm_{j'}}\frac{\pi}{k_{vj}^2}\sum_{J\ell}\sum_{\ell'=|J-j'|}^{|J+j'|}\sum_{m_{\ell'}}\sum_{\bar{J}}$$
$$\times \sum_{\bar{\ell}=|\bar{J}-j'|}^{|\bar{J}+j'|} i^{\ell-\bar{\ell}}T^{J}_{v'j'\ell',vj\ell}T^{\bar{J}*}_{v'j'\ell',vj\bar{\ell}}$$
$$\times (j'm_{j'}\ell'm_{\ell'}|j'\ell'Jm_j)(j'm_{j'}\ell'm_{\ell'}|j'\ell'\bar{J}m_j)$$
$$\times (2\ell+1)^{\frac{1}{2}}(j\ell Jm_j|jm_j\ell 0)(2\bar{\ell}+1)^{\frac{1}{2}}(j\bar{\ell}\bar{J}m_j|jm_j\bar{\ell}0). \qquad (3.34)$$

We now use the orthogonality property of the Clebsch–Gordan coefficients (see eqn 8.1.1(8) of ref. 16 or eqn (3.5.4) of ref. 4)

$$\sum_{m_{\ell'}}\sum_{m_{j'}}(j'm_{j'}\ell'm_{\ell'}|j'\ell'Jm_j)(j'm_{j'}\ell'm_{\ell'}|j'\ell'\bar{J}m_j) = \delta_{J,\bar{J}}. \qquad (3.35)$$

Using this in eqn (3.34) we obtain

$$\bar{\sigma}_{v'j'\leftarrow vj}(E) = \frac{1}{(2j+1)}\sum_{m_j}\frac{\pi}{k_{vj}^2}\sum_{J\ell}\sum_{\ell'=|J-j'|}^{|J+j'|}\sum_{\bar{J}}$$
$$\times \sum_{\bar{\ell}=|\bar{J}-j'|}^{|\bar{J}+j'|} i^{\ell-\bar{\ell}}T^{J}_{v'j'\ell',vj\ell}T^{\bar{J}*}_{v'j'\ell',vj\bar{\ell}}\delta_{J,\bar{J}}$$
$$\times (2\ell+1)^{\frac{1}{2}}(j\ell Jm_j|jm_j\ell 0)(2\bar{\ell}+1)^{\frac{1}{2}}(j\bar{\ell}\bar{J}M|jm_j\bar{\ell}0)$$
$$= \frac{1}{(2j+1)}\sum_{m_j}\frac{\pi}{k_{vj}^2}\sum_{J\ell}\sum_{\ell'=|J-j'|}^{|J+j'|}\sum_{\bar{\ell}=|J-j'|}^{|J+j'|} i^{\ell-\bar{\ell}}T^{J}_{v'j'\ell',vj\ell}T^{J*}_{v'j'\ell',vj\bar{\ell}}$$
$$\times (2\ell+1)^{\frac{1}{2}}(j\ell Jm_j|jm_j\ell 0)(2\bar{\ell}+1)^{\frac{1}{2}}(j\bar{\ell}JM|jm_j\bar{\ell}0). \qquad (3.36)$$

Making use now of the symmetry property of the Clebsch–Gordan coefficients given in eqn (3.32) we can write

$$\bar{\sigma}_{v'j'\leftarrow vj}(E)$$
$$= \frac{1}{(2j+1)}\sum_{m_j}\frac{\pi}{k_{vj}^2}\sum_{J\ell}\sum_{\ell'=|J-j'|}^{|J+j'|}\sum_{\bar{\ell}=|J-j'|}^{|J+j'|} i^{\ell-\bar{\ell}}(2J+1)T^J_{v'j'\ell',vj\ell}T^{J*}_{v'j'\ell',vj\bar{\ell}}$$
$$\times (Jm_jj-m_j|Jj\ell 0)(Jm_jj-m_j|Jj\bar{\ell}0). \quad (3.37)$$

The Clebsch–Gordan orthogonality property of eqn (3.35) can be written in the form

$$\sum_{m_j}(Jm_jj-m_j|Jj\ell 0)(Jm_jj-m_j|Jj\bar{\ell}0) = \delta_{\ell,\bar{\ell}}. \quad (3.38)$$

Using this in eqn (3.37) we obtain

$$\bar{\sigma}_{v'j'\leftarrow vj}(E) = \frac{1}{(2j+1)}\frac{\pi}{k_{vj}^2}\sum_J\sum_{\ell'=|J-j'|}^{|J+j'|}\sum_{\ell=|J-j'|}^{|J+j'|}(2J+1)\left|T^J_{v'j'\ell',vj\ell}\right|^2. \quad (3.39)$$

This cross section now corresponds to one which has been averaged over the initial m_j and summed over the final $m_{j'}$ quantum numbers. The "bar" above the cross section indicates that it has been averaged and summed in this manner. It is sometimes useful to define an "averaged transition probability" known as a "opacity function" $P_J(j',j)$. In terms of the opacity function the cross section can be written as

$$\bar{\sigma}_{v'j'\leftarrow vj}(E) = \frac{\pi}{k_{vj}^2}\sum_J(2J+1)P_J(j',j) \quad (3.40)$$

where

$$P_J(j',j) = \frac{1}{(2j+1)}\sum_{\ell'=|J-j'|}^{|J+j'|}\sum_{\ell=|J-j'|}^{|J+j'|}\left|T^J_{v'j'\ell',vj\ell}\right|^2$$
$$= \frac{1}{(2j+1)}\sum_{\ell'=|J-j'|}^{|J+j'|}\sum_{\ell=|J-j'|}^{|J+j'|}\left|\delta_{v',v}\delta_{j',j}\delta_{\ell',\ell} - S^J_{v'j'\ell',vj\ell}\right|^2. \quad (3.41)$$

The analysis outlined above has been performed using a space-fixed coordinate system and corresponds to the approach initially used in the study of molecular collisions.[10,13] Below we outline an alternative approach, namely that using a body-fixed rather than space-fixed coordinate system.

This approach switches some of the coupling between the scattering channels from potential coupling to coupling arising from the kinetic energy operator.

3.8 Body-Fixed Formulation and Helicities

In the space-fixed formulation used in the preceding section the Z axis was chosen as the direction of the initial relative velocity. This resulted in the simplification that there was no orbital angular momentum component about the Z axis ($m_\ell = 0$). In contrast the final relative velocity is in the $\hat{\boldsymbol{R}}$ direction and the final Z component of the orbital angular momentum ranges over all possible values. Thus there exists a marked difference in the treatment of the initial and final states of the collision system. In a body-fixed coordinate system this difference vanishes as the Z axis is always pointed along the $\hat{\boldsymbol{R}}$ direction and there is a symmetry between the initial and final scattering wavefunctions. This approach leads to a marked simplification in the expressions for the cross sections and is very similar to the helicity scattering representation widely used in nuclear physics.[8,15,17]

The use of body-fixed coordinates in molecular scattering was first discussed by Hirschfelder and Wigner[18] and was later described in more detail by Curtis *et al.*[2,3] and by Pack.[19] Its use in simplifying the form of the differential cross section was demonstrated by Miller.[20,21] A very clear exposition of the theory is given in the work of Schatz and Kuppermann[22] and one of the present authors has also outlined the theory in connection with the theory of photodissociation processes[23,24] (see in particular Appendix B of ref. 23).

Figure 3.3 shows the atom–diatom system together with the angles needed to rotate the axes from the space-fixed to the body-fixed coordinate system. As discussed by Schatz and Kuppermann[22] there are two possible rotations of the coordinate system which can be used to accomplish the space-fixed to body-fixed transformation. We will choose the simpler of these two. Thus the Euler angles[4,5,16] required to rotate the space-fixed axes into the body-fixed ones are just $(\phi_R, \theta_R, 0)$.

To achieve this rotation the space-fixed axes are first rotated about the Z axis by an angle ϕ_R. This takes the original X axis into the X' axis and Y into Y', as shown in Figure 3.3. The axes are then rotated by θ_R about the Y' axis. This then rotates the original Z axis into a position parallel to the $\boldsymbol{R}$ vector. These new rotated axes define the body-fixed coordinate system.

There are different conventions as to how the change in coordinate system is treated. In this book we will use a "passive" rotation approach in which we rotate the coordinate axes and leave the object unchanged. This agrees with the treatment in ref. 16 but differs from the majority of papers cited in this book. This difference leads to a different definition of the important Wigner **D** matrix[4,5,16] and also in the way in which these matrices are used to accomplish the transformation of functions from one reference frame to the other.

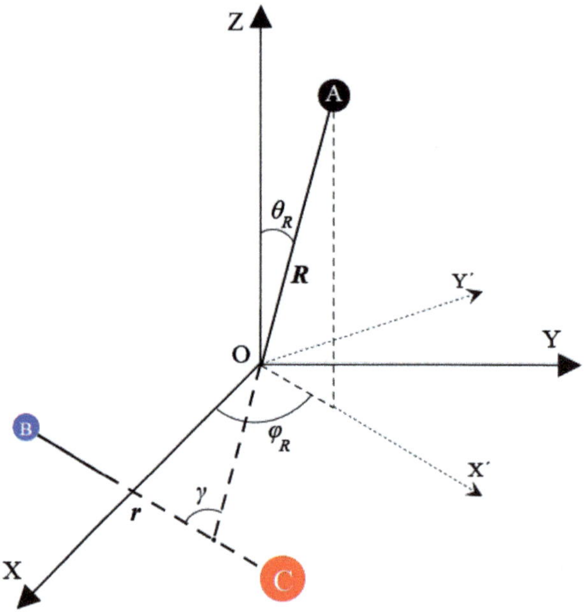

Figure 3.3 Main angles involved in the space-fixed to body-fixed transformation.

Eqn (3.42) below illustrates how the diatomic rotational wavefunction $Y_{j,m_j}(\theta_r, \phi_r)$ is transformed from the space-fixed reference frame to the new body-fixed frame (see ref. 16 eqn 4.1(2))

$$Y_{j,\Omega}(\gamma, \phi_r') = \sum_{m_j} Y_{j,m_j}(\theta_r, \phi_r) D^j_{m_j,\Omega}(\phi_R, \theta_R, 0) \tag{3.42}$$

where (γ, ϕ_r') are the polar and azimuthal angles of the diatomic bond vector $\hat{\boldsymbol{r}}$ in the body-fixed coordinate system.

The **D** matrix is defined as

$$D^J_{m_j,\Omega}(\phi_R, \theta_R, 0) = \mathrm{e}^{-im_j\phi_R} d^J_{m_j,\Omega}(\theta_R) \tag{3.43}$$

where $d^J_{m_j,\Omega}(\theta_R)$ is a small Wigner $\boldsymbol{D}$ matrix and Ω is the helicity quantum number, which corresponds to the quantum number for the body-fixed Z component of the angular momentum.

The detailed formulation of inelastic scattering theory in the body-fixed reference frame is given in Appendix D. The set of coupled differential equations which determine the radial wavefunction in the body-fixed reference frame can be written as (see eqn (D-21) of Appendix D)

$$\left[-\frac{\partial^2}{\partial R^2} - k_{\nu j}^2 + \frac{(J(J+1) + j(j+1) - \Omega^2)}{R^2} + U_{\nu j;\nu j}^{J\Omega}(R)\right] \Phi_{\nu j\Omega}^{J\nu j\Omega}(R,E)$$
$$+ C_{\Omega,\Omega+1}^{J+}\, \Phi_{\nu j\Omega+1}^{J\nu j\Omega}(R,E) + C_{\Omega,\Omega-1}^{J-}\, \Phi_{\nu j\Omega-1}^{J\nu j\Omega}(R,E)$$
$$+ \sum_{\nu',j' \neq \nu,j} U_{\nu j;\nu' j'}^{J\Omega}(R)\Phi_{\nu' j'\Omega}^{J\nu j\Omega}(R,E) = 0 \tag{3.44}$$

where

$$C_{\Omega,\Omega\pm 1}^{J\pm} = -\frac{[J(J+1) - \Omega(\Omega \pm 1)]^{\frac{1}{2}}\,[j(j+1) - \Omega(\Omega \pm 1)]^{\frac{1}{2}}}{R^2}, \tag{3.45}$$

$$U_{\nu j;\nu' j'}^{J\Omega}(R) = \frac{2\mu}{\hbar^2}\left(\frac{2J+1}{4\pi}\right)$$
$$\langle \chi_{\nu j}(r) Y_{j,\Omega}(\hat{\boldsymbol{r}}') D_{M,\Omega}^{J*}(\hat{\boldsymbol{R}})|\,\{V(R,r,\gamma) - V_{\text{diatom}}(r)\}\,|\chi_{\nu' j'}(r) Y_{j',\Omega}(\hat{\boldsymbol{r}}') D_{M,\Omega}^{J*}(\hat{\boldsymbol{R}})\rangle \tag{3.46}$$

and $\Phi_{\nu' j'\Omega'}^{J\nu j\Omega}(R,E)$ is the body-fixed radial wavefunction. We have used the notation $(\hat{\boldsymbol{r}}' = \gamma, \phi_r')$ and $(\hat{\boldsymbol{R}} = \phi_R, \theta_R, 0)$, and as previously the angular brackets $\langle .|\,.\,|.\rangle$ imply integration over all four angles and over the diatomic radial coordinate r.

The potential is diagonal in the helicity quantum number Ω and therefore does not couple different values of Ω, however the $C^{J\pm}$ terms in eqn (3.44), which arise from the orbital angular momentum operator, do couple Ω radial wavefunctions to those with $\Omega \pm 1$.

In Appendix D we show that the body-fixed **T** matrix is related to the space-fixed one through a unitary transformation (eqn (D-41))

$$T_{\nu' j'\Omega',\nu j(\Omega=m_j)}^{J} = \sum_{\ell''\,\ell'} B_{\ell',\Omega'}\, T_{\nu' j'\ell',\nu j\ell''}^{J} B_{\ell'',m_j} \tag{3.47}$$

where the initial m_j quantum number is also equal to the initial helicity quantum number Ω.

The differential cross section is shown to be (see eqn (D-43))

$$\sigma_{\nu' j'\Omega' \leftarrow \nu j m_j}(\theta_R, E) = \frac{1}{4\,k_{\nu j}^2}\left|\sum_J (2J+1) T_{\nu' j'\Omega',\nu j(\Omega=m_j)}^{J} d_{m_j,\Omega'}^{J}(\theta_R)\right|^2. \tag{3.48}$$

This expression for the state-to-state differential inelastic cross section in terms of the body-fixed **T** matrix can be compared with the eqn (3.26) for a very similar detailed differential cross section obtained in a space-fixed formulation. The cross sections differ only in that in eqn (3.48) $m_{j'}$ has been replaced by Ω', which is the quantum number for the Z component of the diatom rotational angular momentum about the scattering direction. We readily see that the present expression is far simpler. This form of the cross section was first made specific in the work of Miller.[20]

The integral cross section, obtained by integrating eqn (3.48) over all angles is (see Appendix D eqn (D-45))

$$\sigma_{v'j'\Omega'\leftarrow vjm_j}(E) = \frac{\pi}{k_{vj}^2}\sum_J (2J+1)\left|T^J_{v'j'\Omega',vjm_j}\right|^2, \tag{3.49}$$

and the integral cross section averaged over the initial m_j and summed over the final Ω' quantum numbers is

$$\bar{\sigma}_{v'j'\leftarrow vj}(E) = \frac{\pi}{k_{vj}^2}\frac{1}{2j+1}\sum_J \sum_{\Omega'=-|\min(J,j')|}^{\Omega'=|\min(J,j')|} \sum_{m_j=-|\min(J,j)|}^{m_j=|\min(J,j)|} (2J+1)\left|T^J_{v'j'\Omega',vjm_j}\right|^2. \tag{3.50}$$

The last two equations for the state-to-state integral cross sections should be compared to the space-fixed expressions of eqn (3.31) and (3.39). We see that the body-fixed expression for the detailed cross section (eqn (3.49)) is far simpler, but the m_j summed and averaged cross sections are of similar complexity.

References

1. B. T. Sutcliffe and J. Tennyson, A general treatment of vibration-rotation coordinates for triatomic molecules, *Int. J. Quantum Chem.*, 1991, **39**, 183.
2. C. F. Curtiss, J. O. Hirschfelder and F. T. Adler, The separation of the rotational coordinates from the n-particle Schroedinger equation, *J. Chem. Phys.*, 1950, **18**, 1638.
3. C. F. Curtiss and F. T. Adler, The scattering of atoms from diatomic molecules, *J. Chem. Phys.*, 1952, **20**, 249.
4. A. R. Edmonds, *Angular Momentum in Quantum Mechanics*, Princeton University Press, Princeton, New Jersey, 1960.
5. R. N. Zare, *Angular Momentum*. John Wiley and Sons, New York, 1988. See also list of corrections at web address: http://web.stanford.edu/group/Zarelab/Errata.pdf.
6. A. Messiah, *Quantum Mechanics*, North Holland Publishing Co., Amsterdam, 1965.
7. L. S. Rodberg and R. M. Thaler, *Introduction to the Quantum Theory of Scattering*, Academic Press, 1st edn, 1967.
8. M. L. Goldberger and K. M. Watson, *Collision Theory*, John Wiley and Sons, 1964.
9. G. G. Balint-Kurti, The theory of rotationally inelastic molecular collisions, ed. A. D. Buckingham and C. A. Coulson, *International Review of Science, Series II*, Butterworth, London, 1975, vol. 1, pp. 286.
10. A. M. Arthurs and A. Dalgarno, The theory of scattering by a rigid rotator, *Proc. R. Soc. London, Ser. A*, 1960, **256**, 540.
11. D. M. Brink and G. R. Satchler, *Angular Momentum*, Clarendon Press, Oxford, 1993.

12. I. C. Percival and M. J. Seaton, *Math. Proc. Cambridge Philos. Soc.*, 1957, **53**, 654.
13. W. A. Lester Jr and R. B. Bernstein, Computational procedure for the close-coupled rotational excitation problem: scattering of diatomic molecules by atoms, *J. Chem. Phys.*, 1968, **48**, 4896.
14. T. y. Wu and T. Omura, *Quantum Theory of Scattering*, Prentice-Hall, 1962.
15. R. G. Newton, *Scattering Theory of Waves and Particle*, David and Charles, 2nd edn, 1982.
16. D. A. Varshalovich, A. N. Moskalev and V. K. Khersonskii, *Quantum Theory of Angular Momentum*, World Scientific, Singapore, 1988.
17. M. Jacob and G. C. Wick, On the general theory of collisions for particles with spin, *Ann. Phys.*, 1959, **7**, 404.
18. J. O. Hirschfelder and E. Wigner, Separation of rotational coordinates from the Schrodinger equation for n particles, *Proc. Natl. Acad. Sci. U. S. A.*, 1935, **21**, 113.
19. R. T. Pack, Space-fixed *vs.* body-fixed axes in atom–diatomic molecule scattering. Sudden approximations, *J. Chem. Phys.*, 1974, **60**, 633.
20. W. H. Miller, Distorted-wave theory for collisions of an atom and a diatomic molecule, *J. Chem. Phys.*, 1968, **49**, 2373.
21. G. Wolken Jr, W. H. Miller and M. Karplus, Theoretical studies of $H + H_2$ rotationally inelastic scattering, *J. Chem. Phys.*, 1972, **56**, 4930.
22. G. C. Schatz and A. Kuppermann, Quantum mechanical reactive scattering for three-dimensional atom plus diatom systems. I. Theory, *J. Chem. Phys.*, 1976, **65**, 4642.
23. G. G. Balint-Kurti and M. Shapiro, Photofragmentation of triatomic molecules. Theory of angular and state distribution of product fragments, *Chem. Phys.*, 1981, **61**, 137.
24. G. G. Balint-Kurti, Wavepacket theory of photodissociation and reactive scattering, *Adv. Chem. Phys.*, 2003, **128**, 249.

CHAPTER 4

Inelastic Scattering: Exact and Approximate Solutions

4.1 Solving the Close-Coupling Equations

The key equations in scattering theory are sets of coupled second order differential equations of the form of eqn (3.16) or (3.42). These are the so-called "close-coupling equations". The variable over which integration must be performed is invariably a radial scattering coordinate. All other variables have vanished from the equation as the wavefunction has been expanded in a basis set involving these variables. There are many standard numerical methods for solving such equations, also over the years several powerful specialized methods have been developed explicitly for scattering problems. These methods have been compared and evaluated in a review article[1] (see also ref. 2). There are also available two extensive suites of computer programs aimed at solving such sets of equations mainly for inelastic scattering problems.[3,4] In this section we will discuss two of the most widely used methods for the numerical solution of the close-coupling equations.

4.1.1 Piecewise Analytic Solutions Method

This method, commonly known as Gordon's method, is described in ref. 5 (see also ref. 6). We outline here its salient points. The coupled differential equations of scattering theory, *e.g.* eqn (3.18) (see also eqn (3.44)) can be written in the form of a matrix equation

$$\left[\mathbf{I}\frac{\partial^2}{\partial R^2} + \mathbf{W}(R)\right]\boldsymbol{\Phi}(R) = 0 \tag{4.1}$$

RSC Theoretical and Computational Chemistry Series No. 7
Theory of Molecular Collisions
By Gabriel G. Balint-Kurti and Alexander P. Palov

Published by the Royal Society of Chemistry, www.rsc.org

where **I**, $\mathbf{W}(R)$ and $\mathbf{\Phi}(R)$ are square matrices. The indices of the matrices are "combined" indices, $i \equiv vj\ell$. **I** is the unit matrix. $\mathbf{W}(R)$ is the "potential" matrix. In the case of eqn (3.18) its elements are

$$W_{i,i}(R) \equiv W_{vj\ell,vj\ell}(R) = k_{vj}^2 - \frac{\ell(\ell+1)}{R^2} - U^J_{vj\ell;vj\ell}(R) \quad \text{diagonal elements,}$$
$$W_{i,j} \equiv W_{vj\ell,v'j'\ell'}(R) = -U^J_{vj\ell;v'j'\ell'}(R) \quad \text{off-diagonal elements.} \tag{4.2}$$

$\mathbf{\Phi}(R)$ is the solution matrix with elements (see eqn (3.18))

$$\Phi_{j,i}(R) \equiv \Phi_{v'j'\ell',vj\ell}(R) = \phi^{Jvj\ell}_{v'j'\ell'}(R,E). \tag{4.3}$$

Each of the columns of $\mathbf{\Phi}(R)$ is an independent solution of the coupled differential equations corresponding to a different initial state of the colliding molecules.

The basic idea of the method is to divide the the scattering coordinate up into segments. At the central point of each segment the potential matrix is diagonalized, this is done by finding its eigenvalues and eigenfunctions using standard numerical methods. Within each segment the scattering equations and the basis set are transformed to the basis of the eigenfunctions of the diagonalized potential corresponding to that segment. A *reference potential* is then defined for each segment in such a way that the uncoupled (*i.e.* diagonalized) equations have known analytic solutions within the segment. In our discussion here we will assume that we use a linear reference potential of the form $W_{i,i}(R) = A + B(R - R_c)$. For this form of potential the two linearly independent solutions of the uncoupled differential equations in each of the channels are standard well known functions, the Airy functions. This means that if the wavefunction is known at the left hand boundary of a segment, the known analytic solutions within the segment permit us to predict the wavefunction at the right hand boundary of the segment. The coupling between the different channels takes place through the the transformation of the basis, and therefore of the wavefunction, from segment to segment. This will be explained in a little more detail below.

Figure 4.1 illustrates the segmentation of the R coordinate for a one-dimensional case.

Let us consider a single segment. We denote the wavefunction or solution matrix at the left hand side of the segment as $\mathbf{\Phi}(R_L)$. We can write the i-th eigenfunction u_{ji} of the potential matrix at the center of the segment as

$$\sum_j W_{ij}(R_c) u_{ji} = \omega_i u_{ji} \tag{4.4}$$

where ω_i is the corresponding eigenvalue and R_c is the value of the radial coordinate at the center of the segment.

We make sure that the eigenvectors are normalized. As $\mathbf{W}(R)$ is a symmetric matrix, its eigenvectors are orthogonal. The matrix $\mathbf{U}^{(k)}$ formed by the

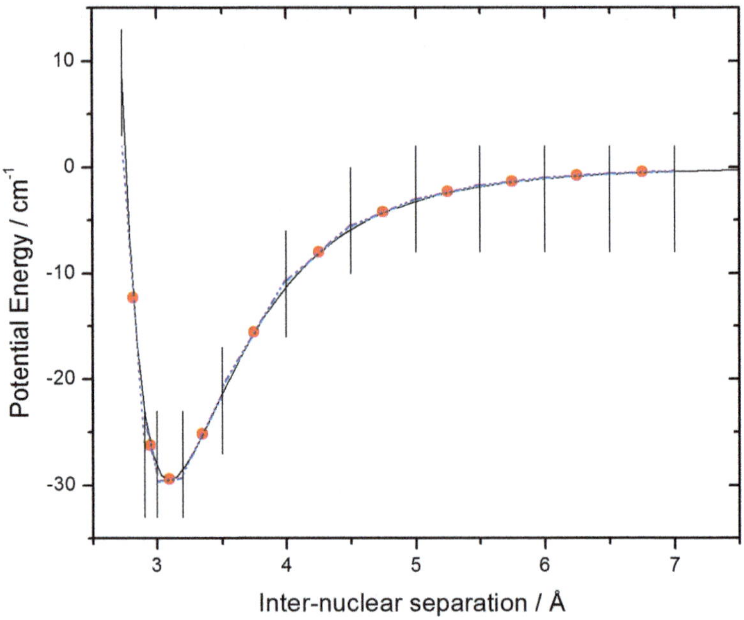

Figure 4.1 Diagram showing potential split into segments. Within each segment the potential is approximated by a straight line. The line touches the exact potential at the center point of the section and has a slope corresponding to that of the exact potential at this point.

column eigenvectors u_{ji} is therefore a unitary matrix. The superscript "(k)" indicates that we are concerned with segment k

$$\mathbf{U}^{(k)}\mathbf{U}^{(k)T} = \mathbf{U}^{(k)T}\mathbf{U}^{(k)} = \mathbf{I}. \tag{4.5}$$

The eigenvalue equation, eqn (4.4), may be written in matrix form as

$$\mathbf{W}(R_c)\mathbf{U}^{(k)} = \boldsymbol{\omega}\,\mathbf{U}^{(k)} \tag{4.6}$$

where $\boldsymbol{\omega}$ is a diagonal matrix with diagonal elements ω_i.

The potential matrix is then transformed to the basis in which it is diagonal at the center of the segment, *i.e.* the basis of eigenfunctions of $\mathbf{W}(R_c)$. But we will need the potential matrix throughout the segment, not just at the central point. Denoting the potential matrix in the new basis by $\tilde{\mathbf{W}}(R)$ and retaining only its diagonal linear terms we can write our reference potential as

$$\begin{aligned}\left[\tilde{\mathbf{W}}(R)\right]_{i,j} &= \left[\mathbf{U}^{(k)T}\mathbf{W}(R)\mathbf{U}^{(k)}\right]_{i,j}\delta_{i,j}\\ &= \left[\omega_j + d_j(R-R_c)\right]\delta_{i,j}\\ &= -\left[x_j\right](d_j)^2\,\delta_{i,j}\end{aligned} \tag{4.7}$$

where d_j is the derivative of $[\tilde{\mathbf{W}}(R)]_{j,j}$ at the center of the segment and we have defined a new variable $x_j = -[\omega_j + d_j(R-R_c)]/(d_j)^2$ which will replace R within the sector.

We now multiply eqn (4.1) on the left by $\mathbf{U}^{(k)T}$ and insert the unit matrix before $\mathbf{\Phi}(R)$ and also using eqn (4.5) we can write

$$\begin{aligned}\mathbf{U}^{(k)T}\left[\mathbf{I}\frac{\partial^2}{\partial R^2}+\mathbf{W}(R)\right]\mathbf{I\Phi}(R) &= \mathbf{U}^{(k)T}\left[\mathbf{I}\frac{\partial^2}{\partial R^2}+\mathbf{W}(R)\right]\mathbf{U}^{(k)}\mathbf{U}^{(k)T}\mathbf{\Phi}(R)\\ &= \left[\mathbf{I}\frac{\partial^2}{\partial R^2}+\left\{\mathbf{U}^{(k)T}\mathbf{W}(R)\mathbf{U}^{(k)}\right\}\right]\left\{\mathbf{U}^{(k)T}\mathbf{\Phi}(R)\right\}\\ &= \left[\mathbf{I}\frac{\partial^2}{\partial R^2}+\tilde{\mathbf{W}}(R)\right]\tilde{\mathbf{\Phi}}(R)\\ &= 0.\end{aligned} \tag{4.8}$$

$\tilde{\mathbf{W}}(R)=\mathbf{U}^{(k)T}\mathbf{W}(R)\mathbf{U}^{(k)}=\omega$ is the diagonal potential matrix in the new basis and $\tilde{\mathbf{\Phi}}(R)$ is the wavefunction matrix in the same basis. The relationship between the wavefunctions in the new and old bases are

$$\begin{aligned}\tilde{\mathbf{\Phi}}(R) &= \mathbf{U}^{(k)T}\mathbf{\Phi}(R) \quad \text{and}\\ \mathbf{\Phi}(R) &= \mathbf{U}^{(k)}\tilde{\mathbf{\Phi}}(R).\end{aligned} \tag{4.9}$$

In the new "diagonalized" basis the "coupled" differential equations are no longer coupled as the potential is now diagonal. We can write the equations for the i-th column of the solution matrix $\tilde{\mathbf{\Phi}}(R)$ as

$$\left[\frac{\partial^2}{\partial R^2}+[\omega_j+d_j(R-R_\mathrm{c})]\right]\tilde{\phi}_{j,i}(R)=0. \tag{4.10}$$

We now transform eqn (4.10) to the new channel variables $x_j=-[\omega_j+d_j(R-R_\mathrm{c})]/(d_j)^2$ so that $\mathrm{d}x_j=-\mathrm{d}R/d_j$. Note that these are different for each of the channels. The transformed differential equations are

$$\left[\frac{\partial^2}{\partial x_j^2}-x_j\right]\tilde{\phi}_{j,i}(x_j)=0. \tag{4.11}$$

We note that from the nature of the definition of the channel variables x_j we can anticipate that problems might occur whenever d_j, the derivative of the diagonal potential, is small or when the potentials are relatively flat. This indeed turns out to be one of the shortcomings of the method.[6]

Eqn (4.11) is the Airy function equation (see ref. 7) and has the two independent solutions $Ai(x_j)$ and $Bi(x_j)$. These are standard functions related to Bessel functions of fractional order.[7]

We assume that we know all the wavefunctions, $(\phi_{\ell,i}(R_\mathrm{L}))$, and their derivatives $\left(\phi'_{\ell,i}(R_\mathrm{L})=\left[\frac{\partial\phi_{\ell,i}(R)}{\partial R}\right]_{R=R_\mathrm{L}}\right)$ at the left boundary of the interval or segment. These we denote as $\mathbf{\Phi}(R_\mathrm{L})$ and $\mathbf{\Phi}'(R_\mathrm{L})$ respectively. The channel wavefunctions and their derivatives in the transformed basis are

$$\begin{aligned}\tilde{\phi}_{j,i}(R_\mathrm{L}) &= \sum_\ell[\mathbf{U}^{(k)T}]_{j,\ell}\phi_{\ell,i}(R_\mathrm{L})\\ &= \sum_\ell u_{\ell,j}\,\phi_{\ell,i}(R_\mathrm{L}),\end{aligned} \tag{4.12}$$

$$\begin{aligned}\tilde{\phi}'_{j,i}(R_\mathrm{L}) &= \sum_\ell [\mathbf{U}^{(k)T}]_{j,\ell}\phi'_{\ell,i}(R_\mathrm{L}) \\ &= \sum_\ell u_{\ell,j}\,\phi'_{\ell,i}(R_\mathrm{L}).\end{aligned} \tag{4.13}$$

In terms of the new x_j channel coordinates we can write

$$\begin{aligned}\tilde{\phi}_{j,i}(x_{j\mathrm{L}}) &= \tilde{\phi}_{j,i}(R_\mathrm{L}), \\ \tilde{\phi}'_{j,i}(x_{j\mathrm{L}}) &= \left[\frac{\partial\tilde{\phi}_{j,i}(x_j)}{\partial x_j}\right]_{x_j=x_{j\mathrm{L}}} \\ &= \sum_\ell u_{\ell,j}\left[\frac{\partial R}{\partial x_j}\frac{\partial\phi_{\ell,i}(R)}{\partial R}\right]_{R=R_\mathrm{L}} \\ &= -d_j\sum_\ell u_{\ell,j}\left[\frac{\partial\phi_{\ell,i}(R)}{\partial R}\right]_{R=R_\mathrm{L}} \\ &= -d_j\tilde{\phi}'_{j,i}(R_\mathrm{L})\end{aligned} \tag{4.14}$$

where $x_{j\mathrm{L}}$ is the new channel coordinate evaluated at the left hand segment boundary, $x_{j\mathrm{L}} = -[\omega_j + d_j(R_\mathrm{L} - R_\mathrm{c})]/(d_\mathrm{j})^2$.

We now need to express the wavefunctions at the left hand boundary of the segment as a linear combination of Airy functions

$$\begin{aligned}\tilde{\phi}_{j,i}(x_{j\mathrm{L}}) &= a\,Ai(x_{j\mathrm{L}}) + b\,Bi(x_{j\mathrm{L}}), \\ \tilde{\phi}'_{j,i}(x_{j\mathrm{L}}) &= a\,Ai'(x_{j\mathrm{L}}) + b\,Bi'(x_{j\mathrm{L}}).\end{aligned} \tag{4.15}$$

There are two equations and two unknowns, a and b. Solving these equations we obtain

$$\begin{aligned}a &= \frac{\tilde{\phi}_{j,i}(x_{j\mathrm{L}})Bi'(x_{j\mathrm{L}}) - \tilde{\phi}'_{j,i}(x_{j\mathrm{L}})Bi(x_{j\mathrm{L}})}{\mathcal{W}}, \\ b &= \frac{Ai(x_{j\mathrm{L}})\tilde{\phi}'_{j,i}(x_{j\mathrm{L}}) - Ai'(x_{j\mathrm{L}})\tilde{\phi}_{j,i}(x_{j\mathrm{L}})}{\mathcal{W}}\end{aligned} \tag{4.16}$$

where $\mathcal{W}$ is the Wronskian of Ai and Bi and is a constant independent of R, $\mathcal{W} = Ai(x_j)Bi'(x_j) - Ai'(x_j)Bi(x_j) = 1/\pi$ (see ref. 7). These equations determine $\tilde{\phi}_{j,i}(x_{j\mathrm{L}})$ in terms of the solutions of the reference potential. The coefficients a and b are completely determined by the wavefunction and its derivative at $R=R_\mathrm{L}$ and by the details of the reference potential in the segment under consideration. As $Ai(x_j)$ and $Bi(x_j)$ are the exact solutions of the reference potential in the segment we may directly write the values of the wavefunctions $\tilde{\phi}_{j,i}(x_j)$ at the right hand boundary of the segment, $R = R_\mathrm{R}$.

$$\begin{aligned}\tilde{\phi}_{j,i}(x_{jR}) &= a\,Ai(x_{jR}) + b\,Bi(x_{jR}), \\ \tilde{\phi}'_{j,i}(x_{jR}) &= a\,Ai'(x_{jR}) + b\,Bi'(x_{jR}).\end{aligned} \tag{4.17}$$

We can now find the channel wavefunctions in the diagonalized basis as a function of R at $R = R_{\mathrm{R}}$ by the inverse of eqn (4.14).

$$\begin{aligned}\tilde{\phi}_{j,i}(R_{\mathrm{R}}) &= \tilde{\phi}_{j,i}(x_{jR}),\\ \tilde{\phi}'_{j,i}(R_{\mathrm{R}}) &= -\frac{1}{d_j}\,\tilde{\phi}'_{j,i}(x_{jR}).\end{aligned} \tag{4.18}$$

Using the inverse of eqn (4.12) and (4.13) to determine the wavefunction and its derivative at the right hand boundary of the segment we obtain

$$\phi_{k,i}(R_{\mathrm{R}}) = \sum_j [\mathbf{U}^{(k)}]_{k,j}\tilde{\phi}_{j,i}(R_{\mathrm{R}}), \tag{4.19}$$

$$\phi'_{k,i}(R_{\mathrm{R}}) = \sum_j [\mathbf{U}^{(k)}]_{k,j}\tilde{\phi}'_{j,i}(R_{\mathrm{R}}). \tag{4.20}$$

We have outlined above how the full set of solutions of the coupled differential equations may be propagated from one side of a segment to the other. The solutions are started off at some small value of R in a "classically" forbidden region of the potential, where both the channel wavefunctions and their derivatives are set to zero. They are then propagated, segment by segment, to a large R value. The reader is referred to the original literature for a more detailed discussion.[1,2,6,8]

For the next section we will find it useful to combine some of the above steps and to determine a propagator such that the wavefunction at the right hand boundary of the segment is given directly by that at the left hand boundary together with the derivative at the left hand boundary. Thus combining eqn (4.14)–(4.18) we obtain

$$\begin{aligned}\tilde{\phi}_{j,i}(R_{\mathrm{R}}) &= \frac{1}{\mathcal{W}}\left[Bi'(x_{j\mathrm{L}})Ai(x_{jR}) - Ai'(x_{j\mathrm{L}})\,Bi(x_{jR})\right]\tilde{\phi}_{j,i}(R_{\mathrm{L}})\\ &\quad - \frac{d_j}{\mathcal{W}}[Ai(x_{j\mathrm{L}})Bi(x_{jR}) - Bi(x_{j\mathrm{L}})\,Ai(x_{jR})]\tilde{\phi}'_{j,i}(R_{\mathrm{L}}),\\ \tilde{\phi}'_{j,i}(R_{\mathrm{R}}) &= -\frac{1}{d_j\mathcal{W}}\left[Bi'(x_{j\mathrm{L}})Ai'(x_{jR}) - Ai'(x_{j\mathrm{L}})\,Bi'(x_{jR})\right]\tilde{\phi}_{j,i}(R_{\mathrm{L}})\\ &\quad + \frac{1}{\mathcal{W}}\left[Ai(x_{j\mathrm{L}})\,Bi'(x_{jR}) - Bi(x_{j\mathrm{L}})\,Ai'(x_{jR})\right]\tilde{\phi}'_{j,i}(R_{\mathrm{L}}).\end{aligned} \tag{4.21}$$

These equations relate the wavefunctions and their derivatives, expressed in the diagonalized basis, at the right hand boundary of the segment to those at the left hand boundary. Using eqn (4.9) we can transform this relationship into one involving the wavefunction and its derivatives expressed in the original untransformed basis

$$\begin{aligned}\mathbf{\Phi}(R^k_{\mathrm{R}}) &= \mathbf{U}^{(k)}\mathbf{\Gamma}^k_1(R^k_{\mathrm{R}}, R^k_{\mathrm{L}})\mathbf{U}^{(k)T}\mathbf{\Phi}(R^k_{\mathrm{L}})\\ &\quad + \mathbf{U}^{(k)}\mathbf{\Gamma}^k_2(R^k_{\mathrm{R}}, R^k_{\mathrm{L}})\mathbf{U}^{(k)T}\mathbf{\Phi}'(R^k_{\mathrm{L}}),\\ \mathbf{\Phi}'(R^k_{\mathrm{R}}) &= \mathbf{U}^{(k)}\mathbf{\Gamma}^k_3(R^k_{\mathrm{R}}, R^k_{\mathrm{L}})\mathbf{U}^{(k)T}\mathbf{\Phi}(R^k_{\mathrm{L}})\\ &\quad + \mathbf{U}^{(k)}\mathbf{\Gamma}^k_4(R^k_{\mathrm{R}}, R^k_{\mathrm{L}})\mathbf{U}^{(k)T}\mathbf{\Phi}'(R^k_{\mathrm{L}}).\end{aligned} \tag{4.22}$$

$\mathbf{U}^{(k)T}$ transforms the wavefunction to the diagonalized basis of segment k (see eqn (4.8) and (4.9), while $\mathbf{U}^{(k)}$ transforms it back to the standard basis. The matrices $\mathbf{\Gamma}_1^k(R_\mathrm{R}^k, R_\mathrm{L}^k)$ *etc.* are diagonal matrices and propagate the wavefunction in the diagonalized basis from the left hand boundary of segment k to the right hand boundary of the segment

$$\begin{aligned}
\mathbf{\Gamma}_1^k(R_\mathrm{R}^k, R_\mathrm{L}^k) &= \frac{1}{\mathcal{W}}\left[Bi'(x_{j\mathrm{L}})Ai(x_{jR}) - Ai'(x_{j\mathrm{L}})\,Bi(x_{jR})\right] \\
\mathbf{\Gamma}_2^k(R_\mathrm{R}^k, R_\mathrm{L}^k) &= -\frac{d_j}{\mathcal{W}}\left[Ai(x_{j\mathrm{L}})Bi(x_{jR}) - Bi(x_{j\mathrm{L}})\,Ai(x_{jR})\right] \\
\mathbf{\Gamma}_3^k(R_\mathrm{R}^k, R_\mathrm{L}^k) &= -\frac{1}{d_j\mathcal{W}}\left[Bi'(x_{j\mathrm{L}})Ai'(x_{jR}) - Ai'(x_{j\mathrm{L}})\,Bi'(x_{jR})\right] \\
\mathbf{\Gamma}_4^k(R_\mathrm{R}^k, R_\mathrm{L}^k) &= \frac{1}{\mathcal{W}}\left[Ai(x_{j\mathrm{L}})\,Bi'(x_{jR}) - Bi(x_{j\mathrm{L}})\,Ai'(x_{jR})\right].
\end{aligned} \tag{4.23}$$

4.1.2 Log-Derivative Method

A problem with Gordon's method, noted in the original paper,[6] is concerned with starting the integration or solution process. The wavefunctions must be started in the classically forbidden region where, for the "regular" solution in which we are interested, the wavefunction and its derivative tend to zero as R tends to zero. When the wavefunction and its derivative is set to zero somewhere deep inside the classically forbidden region, it grows exponentially as R increases. As this happens the independence of the different solutions is compromised and steps have to be taken to reimpose their mutual independence.

In the log-derivative method this problem is completely eliminated.[2,8–10] In this sub-section we will outline the basic ideas of the method. The log-derivative matrix of the wavefunction is

$$\mathbf{Y} = \mathbf{\Phi}'(R)\mathbf{\Phi}^{-1}(R). \tag{4.24}$$

Differentiating eqn (4.24) and utilizing eqn (4.1) we can write the following equation for the log-derivative[†]

$$\mathbf{Y}' + \mathbf{W} + \mathbf{Y}^2 = 0. \tag{4.25}$$

The wavefunction $\mathbf{\Phi}(R)$ should tend to zero as the scattering coordinate is decreased progressively further into the classically forbidden region. At the same time the log-derivative tends to infinity. In practical terms it turns out that choosing some very large value for the log-derivative in this region is a satisfactory way of starting off the calculation.[2,10] Following the procedures of sub-Section 4.1.1 it is possible to derive equations similar to eqn (4.22) and (4.23) but for the propagation of the log-derivative matrix, rather than for that of the wavefunction matrix. The overall recommendation for the solution of the coupled differential equations[2] is to use a numerical method

---

[†]Note that the sign of **W** used here differs from that used in ref. 2 and 10 but is the same as that used in ref. 9

for the direct solution of the the log-derivative equations[10] (*i.e.* without fitting the potential to straight lines within each segment) in the inner region where the potential varies a lot, and to use the log-derivative method, but now fitting straight lines to the potentials within segments and using the piecewise analytic solutions to propagate the log-derivative matrix across segments.

4.1.3 Comparison of Theory and Experiment

He + N_2

Figure 4.2 shows a comparison between theory and experiment for total and rotationally inelastic differential cross sections for the He + N_2 system.[11] The experiments were some of the first crossed molecular beam experiments carried out for rotationally inelastic scattering. There is good general agreement between the calculated and experimental results. Complete agreement is not expected as the calculations used a model interaction potential. Both the total and inelastic differential cross sections show marked oscillations, quite similar in nature to those occurring for Ne+Ne (see Chapter 2).

He + ND_3

Figure 4.3 shows a recent comparison of experiment and theory for the rotationally inelastic differential cross section of He + ND_3 at a collision energy of 430 ± 40 cm^{-1}, starting in a well defined initial state of ND_3 and

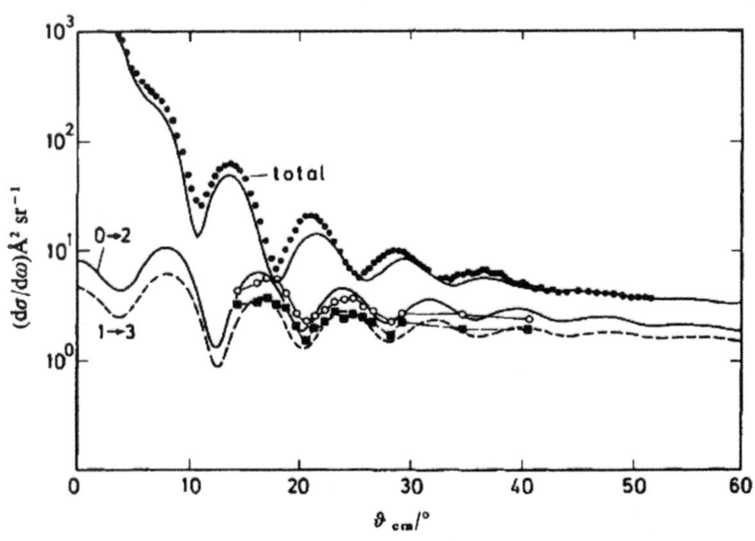

Figure 4.2 Comparison of experimental and computed differential cross sections for He + N_2. The smooth curves are the results of close-coupling calculations while the circles and squares are from experimental measurements. The figure is reproduced with permission from ref. 11.

being scattered into different final rotational states of negative parity.[12] In view of the complexity of the system the agreement between experiment and theory is very impressive and demonstrates the high level of accuracy and sophistication that is now achievable in theoretical calculations.

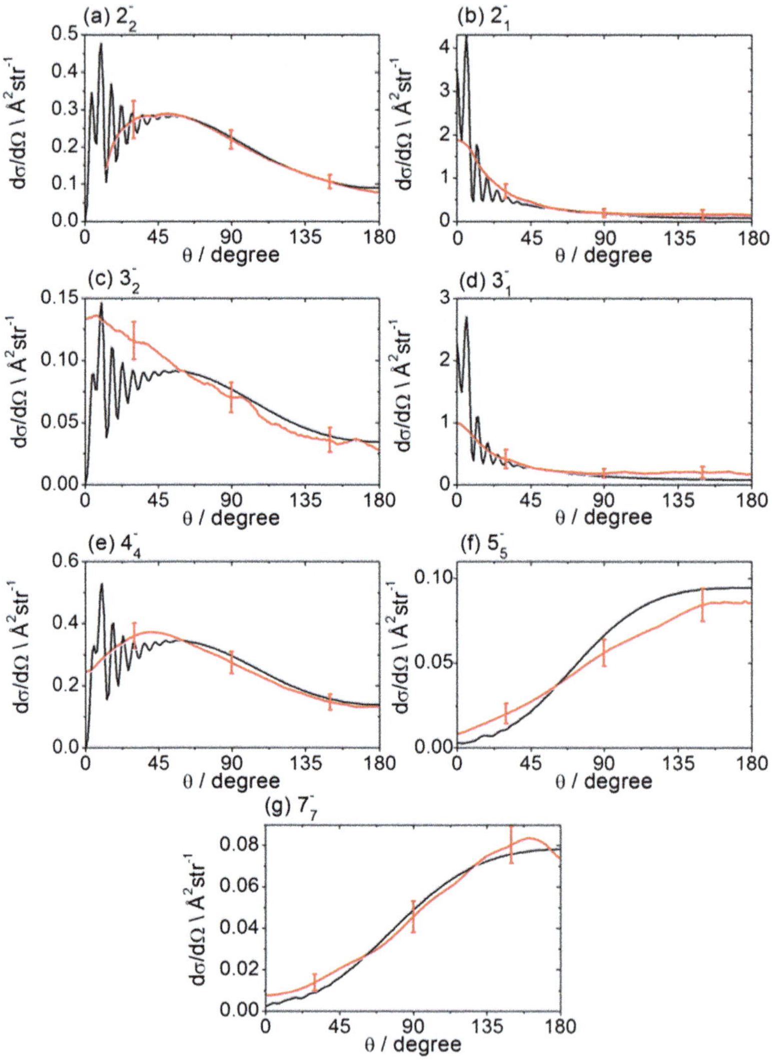

Figure 4.3 Comparison of experimental and computed differential state-to-state differential cross sections cross sections for He + ND_3. The ND_3 is initially in its ground vibrational level and in a prepared initial rotational state (1_1^-). The figure shows both experimental (red line with error bars) and computed (black line) differential cross sections for scattering into other rotational levels within the same vibrational manifold. Figure reproduced from ref. 12 with permission from the PCCP Owner Societies.

4.2 Approximate Methods

4.2.1 Neglect of Coupling

A major problem with solving the quantum mechanical equations of molecular scattering theory (see eqn (3.16) and (3.42)) is the very large number of fragment quantum states or coupled channels that arise.[13–18] The most obvious way of simplifying the process of solving these equations is to break them up into groups of coupled differential equations with each group containing fewer equations, *i.e.* by somehow reducing the coupling terms. This is indeed the motivation behind using the total angular momentum representation (see Appendix D sub-Section D.4) and also behind the use of the parity symmetry in the body-fixed formulation (see Section 3.8). Pack[15] pointed out that, in this respect, the body-fixed representation had some advantages as the helicity quantum numbers are coupled only through the orbital angular momentum operator and potential energy matrix elements couple only the rotational–vibrational quantum numbers, while in the space-fixed representation all the coupling is contained in the potential energy matrix.

The simplest approximate method[17,18] is to ignore the coupling between the different helicity quantum numbers, *i.e.* to set term $C^{J\pm}_{\Omega,\Omega\pm1}$ in eqn (3.42) to zero. This approximation retains all the channels, but eliminates a lot of coupling terms and therefore drastically reduces the size of the coupled differential equations which need to be solved. It is called "Centrifugal Decoupling" or "Helicity Decoupling". Figure 4.4 shows a comparison of exact close-coupling and Helicity Decoupling calculations to the calculations of opacity functions for Ar + N_2 scattering. The approximation is seen to give excellent results for this simple case.

In the "j_z-conserving Coupled States Approximation" or more briefly the "Coupled States Approximation", further simplifications are made[16,20] and all the diagonal elements of the $\hat{\ell}^2$ operator are assumed equal. The approximation effectively replaces the term $\dfrac{J(J+1)+j(j+1)-\Omega^2}{R^2}$ by $\dfrac{J(J+1)}{R^2}$ and the helicity quantum number Ω becomes redundant and can be ignored. Unlike Helicity Decoupling, this approximation not only reduces the number of coupling terms, it actually eliminates many of the channels. The channels are no longer labeled by j, Ω (or j, ℓ) but only by j. Figure 4.5 shows an application of this method to the calculation of differential vibrationally inelastic cross sections scattering for He + H_2 scattering. This system processes a relatively small anisotropy and the agreement between the exact and the approximate calculations is impressive.

The Infinite Order Sudden (IOS) approximation[21–23] is designed for high energy collisions and sets all the rotational energy levels equal to zero. It is most conveniently discussed within the body-fixed formalism. As all the rotational levels are assumed to be of equal energy, the rotational basis functions can be subjected to any arbitrary unitary transformation, without changing

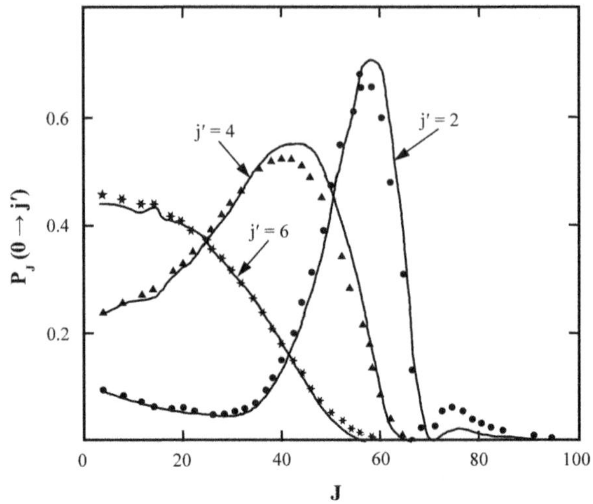

Figure 4.4 Comparison of exact and Helicity Decoupled partial cross sections for a model of Ar + N_2 rotationally inelastic scattering. The dots correspond to the exact close-coupling calculations while the solid line corresponds to the approximate Helicity Decoupled calculations. Reprinted from M. Shapiro and M. Tamir, *Chem. Phys.*, **13**, 215 with permission from Elsevier.[19]

the wave numbers k_{vj} which appear in the coupled differential equations, eqn (3.44). The consequence of this is that we do not perform the integration over angular coordinates so as to create matrix elements of the potential, but rather solve the scattering equations for fixed angles and take the matrix elements of the fixed-angle **S** over the angular basis functions after they have been evaluated over a sufficiently dense grid of angles.[23] If only rotationally inelastic transitions are being computed, then the differential equation that needs to be solved for each fixed angle is just of the same form as that encountered in atom–atom elastic scattering (see Chapter 2). In the more general case where vibrational inelasticity is also considered, the fixed angle equations have the dimensionality just of the number of vibrational levels included in the calculations. The IOS approximation leads to a huge reduction in the number of channels that need to be included in each of the coupled differential equations.

Figure 4.6 compares exact close-coupling and IOS approximation calculations of the opacity function for rotationally inelastic scattering of Ar — TlF. The numbers along the top of the figure give the number of channels required in the exact close-coupling calculations to achieve convergence at the different values of J, which are shown along the lower axis. We see that for large values of J fewer channels are required. This is due to the centrifugal potential which effectively keep the collision partners further apart for larger values of J. We also see that when many channels are required in the

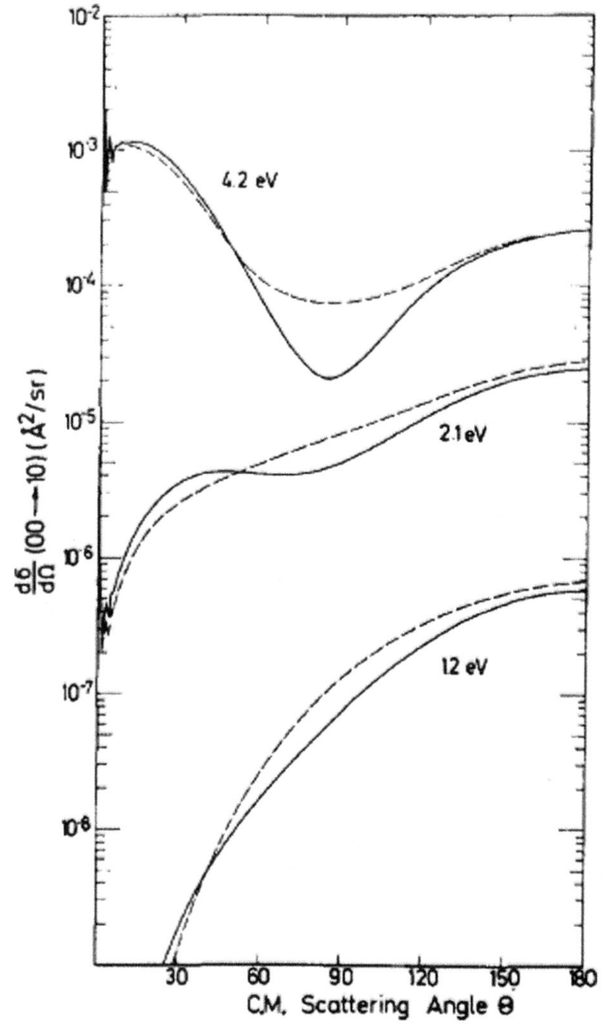

Figure 4.5 Comparison of exact close-coupling and j_z-conserving Coupled States Approximation calculations of the differential vibrationally inelastic scattering cross section for He + H_2. Dashed line: exact close-coupling calculations. Solid line: j_z-conserving Coupled States Approximation calculations. Reproduced with permission from P. McGuire, *J. Chem. Phys.*, **62**, 525. Copyright 1975. AIP Publishing LLC.[20] DOI: 10.1063/1.430453.

close-coupling calculations, the IOS approximation performs much better because of the greater degree of coupling between the channels.

4.2.2 Born and Distorted Wave Approximations

The integral equation solution to the scattering problem, outlined in Appendix E, lends itself naturally to a hierarchy of approximations which we discuss below.

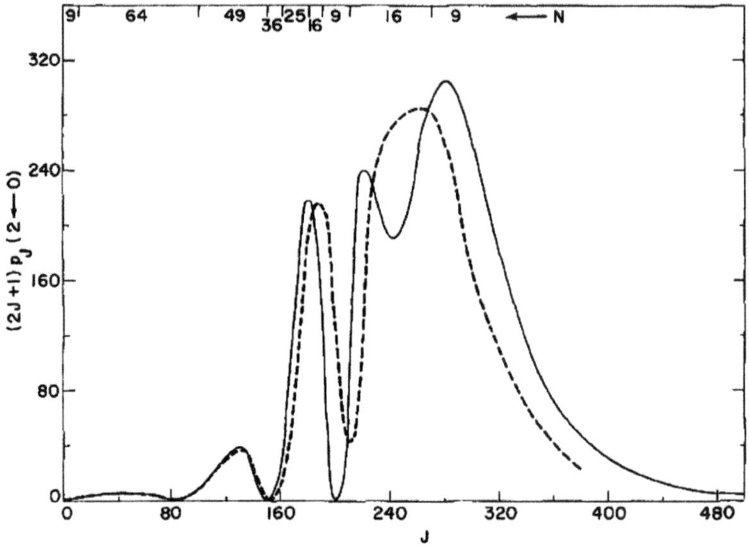

Figure 4.6 Comparison of exact close-coupling and IOS Approximation calculations of the opacity function for rotationally inelastic scattering of the Ar–TlF collision system. Solid line: exact close-coupling calculations. Dashed line: IOS calculations. Reproduced with permission from T. P. Tsien, G. A. Parker and R. T. Pack, *J. Chem. Phys.*, **59**, 5373. Copyright 1973. AIP Publishing LLC.[21] DOI: 10.1063/1.1679884.

Born Approximation

The integral equations, analogous to the set of coupled differential equations eqn (3.16) are (see eqn (E-17))

$$\phi^{vj\ell}_{v'j'\ell'}(R) = \hat{j}_\ell(k_{vj}R)\delta_{v,v'}\delta_{j,j'}\delta_{\ell,\ell'} - \int_0^\infty G_{v'j'\ell'}(R,R') \times \sum_{v'',j'',\ell''} U_{v'j'\ell';v''j''\ell''}(R')\phi^{vj\ell}_{v''j''\ell''}(R')\mathrm{d}R'. \quad (4.26)$$

Note that we have simplified the notation here by omitting the argument E from the wavefunctions and the Green's function (*i.e.* writing $\phi^{vj\ell}_{v'j'\ell'}(R)$ instead of $\phi^{vj\ell}_{v'j'\ell'}(R,E)$). Note also that the Green's function, $G_{vj\ell}(R,R')$, depends on the vibrational quantum number v only through the wave number k_{vj}, where

$$\begin{aligned} G_{vj\ell}(R,R') &= u^1_{vj\ell}(R)\,u^2_{vj\ell}(R') \quad \text{for} \quad R<R', \\ G_{vj\ell}(R,R') &= u^2_{vj\ell}(R)\,u^1_{vj\ell}(R') \quad \text{for} \quad R>R' \end{aligned} \quad (4.27)$$

and $u^1_{vj\ell}(R)$ and $u^2_{vj\ell}(R)$ are solutions of the homogeneous equation derived from eqn (3.18) by setting all the terms containing the potential to zero. They have the following asymptotic behavior

$$u^1_{\nu j\ell}(R) = \hat{j}_\ell(k_{\nu j}R) \widetilde{_{R\to\infty}} \sin\left(k_{\nu j}R - \frac{\ell\pi}{2}\right)$$

$$u^2_{\nu j\ell}(R) = -\frac{1}{k_{\nu j}}\left[\hat{n}_\ell(k_{\nu j}R) \; - i\,\hat{j}_\ell(k_{\nu j}R)\right] \widetilde{_{R\to\infty}} \frac{1}{k_{\nu j}} e^{i(k_{\nu j}R - \frac{\ell\pi}{2})} \tag{4.28}$$

where $\hat{j}_\ell(kr)$ and $\hat{n}_\ell(kr)$ are the Riccati–Bessel functions (see eqn (2.12)).

The Born approximation is obtained by assuming that the collision has only a minimal effect upon the scattering wavefunction and therefore justifies substituting the first term on the right hand side of eqn (4.26) for the final term $\phi^{\nu j\ell}_{\nu'' j'' \ell''}(R')$ to obtain

$$\begin{aligned}\phi^{\nu j\ell}_{\nu' j' \ell'}(R) = {} & \hat{j}_\ell(k_{\nu j}R)\delta_{\nu,\nu'}\delta_{j,j'}\delta_{\ell,\ell'} \\ & - \int_0^\infty G_{\nu' j' \ell'}(R, R')U_{\nu' j' \ell';\nu j\ell}(R')\hat{j}_\ell(k_{\nu j}R')\mathrm{d}R'.\end{aligned} \tag{4.29}$$

This is the Born (or first Born) approximation. We could form a "Born series" by substituting the whole of the right hand side of eqn (4.26) for the wavefunction $\phi^{\nu j\ell}_{\nu'' j'' \ell''}(R')$ at the far right of the equation and continuing to iterate this process. Each term in the series would then be a successive term in the Born series.[24–27]

Let us now use eqn (4.27) and (4.28) in eqn (4.29) to find the asymptotic form of the Born wavefunction and hence, by comparison to eqn (3.21) find the Born approximation to the **S** matrix

$$\begin{aligned}\phi^{\nu j\ell}_{\nu' j' \ell'}(R) \widetilde{_{R\to\infty}} & \frac{e^{i\left(k_{\nu j}R - \frac{\ell\pi}{2}\right)} - e^{-i\left(k_{\nu j}R - \frac{\ell\pi}{2}\right)}}{2i}\delta_{\nu,\nu'}\delta_{j,j'}\delta_{\ell,\ell'} \\ & - \frac{1}{k_{\nu' j'}} e^{i\left(k_{\nu' j'}R - \frac{\ell'\pi}{2}\right)} \int_0^\infty \hat{j}_{\ell'}(k_{\nu' j'}R')U_{\nu' j' \ell';\nu j\ell}(R')\hat{j}_\ell(k_{\nu j}R')\mathrm{d}R' \\ \widetilde{_{R\to\infty}} & -\frac{1}{2i}\left\{e^{-i\left(k_{\nu j}R - \frac{\ell\pi}{2}\right)}\delta_{\nu,\nu'}\delta_{j,j'}\delta_{\ell,\ell'} - e^{i\left(k_{\nu j}R - \frac{\ell\pi}{2}\right)}\left[\delta_{\nu,\nu'}\delta_{j,j'}\delta_{\ell,\ell'}\right.\right. \\ & \left.\left. - 2i\frac{1}{k_{\nu' j'}}\int_0^\infty \hat{j}_{\ell'}(k_{\nu' j'}R')U_{\nu' j' \ell';\nu j\ell}(R')\hat{j}_\ell(k_{\nu j}R')\mathrm{d}R'\right]\right\}.\end{aligned} \tag{4.30}$$

Comparing this with eqn (3.21) we see that the Born approximation to the **S** matrix may be written as

$$S^J_{\nu' j' \ell',\nu j\ell} = \delta_{\nu,\nu'}\delta_{j,j'}\delta_{\ell,\ell'} - 2i\frac{1}{\sqrt{k_{\nu' j'}k_{\nu j}}}\int_0^\infty \hat{j}_{\ell'}(k_{\nu' j'}R')U_{\nu' j' \ell';\nu j\ell}(R')\hat{j}_\ell(k_{\nu j}R')\mathrm{d}R'. \tag{4.31}$$

The Born approximation has a very limited range of validity and has not been extensively used. It is however the lowest in a hierarchy of related methods and, as such, forms a valuable building block to the theories discussed below.

Distorted Wave Approximation

In the Born approximation the wavefunctions used to build the Green's function, and which then appear in the key equations for the **S** matrix (see eqn (4.31)), are the solutions of the radial Schrödinger equation without any potential. In the distorted wave approximation the basic radial wavefunctions are the solutions of the one-dimensional radial equations which include the "distorting" diagonal channel potentials but do not include the coupling to other channels. The relevant integral equations have been discussed in Appendix E (see Section E.1) and are

$$\phi^{vj\ell}_{v'j'\ell'}(R) = u^{1}_{v'j'\ell'}(R)\delta_{v,v'}\delta_{j,j'}\delta_{\ell,\ell'} - \int_0^\infty G_{v'j'\ell'}(R,R') \times \sum_{v''j''\ell''\neq v'j'\ell'} U_{v'j'\ell';v''j''\ell''}(R')\phi^{vj\ell}_{v''j''\ell''}(R')\mathrm{d}R'. \tag{4.32}$$

The Green's function is defined in eqn (4.27), as before, but the definition of its basic constituent one-dimensional radial wavefunctions has now changed such that they are the two independent solutions of the one-dimensional Schrödinger equation (see Appendix E, eqn (E-20) and eqn (3.18))

$$\left[-\frac{\partial^2}{\partial R^2} + \frac{\ell(\ell+1)}{R^2} - k^2_{vj} + U^J_{vj\ell;vj\ell}(R)\right] u_{vj\ell}(R) = 0 \tag{4.33}$$

obeying the asymptotic boundary conditions

$$u^1_{vj\ell}(R) \widetilde{_{R\to\infty}} \sin\left(k_{vj}R - \frac{\ell\pi}{2} + \eta_\ell\right)$$
$$u^2_{vj\ell}(R) \widetilde{_{R\to\infty}} \frac{1}{k_{vj}} \mathrm{e}^{i\left(k_{vj}R - \frac{\ell\pi}{2} + \eta_\ell\right)}. \tag{4.34}$$

These functions and the associated phase shifts must be obtained by numerical integration of eqn (4.33) or approximated using semiclassical theory (see Appendix F). The distorted wave approximation is now obtained in the same manner as was used in the formulation of the Born approximation (see eqn (4.29) above) by substituting the first term on the right hand side of eqn (4.32) for $\phi^{vj\ell}_{v''j''\ell''}(R')$ to yield

$$\phi^{vj\ell}_{v'j'\ell'}(R) = u^1_{v'j'\ell'}(R)\delta_{v,v'}\delta_{j,j'}\delta_{\ell,\ell'} - (1 - \delta_{v,v'}\delta_{j,j'}\delta_{\ell,\ell'}) \times \int_0^\infty G_{v'j'\ell'}(R,R')U_{v'j'\ell';vj\ell}(R')u^1_{vj\ell}(R')\mathrm{d}R'. \tag{4.35}$$

We can obtain the asymptotic form of the radial wavefunctions by substituting eqn (4.34) and (4.27) into eqn (4.35)

$$\begin{aligned}
\phi^{vj\ell}_{v'j'\ell'}(R) \widetilde{_{R\to\infty}} & \frac{1}{2i}\left\{ e^{i\left(k_{vj}R-\frac{\ell\pi}{2}+\eta_\ell\right)} - e^{-i\left(k_{vj}R-\frac{\ell\pi}{2}+\eta_\ell\right)}\right\}\delta_{v,v'}\delta_{j,j'}\delta_{\ell,\ell'} \\
& - \frac{e^{i\left(k_{v'j'}R-\frac{\ell'\pi}{2}+\eta'_\ell\right)}}{k_{v'j'}}\left(1-\delta_{v,v'}\delta_{j,j'}\delta_{\ell,\ell'}\right) \\
& \times \int_0^\infty u^1_{v'j'\ell'}(R')U_{v'j'\ell';vj\ell}(R')u^1_{vj\ell}(R')\mathrm{d}R' \\
\widetilde{_{R\to\infty}} & -\frac{e^{-i\eta_\ell}}{2i}\left\{ e^{-i\left(k_{vj}R-\frac{\ell\pi}{2}\right)} - e^{i\eta_\ell}e^{i\eta_{\ell'}}e^{i\left(k_{v'j'}R-\frac{\ell'\pi}{2}\right)}\left[\delta_{v,v'}\delta_{j,j'}\delta_{\ell,\ell'}\right.\right. \\
& \left.\left. - \frac{2i\left(1-\delta_{v,v'}\delta_{j,j'}\delta_{\ell,\ell'}\right)}{k_{v'j'}}\int_0^\infty u^1_{v'j'\ell'}(R')U_{v'j'\ell';vj\ell}(R')u^1_{vj\ell}(R')\mathrm{d}R'\right]\right\}.
\end{aligned} \tag{4.36}$$

Comparing this with eqn (3.21) we can write the following expression for the **S** matrix

$$\begin{aligned}
S^J_{v'j'\ell',vj\ell} = e^{i\eta_{\ell'}} & \left[\delta_{v,v'}\delta_{j,j'}\delta_{\ell,\ell'} - \frac{2i\left(1-\delta_{v,v'}\delta_{j,j'}\delta_{\ell,\ell'}\right)}{\sqrt{k_{v'j'}k_{vj}}}\right. \\
& \left. \times \int_0^\infty u^1_{v'j'\ell'}(R')U_{v'j'\ell';vj\ell}(R')u^1_{vj\ell}(R')\mathrm{d}R'\right] e^{i\eta_\ell}.
\end{aligned} \tag{4.37}$$

Note the term $(1-\delta_{v,v'}\delta_{j,j'}\delta_{\ell,\ell'})$ which precedes the integral in the distorted wave approximation. This term excludes the diagonal term from the **S** matrix. It reflects the exclusion of this term in the summation over the potential terms in eqn (3.18) and the resulting form of the summation in the distorted wave wavefunction (eqn (4.32)).

Figure 4.7 shows a comparison of exactly computed and distorted wave approximation differential rotationally inelastic cross sections for $H + H_2$ scattering. The Born and distorted wave approximations work best for weak interactions and high collision energies. In the case shown the approximation yields results in good qualitative agreement with the exact calculations, but as we will see this is not the general case.

Exponential Born and Exponential Distorted Wave Approximations

A failing of both the Born and the distorted wave approximations is that they yield **S** matrices that are not unitary (see Section 3.6). This is a reflection of the fact that the approximations fail to conserve particle flux. The solution to this problem was first discussed by Percival[29] and Seaton.[30] We will concentrate in this book on just one way of overcoming this shortcoming of the Born and distorted wave approximations.

If **A** is a Hermitian matrix then the exponential form

$$\mathbf{S} = e^{i\mathbf{A}} \tag{4.38}$$

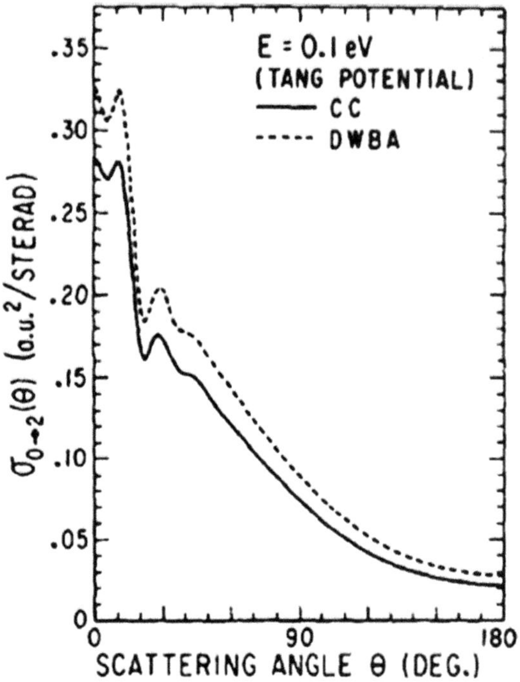

Figure 4.7 Comparison of exact close-coupling and distorted wave approximation calculations of the differential rotationally inelastic cross section for H + H_2 scattering. Solid line: exact close-coupling calculations. Dashed line: distorted wave approximation calculations (referred to as distorted wave Born in the paper). Reproduced with permission from B. H. Choi and K. T. Tang, *J. Chem. Phys.*, **63**, 1783. Copyright 1975. AIP Publishing LLC.[28] DOI: 10.1063/1.431554.

guarantees that the **S** matrix will be unitary. We can consider expanding the matrix **A** in a power series of the potential and if we compare the resulting expansion with the Born or distorted wave series we can identify the terms in the series. This leads to a "re-summation" of the series in a form guaranteed to be unitary.[26] The distorted wave approximation for the **S** matrix (eqn (4.37)) has the form

$$\mathbf{S} = \mathrm{e}^{i\eta}[\mathbf{I} + i\mathbf{D}]\mathrm{e}^{i\eta} \tag{4.39}$$

where η is a diagonal matrix and the elements of **D** can be identified by comparison with eqn (4.37).

In the exponential distorted wave approximation this form is replaced by the intrinsically unitary form

$$\mathbf{S} = \mathrm{e}^{i\eta}\mathrm{e}^{i\mathbf{D}}\mathrm{e}^{i\eta}. \tag{4.40}$$

Figure 4.8 compares exact and approximate computed opacity functions for rotationally inelastic scattering in a model atom–diatom system.[31] The definition for the opacity function is taken from ref. 32 and differs

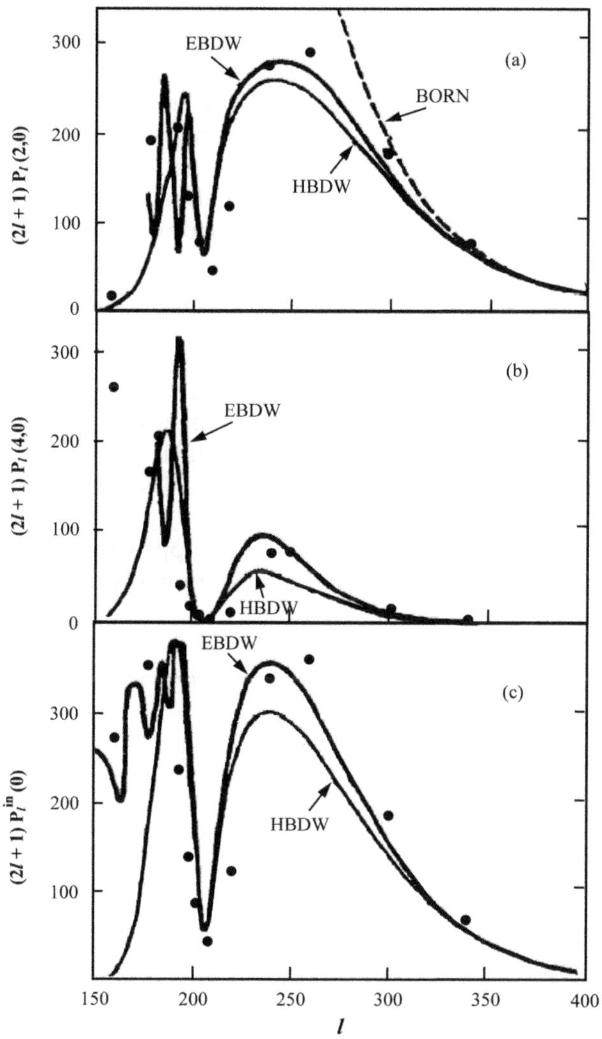

Figure 4.8 Comparison of exact close-coupling, Born approximation and exponential Born distorted wave approximation calculations of the opacity function for three rotationally inelastic transitions in a model atom–diatom collision.[31] The top panel (a) shows the opacity factor for $j=0 \rightarrow j=2$ transitions and the second panel (b) for $j=0 \rightarrow j=4$ collisions. The lowest panel (c) shows the opacity function for the sum of all transitions out of the $j=0$ level. Dots: exact close-coupling calculations. The solid and dashed lines are labeled with the approximations used (see text). Reprinted from G. G. Balint-Kurti and R. D. Levine, *Chem. Phys. Letts.*, 1970, 7, 107 with permission from Elsevier.[31]

from that given in eqn (3.41). The line marked EBDW is the exponential Born distorted wave approximation. This method corresponds to the use of eqn (4.40) with the phase shifts being computed exactly (as in the distorted wave approximation) and the integrals required to make up the **D** matrix being evaluated using the Born approximation (see eqn (4.31)). We will not discuss here the line marked HBDW in the figure. It corresponds to an alternative method for making the **S** matrix unitary. The exact computations are shown as large dots. The top panel of the figure (a) shows that the Born approximation fails completely at low orbital angular momentum quantum numbers. The failure encompassing almost the entire important range of the opacity function. The unitarized EBDW results, involving only slightly more work, in contrast give very reasonable results. The second panel (b) of the figure does not have any line for the Born approximation result. This is because the Born approximation predicts a zero value for this opacity. In contrast the EBDW results are again very reasonable. The bottom panel (c) shows the opacity for the sum of all transitions emanating from the lowest $j = 0$ rotational level. We see that the EBDW method agrees well with the exact calculations despite the complete failure of the underlying Born approximation.

Figure 4.9 compares exactly computed and approximate partial cross sections, plotted as a function of total angular momentum quantum number,

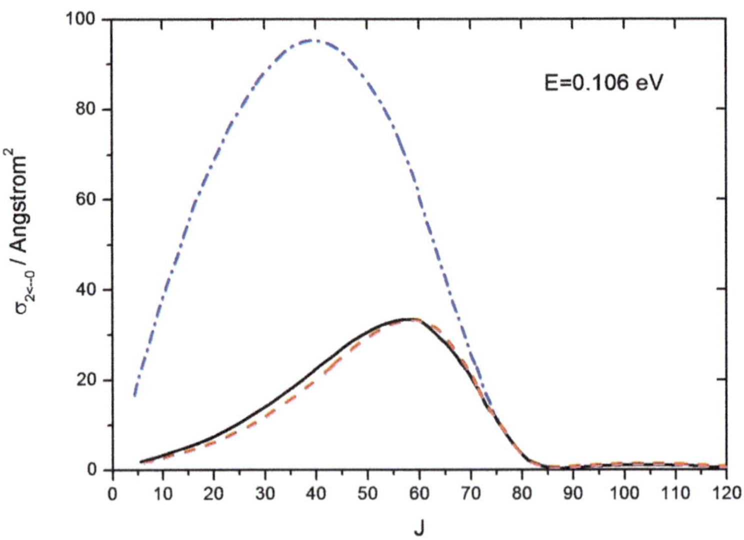

Figure 4.9 Comparison of exact close-coupling, semiclassical distorted wave and exponential semiclassical distorted wave calculations of $j = 0 \rightarrow j = 2$ rotationally inelastic partial cross sections for a model of Ne + N_2 scattering.[33] Solid line: exact close-coupling calculations, dashed line (close to solid line): ESCDW, and dot-dashed line: SCDW (see text). The figure is reproduced with permission from S. Bosanac and G. G. Balint-Kurti, *Mol. Phys.*, 1975, **29**, 1797, published by Taylor & Francis Ltd.[33]

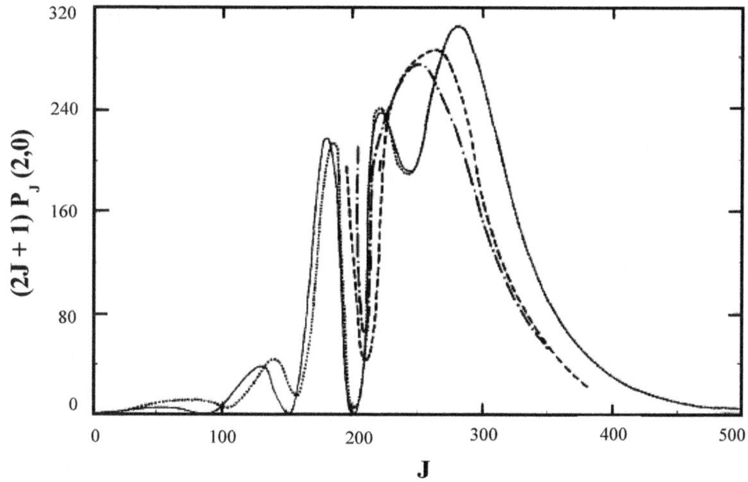

Figure 4.10 Comparison of exact close-coupling (solid line), Exponential Born Distorted wave (EBDW, dot-dashed line) and Infinite Order Sudden (IOS, dashed line) calculations of $j = 0 \rightarrow j = 2$ rotationally inelastic partial cross sections for a model of Ar + TlF scattering. Reproduced with permission from T. P. Tsien, G. A. Parker and R. T. Pack, *J. Chem. Phys.*, **59**, 5373. Copyright 1973. AIP Publishing LLC.[21] DOI: 10.1063/1.1679884.

for Ne + N_2 scattering on a model potential energy surface. The approximate calculations are computed using the distorted wave approximation (eqn (4.37)) and the corresponding unitarized form (eqn (4.40)) but the distorted channel wavefunctions have now been computed using a semiclassical approximation (see Appendix F). We refer to these approximations as the semiclassical distorted wave (SCDW) and the exponential semiclassical distorted wave (ESCDW) approximations. We see immediately that the SCDW approximation greatly overestimates the magnitude of the partial cross sections, while the unitarized form (ESCDW) predicts almost exactly the correct value.

Figure 4.10 shows a comparison of the exact and three different approximate methods for the calculation of the partial $j = 0 \rightarrow j = 2$ inelastic cross section for a model of the Ar + TlF system. Of all the approximations only the Exponential Born Distorted Wave (EBDW) reproduces all the main features of the exact close-coupling results. For total angular momentum quantum numbers higher than $J = 250$ the exact and the EBDW results are indistinguishable.

References

1. L. D. Thomas, M. H. Alexander, B. R. Johnson, W. A. Lester Jr, J. C. Light, K. D. McLenithan, G. A. Parker, M. J. Redmon, T. G. Schmalz, D. Secrest and R. B. Walker, Comparison of numerical methods for solving the second-order differential equations of molecular scattering theory, *J. Comp. Phys.*, 1981, **41**, 407.

2. M. H. Alexander and D. E. Manolopoulos, A stable linear reference potential algorithm for solution of the quantum close-coupled equations in molecular scattering theory, *J. Chem. Phys.*, 1987, **86**, 2044.
3. J. M. Hutson and S. Green, MOLSCAT computer code, version 14 (1994), distributed by Collaborative Computational Project No. 6 of the Engineering and Physical Sciences Research Council (UK). Available from the web site: http://www.giss.nasa.gov/tools/molscat/.
4. M. H. Alexander, D. E. Manolopoulos, H. J. Werner, with contributions from P. F. Vohralik B. Follmeg, D. Lemoine, G. Corey, B. Johnson, T. Orlikowski, A. Berning, A. Degli-Esposti, C. Rist, P. Dagdigian, B. Pouilly, G. van der Sanden and M. Yang, HIBRIDON is a package of programs for the time-independent quantum treatment of inelastic collisions and photodissociation. Available from the web site: http://www2.chem.umd.edu/groups/alexander/hibridon/hib43/hibhelp.html.
5. R. G. Gordon, New method for constructing wavefunctions for bound states and scattering,*J. Chem. Phys.*, 1969, **51**, 14.
6. R. G. Gordon, Quantum scattering using piecewise analytic solutions, ed. B. Alder, S. Fernbach and M. Rotenberg, *Atomic and Molecular Scattering*, volume 10 of *Methods in Computational Physics: Advances in Research and Applications*, Elsevier, 1971, pp. 81.
7. M. Abromowitz and I. A. Stegun, *Handbook on Mathematical Functions, Applied Mathematics Series 55*, National Bureau of Standards, 10th edn, 1972.
8. M. H. Alexander, Hybrid quantum scattering algorithms for long-range potentials,*J. Chem. Phys.*, 1984, **81**, 4510.
9. B. R. Johnson, The multichannel log-derivative method for scattering calculations,*J. Comp. Phys.*, 1973, **13**, 445.
10. D. E. Manolopoulos, An improved log derivative method for inelastic scattering,*J. Chem. Phys.*, 1986, **85**, 6425.
11. M. Faubel, K.-H. Kohl, J. P. Toennies, K. T. Tang and Y. Y. Yung, *Faraday Discuss. Chem. Soc.*, 1982, **73**, 205.
12. O. Tkac, A. K. Saha, J. Onvlee, C.-H. Yang, G. Sarma, C. K. Bishwakarma, S. Y. T. van de Meerakker, A. van der Avoird, D. H. Parker and A. J. Orr-Ewing, State-to-state resolved differential cross sections for rotationally inelastic scattering of ND3 with He, *Phys. Chem. Chem. Phys.*, 2014, **16**, 477.
13. H. Rabitz, Effective potentials in molecular collisions, *J. Chem. Phys.*, 1972, **57**, 1718.
14. G. Zarur and H. Rabitz, Rotationally inelastic scattering with effective potentials,*J. Chem. Phys.*, 1973, **59**, 943.
15. R. T. Pack, Space-fixed *vs.* body-fixed axes in atom–diatomic molecule scattering. sudden approximations,*J. Chem. Phys.*, 1974, **60**, 633.
16. P. McGuire and D. J. Kouri, Quantum mechanical close coupling approach to molecular collisions. j_z-Conserving coupled states approximation,*J. Chem. Phys.*, 1974, **60**, 2488.

17. M. Tamir and M. Shapiro, The approximate conservation of P-helicity in rotational excitation: a new decoupling scheme, *Chem. Phys. Lett.*, 1975, **31**, 166.
18. R. B. Walker and J. C. Light, Body-fixed equations for atom–molecule scattering: exact and centrifugal decoupling methods, *Chem. Phys.*, 1975, **7**, 84.
19. M. Shapiro and M. Tamir, Rotational excitation in the Ar + N_2 system: a computational study of opacity functions and differential cross sections by the p-helicity decoupling approximation. *Chem. Phys.*, 1976, **13**, 215.
20. P. McGuire, Coupled-states approach for elastic and for rotationally and vibrationally inelastic atom–molecule collisions,*J. Chem. Phys.*, 1975, **62**, 525.
21. T. P. Tsien, G. A. Parker and R. T. Pack, Rotationally inelastic molecular scattering: computational tests of some simple solutions of the strong coupling problem,*J. Chem. Phys.*, 1973, **59**, 5373.
22. R. T Pack, Close coupling test of classical and semiclassical cross sections for rotationally inelastic Ar–N_2 collisions,*J. Chem. Phys.*, 1975, **62**, 3143.
23. D. Secrest, Theory of angular momentum decoupling approximations for rotational transitions in scattering,*J. Chem. Phys.*, 1975, **62**, 710.
24. A. Messiah, *Quantum Mechanics*, North Holland Publishing Co., Amsterdam, 1965.
25. N. F. Mott and H. S. W. Massey, *The Theory of Atomic Collisions*, Clarendon Press, Oxford, 2nd edn, 1949.
26. G. G. Balint-Kurti, The theory of rotationally inelastic molecular collisions, in *International Review of Science, Series II*, ed. A. D. Buckingham and C. A. Coulson, Butterworth, London, 1975, vol. 1, pp. 286.
27. R. D. Levine, On the partial summation of the born series, *J. Phys. B*, 1969, **2**, 839.
28. B. H. Choi and K. T. Tang, Inelastic collisions between an atom and a diatomic molecule. II. H + H_2 rotational excitation,*J. Chem. Phys.*, 1975, **63**, 1783.
29. I. C. Percival, A variational principle for scattering phases, *Proc. Phys. Soc.*, 1960, **76**, 206.
30. M. J. Seaton, Strong coupling in optically allowed atomic transitions produced by electron impact, *Proc. Phys. Soc.*, 1961, **77**, 174.
31. G. G. Balint-Kurti and R. D. Levine, Opacity analysis of inelastic molecular collisions: computational studies of the exponential born approximation for rotational excitation, *Chem. Phys. Lett.*, 1970, **7**, 107.
32. S. Bosanac and G. G. Balint-Kurti, Cross sections for the rotationally inelastic scattering of Ne + N2: application of the exponential semiclassical distorted wave approximation, *Mol. Phys.*, 1975, **29**, 1797.
33. R. D. Levine and R. B. Bernstein, Opacity analysis of inelastic molecular collisions. IV. Statistical aspects of rotational excitation theory, *J. Chem. Phys.*, 1970, **53**, 686.

CHAPTER 5

Rate Constants, Cross Sections and Reactive Scattering

Let us consider the bimolecular reaction

$$\mathrm{F} + \mathrm{HD} \rightarrow \mathrm{HF} + \mathrm{D}. \tag{5.1}$$

The rate of this reaction in a standard physical chemistry text book would be given as

$$\frac{\mathrm{d[F]}}{\mathrm{d}t} = -k(T)[\mathrm{F}][\mathrm{HD}] \tag{5.2}$$

where $k(T)$ is the temperature dependent rate constant and the terms in square brackets, such as [F], are concentrations in moles per cubic decimeter.

In this chapter we will discuss the connection between the rate constant and various types of cross sections. We will also survey some of the key experimental methods used to measure the rate constants and cross sections involved. Rate constants may also be defined for inelastic, non-reactive processes. The formalism is exactly the same, but we will concentrate on chemical reactions in this chapter.

5.1 Cross Sections and Rates of Chemical Reactions

We now need to consider reaction (5.1) in greater detail. The atoms and molecules will be present in different quantum states, and these will react at different rates. So we will replace reaction (5.1) by

$$\mathrm{F}_i + \mathrm{HD}_j \rightarrow \mathrm{HF}_k + \mathrm{D}_\ell \tag{5.3}$$

where the subscripts i, j, k, ℓ enumerate the different possible quantum states of the atoms and molecules involved.

RSC Theoretical and Computational Chemistry Series No. 7
Theory of Molecular Collisions
By Gabriel G. Balint-Kurti and Alexander P. Palov

Published by the Royal Society of Chemistry, www.rsc.org

In an experiment, carried out in bulk rather than beam conditions, the number of molecules of the reagent F consumed per unit volume $n(\mathrm{F})$ per second is[1,2]

$$\begin{aligned}\frac{\mathrm{d}n(\mathrm{F})}{\mathrm{d}t} &= \frac{\mathrm{d}\left(\sum_i n(\mathrm{F}_i)\right)}{\mathrm{d}t}\\ &= \int_0^\infty \mathrm{d}\mathrm{v}\sum_{ij,k\ell} P_iP_jP(\mathrm{v})\sigma_{k\ell\leftarrow ij}(E)\,\mathrm{v}\,n(\mathrm{F})n(\mathrm{HD})\end{aligned} \tag{5.4}$$

where $n(\mathrm{F}_i)$ is the number of atoms of F in quantum state i per unit volume, v is the relative velocity of the reacting fragments, $P(\mathrm{v})$ is the probability of them possesing this velocity, P_i and P_j are the probabilities of finding these quantum states and $\sigma_{k\ell\leftarrow ij}(E)$ is the integral reaction cross section for the reaction. E is the total energy of the system, $E = \frac{\mu \mathrm{v}^2}{2} + \varepsilon_i + \varepsilon_j$, where ε_i and ε_j are the energies of quantum states i and j of the reactants. $n(\mathrm{HD})$ is the number of molecules of HD per unit volume.

As we are dealing with a bulk phase experiment under equilibrium conditions, the various probabilities are given by their Boltzmann factors[3]

$$P_i = \frac{\mathrm{e}^{-\varepsilon_i/k_\mathrm{B}T}}{\sum_i \mathrm{e}^{-\varepsilon_i/k_\mathrm{B}T}} = \frac{\mathrm{e}^{-\varepsilon_i/k_\mathrm{B}T}}{q_{\mathrm{F}}} \tag{5.5}$$

where $q_{\mathrm{F}} = \sum_i \mathrm{e}^{-\varepsilon_i/k_\mathrm{B}T}$ is the "internal" partition function for the F atom, (*i.e.* the partition function excluding the translational partition function) and k_B denotes Boltzmann's constant. A similar expression can be written for the probability of the HD molecule being in the state j.

The distribution of relative velocities v is given by the Maxwell–Boltzmann distribution

$$P(v) = 4\pi\left(\frac{\mu}{2\pi k_\mathrm{B}T}\right)^{\frac{3}{2}} \mathrm{v}^2\mathrm{e}^{-\mu\mathrm{v}^2/2k_\mathrm{B}T}. \tag{5.6}$$

Placing these probabilities into eqn (5.4) we obtain

$$\begin{aligned}\frac{\mathrm{d}n(\mathrm{F})}{\mathrm{d}t} &= \left[\frac{4\pi}{q_{\mathrm{F}}q_{\mathrm{HD}}}\left(\frac{\mu}{2\pi k_\mathrm{B}T}\right)^{\frac{3}{2}}\sum_{ij,k\ell}\int_0^\infty \mathrm{d}\mathrm{v}\,\mathrm{v}^3\mathrm{e}^{-(\varepsilon_i+\varepsilon_j+\mu\mathrm{v}^2/2)/k_\mathrm{B}T}\sigma_{k\ell\leftarrow ij}(E)\right]\\ &\quad n(\mathrm{F})n(\mathrm{HD})\\ &= \left[\frac{4\pi}{q_{\mathrm{F}}q_{\mathrm{HD}}}\left(\frac{\mu}{2\pi k_\mathrm{B}T}\right)^{\frac{3}{2}}\sum_{ij,k\ell}\int_0^\infty \mathrm{d}\mathrm{v}\,\mathrm{v}^3\mathrm{e}^{-E/k_\mathrm{B}T}\sigma_{k\ell\leftarrow ij}(E)\right]n(\mathrm{F})n(\mathrm{HD}).\end{aligned} \tag{5.7}$$

We see that eqn (5.7) is of the same form as eqn (5.2) and we can identify the reaction rate constant as the term in square brackets on the right hand side

$$k(T) = \frac{4\pi}{q_{\mathrm{F}}q_{\mathrm{HD}}}\left(\frac{\mu}{2\pi k_\mathrm{B}T}\right)^{\frac{3}{2}}\sum_{ij,k\ell}\int_0^\infty \mathrm{d}\mathrm{v}\,\mathrm{v}^3\mathrm{e}^{-E/k_\mathrm{B}T}\sigma_{k\ell\leftarrow ij}(E). \tag{5.8}$$

Now, however, the rate constant has the units of molecules per unit volume per unit time, rather than moles.

This is an overall rate constant. We may ask more detailed questions, such as what is the rate of the reaction of F in quantum state i with HD in quantum state j to produce HF in quantum state k and D in quantum state ℓ? To obtain this detailed rate constant we just omit the four fold summation on the right hand side of eqn (5.8)

$$k_{k\ell\leftarrow ij}(T) = \frac{4\pi}{q_{\mathrm{F}}q_{\mathrm{HD}}}\left(\frac{\mu}{2\pi k_{\mathrm{B}}T}\right)^{\frac{3}{2}}\int_0^\infty \mathrm{d}v\, v^3 \mathrm{e}^{-E/k_{\mathrm{B}}T}\sigma_{k\ell\leftarrow ij}(E). \tag{5.9}$$

Let us now try to recast the expression for the overall thermal rate constant into an alternative form. First of all let us remember that the inelastic cross section is a function of the **T** matrix, see eqn (3.48), or equivalently eqn (D-45). In the next chapter we will show that the integral reaction cross section can be represented by an expression similar to that of eqn (3.48). We now replace the **T** matrix by the **S** matrix in the expression. This is permitted as we refer here specifically to reactive collisions, so the initial channel is by definition different from the final channel. From eqn (3.22) we see that $S^J_{f,i} = -T^J_{f,i}$ when $i \neq f$. From the definition $E = \frac{\mu v^2}{2} + \varepsilon_i + \varepsilon_j$ also using $\mathrm{d}E = \mu v \mathrm{d}v$ we can substitute in eqn (5.9) to obtain

$$k(T) = \frac{4\pi}{q_{\mathrm{F}}q_{\mathrm{HD}}}\left(\frac{\mu}{2\pi k_{\mathrm{B}}T}\right)^{\frac{3}{2}}\sum_{ij\Omega,k\ell\Omega'}\int_0^\infty \frac{1}{\mu}\mathrm{d}E v^2 \mathrm{e}^{-E/k_{\mathrm{B}}T}\frac{\pi}{k^2_{ij}} \times \sum_J (2J+1)\left|S^J_{k\ell\Omega',ij\Omega}(E)\right|^2 \tag{5.10}$$

where we have included a summation over Ω and Ω' as we need to sum over all initial and final quantum states.

As the wavenumber is related to the relative velocity by the equation $k_{ij}\hbar = \mu v$ we can write

$$k(T) = \frac{1}{q_{\mathrm{F}}q_{\mathrm{HD}}}\left(\frac{h^2}{2\pi\mu k_{\mathrm{B}}T}\right)^{\frac{3}{2}}\frac{1}{h}\int_0^\infty \mathrm{d}E \mathrm{e}^{-E/k_{\mathrm{B}}T} \times \sum_{ij\Omega,k\ell\Omega'}\sum_J (2J+1)\left|S^J_{k\ell\Omega',ij\Omega}(E)\right|^2. \tag{5.11}$$

$\left(\frac{2\pi\mu k_{\mathrm{B}}T}{h^2}\right)^{\frac{3}{2}}$ is the translational partition function per unit volume for the relative motion of the reactants.[3] Using Q_{R} to denote the product of the two internal reactant partition functions with this translational partition function, *i.e.* $Q_{\mathrm{R}}(T) = q_{\mathrm{F}} q_{\mathrm{HD}} \left(\frac{2\pi\mu k_{\mathrm{B}} T}{h^2}\right)^{\frac{3}{2}}$ we can write

$$k(T) = \frac{1}{Q_{\mathrm{R}}(T)h} \int_0^\infty \mathrm{d}E\, \mathrm{e}^{-E/k_{\mathrm{B}}T} \left\{ \sum_{ij\Omega,k\ell\Omega'} \sum_J (2J+1) \left| S^J_{k\ell\Omega',ij\Omega}(E) \right|^2 \right\}. \tag{5.12}$$

$Q(T)$ is sometimes referred to as the "total partition function per unit volume for the reactants".[4,5] This, however, is not completely correct as the translational partition function for the motion of the center-of-mass has been omitted. The summation in curly brackets is called "the cumulative reaction probability", $N(E) = \sum_{ij\Omega,k\ell\Omega'} \sum_J (2J+1) \left| S^J_{k\ell\Omega',ij\Omega}(E) \right|^2$. Using this notation we can write

$$k(T) = \frac{1}{Q_{\mathrm{R}}(T)h} \int_0^\infty \mathrm{d}E\, \mathrm{e}^{-E/k_{\mathrm{B}}T} N(E). \tag{5.13}$$

This expression for the thermal rate constant is exact and has formed the basis for several treatments that aim to compute the rate constant directly without first computing all the state-to-state reaction **S** matrix elements and cross sections.[4-8]

5.1.1 Detailed Balance

An expression for the detailed rate constant in terms of the **S** matrix can be obtained from eqn (5.12) by omitting the summations over the quantum states of the reactants (i, j) and products (k, ℓ)

$$k_{k\ell\leftarrow ij}(T) = \frac{1}{Q_{\mathrm{R}}(T)h} \int_0^\infty \mathrm{d}E\, \mathrm{e}^{-E/k_{\mathrm{B}}T} \left\{ \sum_{\Omega,\Omega'} \sum_J (2J+1) \left| S^J_{k\ell\Omega',ij\Omega}(E) \right|^2 \right\}, \tag{5.14}$$

and the detailed rate constant for the reverse reaction may be written as

$$k_{ij\leftarrow k\ell}(T) = \frac{1}{Q_{\mathrm{P}}(T)h} \int_0^\infty \mathrm{d}E\, \mathrm{e}^{-E/k_{\mathrm{B}}T} \left\{ \sum_{\Omega,\Omega'} \sum_J (2J+1) \left| S^J_{ij\Omega,k\ell\Omega'}(E) \right|^2 \right\}. \tag{5.15}$$

The **S** matrix is symmetric (see Section 3.6), so $S^J_{k\ell\Omega',ij\Omega}(E) = S^J_{ij\Omega,k\ell\Omega'}(E)$. From eqn (5.14) and (5.15) it follows that

$$\frac{k_{k\ell\leftarrow ij}(T)}{k_{ij\leftarrow k\ell}(T)} = \frac{Q_{\mathrm{P}}(T)}{Q_{\mathrm{R}}(T)} = K(T) \tag{5.16}$$

where $Q_R(T)$ and $Q_P(T)$ are the partition functions for the reactants and products respectively evaluated using the same energy origin and $K(T)$ is the equilibrium constant for reaction (5.3).

5.2 Determination of Detailed Rate Constants

Figure 5.1 shows a schematic view of an apparatus used by the Polanyi group[9] for measuring detailed rate constants (see also Polanyi's Nobel Prize lecture[10]). The method used is called infrared chemiluminescence. The reactants are introduced into an evacuated chamber maintained at a very low pressure. The products of the reaction are produced in a variety of vibrational–rotational energy levels. Convex mirrors are mounted at the ends of the cavity. One of the mirrors has a small hole to permit the extraction and analysis of any infrared radiation (IR) emitted. The excited states of the product molecules, produced by the experiment, emit IR radiation. The detailed spectrum of this radiation is then measured and through its analysis the relative portion of the molecules produced in each of the excited vibrational–rotational levels can be determined. The scheme that the experimentalists use to determine the quantum state distribution of the newly produced products is called the "arrested relaxation" method. This consists basically of ensuring that immediately after the reaction the products find themselves in a very low pressure environment (a near vacuum) so that they spontaneously emit infrared radiation before experiencing any significant number of collisions. Initial considerations would suggest that the relative population in the lowest vibrational state of the products could not be determined in this

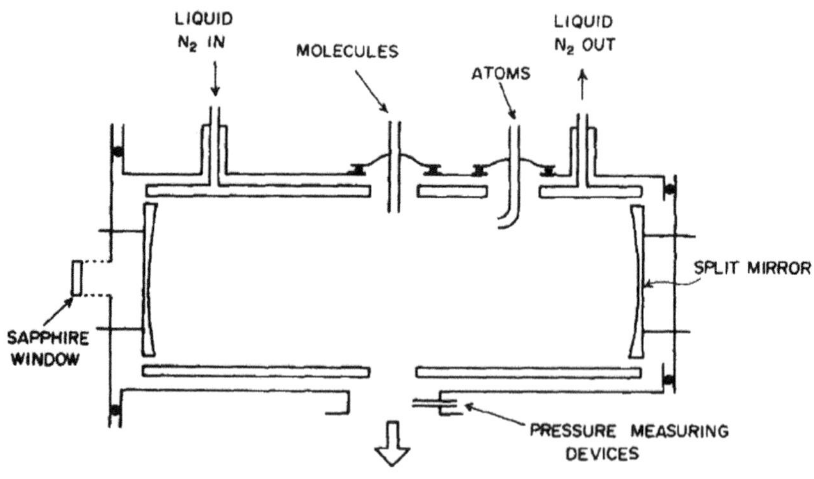

Figure 5.1 Schematic diagram of experimental apparatus used to determine detailed reaction rate constant using the techniques of "infrared chemiluminescence" and "arrested relaxation". Reproduced with permission from D. H. Maylotte, J. C. Polanyi and K. B. Woodall, *J. Chem. Phys.*, **57**, 1547. Copyright 1972, AIP Publishing LLC:[9] DOI: 10.1063/1.1678436.

way, as there will be no spontaneous IR emission from this level. However by taking account of self absorption of the emitted radiation and of the different intensities of isotopic lines detailed observation and modeling of the spectra have enabled the group to extract even this information from their observations.[11]

The results of Polanyi's experiments were often presented as "triangle plots". Figure 5.2 shows an example of such a plot. The energy in rotational motion of the products is plotted along the horizontal axis (abscissa) while the energy in vibration is plotted on the vertical axis (ordinate). The dashed lines joining the two axes are lines of constant translational energy. The contour plot shows the relative values of the detailed rate constant for producing products in the various available product quantum states. The plot shows that the most probable vibrational state of the HF product in the reaction is $v' = 2$.

The triangle plots demonstrate that the quantum state distribution in the products of a chemical reaction may be inverted, with a greater population in excited vibrational levels than in the lowest level. This immediately suggests the possibility of forming an IR laser system based on the chemical reaction. Kasper and Pimentel[12] demonstrated a chemical laser based on the Cl + H_2 reaction in 1965.

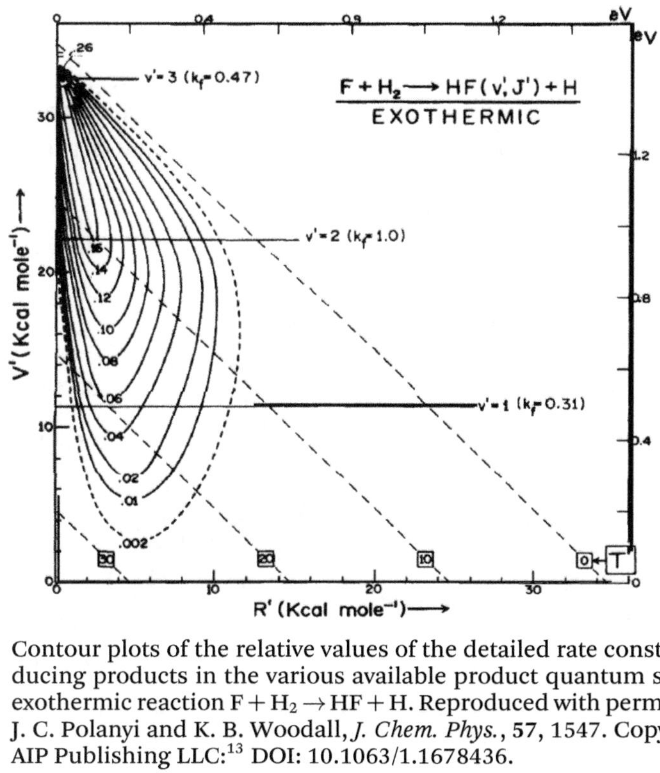

Figure 5.2 Contour plots of the relative values of the detailed rate constant for producing products in the various available product quantum states in the exothermic reaction $F + H_2 \rightarrow HF + H$. Reproduced with permission from J. C. Polanyi and K. B. Woodall, *J. Chem. Phys.*, **57**, 1547. Copyright 1972, AIP Publishing LLC:[13] DOI: 10.1063/1.1678436.

5.3 Measurement of Differential Reactive Cross Sections

5.3.1 The Alkali Age

Figure 5.3 shows a sketch of the crossed molecular beam apparatus used in some early "alkali age" experiments of the Herschbach group.[14,15] The alkali age was so called because of the surface ionization detectors used, which could detect only alkali atoms, and could distinguish between them and the product alkali halides. It was only at a later stage that the design of the experimental apparatus improved to permit the detection of any product molecule using mass spectrometric techniques.[16]

One of the first systems examined was

$$K + CH_3I \rightarrow KI + CH_3. \tag{5.17}$$

Figure 5.4 shows the laboratory angle dependence of the scattered KI product as measured in two different experiments[14,17] while Figure 5.5 shows the Newton diagram for the most probable velocities in the reactant beams. The role of a Newton diagram is to relate the laboratory velocities and scattering angles to those in the center-of-mass reference frame (see Section 1.3).

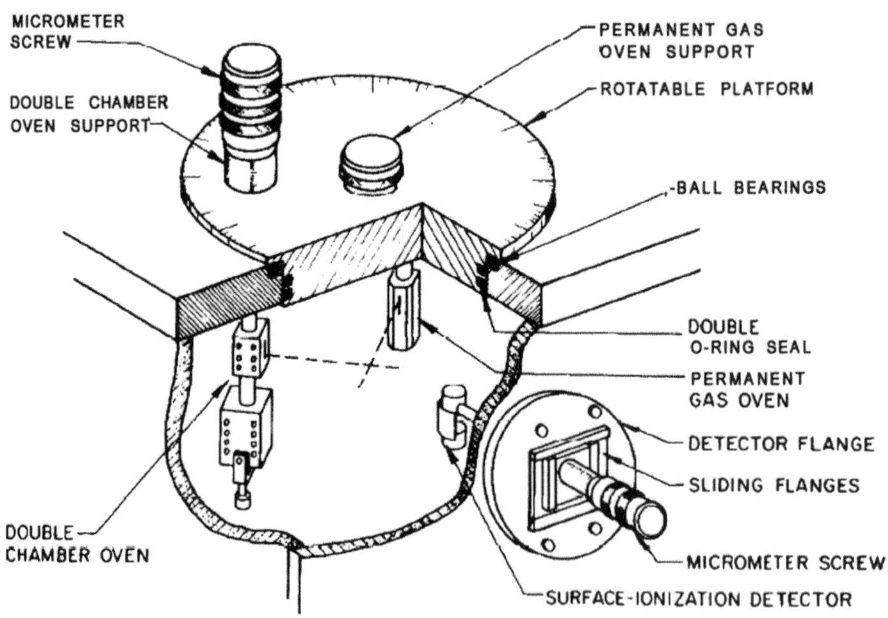

Figure 5.3 Illustrative drawing of crossed molecular beam apparatus used in early "alkali age" experiments of Herschbach group.[14,15] Figure reproduced with permission from ref. 14.

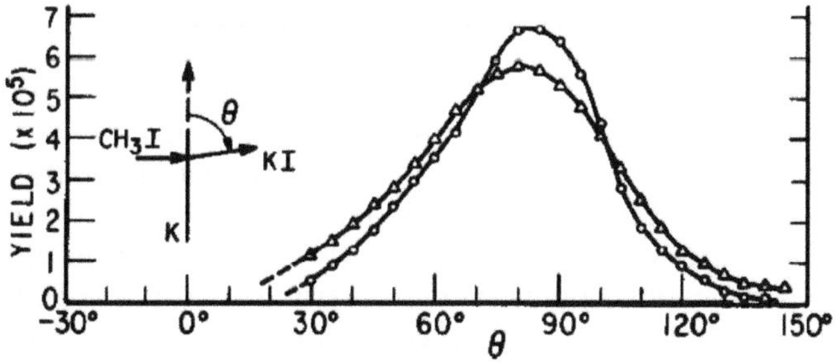

Figure 5.4 Laboratory angle dependence of the scattered KI product in the crossed molecular beam reaction of $K + CH_3I \rightarrow KI + CH_3$. The data is taken from two separate experiments. Figure reproduced with permission from D. R. Herschbach's Nobel prize lecture.[15] © The Nobel Foundation 1986.

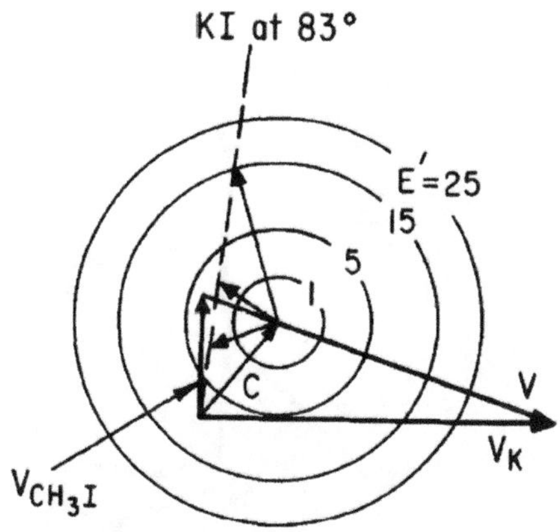

Figure 5.5 Newton diagram for the $K + CH_3I \rightarrow KI + CH_3$ crossed molecular beam experiment drawn using most probable initial reactant velocities. Figure reproduced with permission from D. R. Herschbach's Nobel prize lecture.[15] © The Nobel Foundation 1986.

The Newton diagram is drawn using the most probable initial velocities in the molecular beams. A dotted line is drawn at the laboratory scattering angle of 83°, which is the angle observed to be the maximum of the KI scattering. The circles represent the end positions of possible KI product velocity vectors. Each circle corresponds to a fixed product relative kinetic energy (marked in kcal mol^{-1}). We see from the diagram that many different product relative kinetic energies may result in scattering into the same final laboratory angle. The fact that the most probable KI laboratory scattering angle is observed to be at an angle of 83° clearly indicates that the KI products are scattered "backwards" in the center-of-mass reference frame. Note that the "forward" direction is taken to be that of the initial velocity of the K atom in the center-of-mass frame. Because of this reaction (5.17) is called a "rebound" reaction.

5.3.2 The Modern Age—Universal Detectors

The initial crossed molecular beam reactive scattering experiments were limited to reactions involving alkali atoms. This is because of the very efficient surface ionization detectors used which could detect alkali atoms and their compounds and could also distinguish between them.[18–20] The introduction of universal detectors greatly extended the reactions which could be studied in crossed molecular beams. The universal detectors were based on mass spectrometers with electron impact ionization. The use of this form of detection required the creation of much greater vacuums in the interaction region of the apparatus.[21] Figure 5.6 shows a schematic of a modern crossed molecular beam machine. The apparatus illustrated in the figure, besides being able to detect any atomic or molecular product, also selects the velocity of

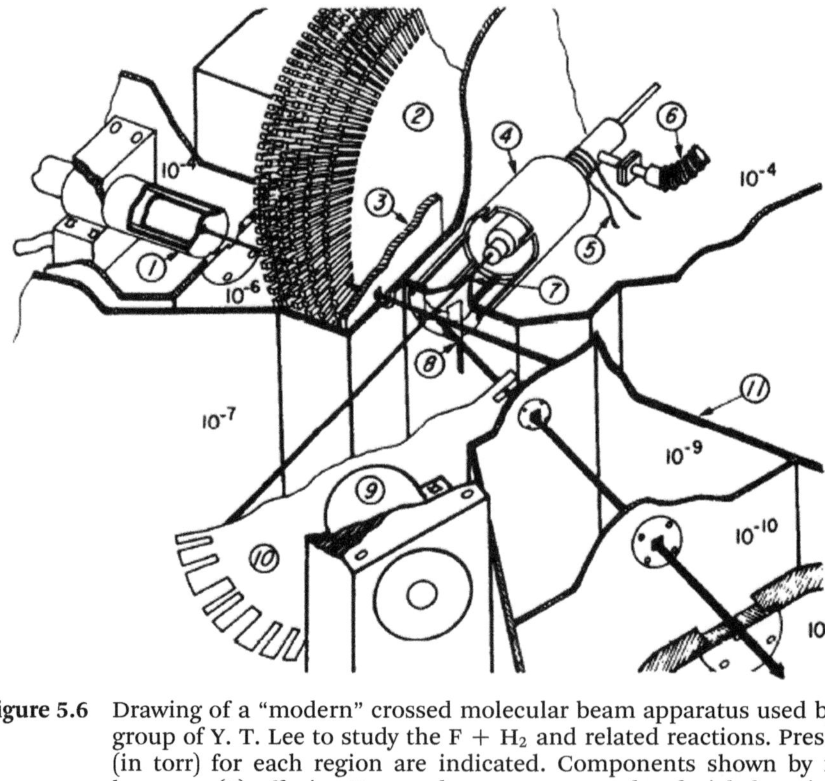

Figure 5.6 Drawing of a "modern" crossed molecular beam apparatus used by the group of Y. T. Lee to study the $F + H_2$ and related reactions. Pressures (in torr) for each region are indicated. Components shown by numbers are: (1) effusive F atom beam source made of nickel, resistively heated; (2) velocity selector; (3) liquid nitrogen cooled cold trap; (4) D_2 or H_2 beam source, supersonic expansion; (5) heater; (6) liquid nitrogen feed line; (7) skimmer; (8) tuning fork chopper; (9) synchronous motor; (10) cross correlation chopper for time-of-flight velocity analysis; (11) ultrahigh vacuum, triply differentially pumped, mass spectrometric detector chamber. Figure reproduced with permission from Y. T. Lee's Nobel prize lecture.[21] © The Nobel Foundation 1986.

one of the initial molecular beams and that of the detected products. The velocity selection provides a far more detailed picture of the scattering process than previously and enables a more reliable theoretical modeling to be undertaken.

Figure 5.7 shows the results obtained for the reaction $F + H_2(j = 0) \rightarrow HF + H$.[22] Because of the ability to select the initial and product beam velocities both the collision and the total energies of the colliding molecules can be specified within close limits. For the figure shown, the collision energy is 1.84 kcal mol^{-1}. The figure shows the angle–velocity distribution of product HF molecules in the center-of-mass reference frame. The dotted circles in the figure indicate the maximum velocity attainable by the HF product in the vibrational levels indicated. A very interesting features of the of the figure

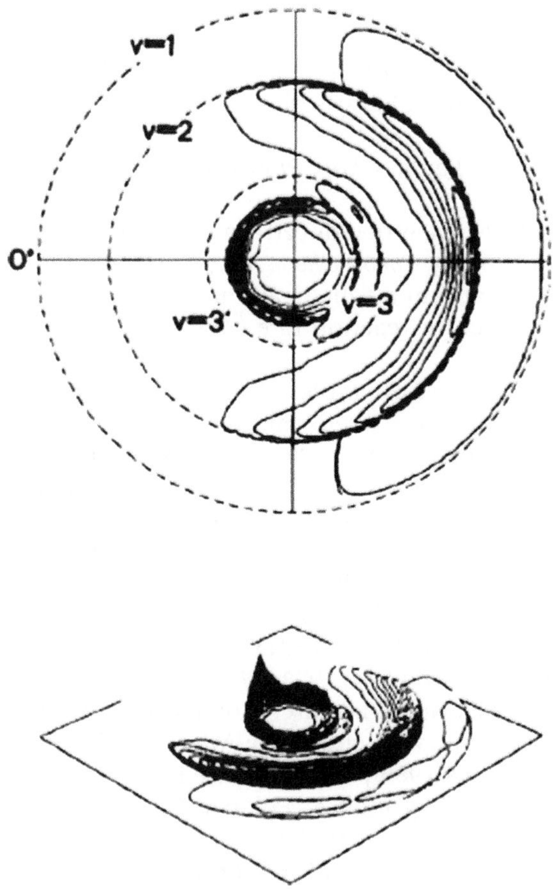

Figure 5.7 Center-of-mass velocity flux contour map for F + *para*-H_2, at collision energy of 1.84 kcal mol^{-1}, shown in three-dimensional perspective. Figure reproduced with permission from Y. T. Lee's Nobel prize lecture.[21] (See also ref. 22). © The Nobel Foundation 1986.

is the forward scattering peak observed for HF($\nu = 3$) as compared with the backward scattering observed for the lower vibrational levels. This together with the sharp energy variation of the scattering patterns was taken to indicate the important contribution that resonances or temporary quasi-bound states played in the dynamics of the collision.

Since the introduction of mass spectrometric detectors in 1969 many new detection methods based on spectroscopy and the use of lasers have been devised. These are discussed in Chapter 9. The reader is referred to an excellent recently written book which reviews the current state of experimental and theoretical research in molecular reaction dynamics.[23]

References

1. W. Kauzmann, *Kinetic Theory of Gases*, W. A. Benjamin Inc., 1st edn, 1966.
2. M. A. Eliason and J. O. Hirschfelder, General collision theory treatment for the rate of bimolecular, gas phase reactions, *J. Chem. Phys.*, 1959, **30**, 1426.
3. P. Atkins and J. de Paula, *Atkins' Phys. Chem.*, Oxford University Press, 9th edn, 2009.
4. J. M. Bowman, Overview of reduced dimensionality quantum approaches to reactive scattering, *Theor. Chem. Acc.*, 2002, **108**, 125.
5. W. H. Miller, Quantum mechanical transition state theory and a new semiclassical model for reaction rate constants, *J. Chem. Phys.*, 1974, **61**, 1823.
6. T. Seideman and W. H. Miller, Calculation of the cumulative reaction probability via a discrete variable representation with absorbing boundary conditions, *J. Chem. Phys.*, 1992, **96**, 4412.
7. U. Manthe and W. H. Miller, The cumulative reaction probability as eigenvalue problem, *J. Chem. Phys.*, 1993, **99**, 3411.
8. D. H. Zhang and J. C. Light, The cumulative reaction probability for the H_2 + OH reaction, *J. Chem. Phys.*, 1997, **106**, 551.
9. D. H. Maylotte, J. C. Polanyi and K. B. Woodall, Energy distribution among reaction products. IV. X + HY (X [equivalent] Cl, Br; Y [equivalent] Br, I), Cl + DI, *J. Chem. Phys.*, 1972, **57**, 1547.
10. J. C. Polanyi, 1986. Nobel Lecture: "Some Concepts in Reaction Dynamics". Available from web site: http://www.nobelprize.org/nobel_prizes/chemistry/laureates/1986/polanyi-lecture.html.
11. F. D. Findlay, J. R. Airey and J. C. Polanyi, The hydrogen–chlorine system in the mm pressure range II. The vibrational ground state studied by self-absorption, *Can. J. Chem.*, 1964, **42**, 2193.
12. J. V. V. Kasper and G. C. Pimentel, HCl chemical laser, *Phys. Rev. Lett.*, 1965, **14**, 352.
13. J. C. Polanyi and K. B. Woodall, Energy distribution among reaction products. VI. F + H_2, D_2, *J. Chem. Phys.*, 1972, **57**, 1574.

14. D. R. Herschbach, Reactive collisions in crossed molecular beams, *Discuss. Faraday Soc.*, 1962, **33**, 149.
15. D. R. Herschbach, 1986. Nobel Lecture: "Molecular Dynamics of Elementary Chemical Reactions". Available from web site: http://www.nobelprize.org/nobel_prizes/chemistry/laureates/1986/herschbach-lecture.html.
16. J. D. McDonald, P. R. LeBreton, Y. T. Lee and D. R. Herschbach, Molecular beam kinetics: reactions of deuterium atoms with halogen molecules, *J. Chem. Phys.*, 1972, **56**, 769.
17. D. R. Herschbach, G. H. Kwei and J. A. Norris, Reactive scattering in crossed molecular beams. K atoms with CH_3I and C_2H_5I, *J. Chem. Phys.*, 1961, **34**, 1842.
18. E. H. Taylor and S. Datz, Study of chemical reaction mechanisms with molecular beams. The reaction of K with HBr, *J. Chem. Phys.*, 1955, **23**, 1711.
19. S. Datz and E. H. Taylor, Ionization on platinum and tungsten surfaces. I: The alkali metals, *J. Chem. Phys.*, 1956, **25**, 389.
20. S. Datz and E. H. Taylor, Ionization on platinum and tungsten surfaces. II: The potassium halides, *J. Chem. Phys.*, 1956, **25**, 395.
21. Y. T. Lee, Nobel Lecture: "Molecular Beam Studies of Elementary Chemical Processes", 1986, available from web site: http://www.nobelprize.org/nobel_prizes/chemistry/laureates/1986/lee-lecture.html.
22. D. M. Neumark, A. M. Wodtke, G. N. Robinson, C. C. Hayden and Y. T. Lee, Molecular beam studies of the F + H_2 reaction, *J. Chem. Phys.*, 1985, **82**, 3045.
23. *Tutorials in Molecular Reaction Dynamics*, ed. M. Brouard and C. Vallance, Royal Society of Chemistry, Cambridge, UK, 2011.

CHAPTER 6

Time-Independent Quantum Theory of Reactive Scattering

The key problem in the quantum theory of reactive scattering is that the most suitable coordinate system for treating the reactants differs from that which is best suited to the description of the products. In a reaction of the type

$$A + BC \rightarrow AB + C \tag{6.1}$$

the reactants are best described by a Jacobi coordinate system in which the $\boldsymbol{R}$ coordinate is the vector connecting the center-of-mass of the diatomic molecule BC to the atom A (see Chapter 3, Figure 3.2 and Appendices C and G). The products however are best described by a different Jacobi coordinate system, namely one in which the $\boldsymbol{R}$ coordinate is the vector connecting the center-of-mass of the diatomic molecule AB to the atom C.

The first explorations of quantum reactive scattering were undertaken on model systems in which all the atoms were confined to lie on a line. Even for this highly simplified model system the coordinate problem is of central importance. To handle the problem in this simple case, a coordinate system called "natural reaction coordinates" was developed. These coordinates change in a continuous manner from reactant to product Jacobi coordinates as the reaction progresses.[1–5] This approach has not proved to be useful for exact three-dimensional quantum calculations and we will not discuss it further.

In 1976 Schatz and Kuppermann[6,7] performed the first exact quantum reactive scattering calculation on a chemically reactive system (see also the calculations of Elkowitz and Wyatt[5]). They used a procedure which required them to solve the Schrödinger equation in all the three reaction channels

RSC Theoretical and Computational Chemistry Series No. 7
Theory of Molecular Collisions
By Gabriel G. Balint-Kurti and Alexander P. Palov

Published by the Royal Society of Chemistry, www.rsc.org

(see Figure 3.1). The configuration space within each channel was divided into three regions and different coordinates were used in each region. The wavefunction was joined smoothly at the boundaries of each region. This required that both the wavefunctions themselves and their first derivatives should be continuous at a chosen dividing or matching plane. In the innermost region the wavefunctions for each of the arrangement channels were matched smoothly to each other. These seminal calculations represented an immense leap forward in the efforts to solve the reactive scattering problem and introduced many methods and concepts which continue to be used in current calculations.

There are in principle many different methods of solving the Schrödinger equation for reactive scattering processes. The three most useful ones may be identified as the use of variational method techniques,[8] the **R** matrix propagation method of Light and co-workers[9,10] and the use of hyperspherical coordinates.[11] Of these the last one has emerged as the most generally useful method and consequently the discussion here will focus on it.

6.1 Potential Energy Surfaces

At this stage of the book it is necessary to say something about the interaction potential V. Until now we have just assumed that it was a simple function of the internal coordinates and have not discussed it further. A potential energy surface is defined within the Born–Oppenheimer approximation.[12–15] This states that we should fix the geometry of the nuclei and solve the Schrödinger equation for the motion of the electrons only. The calculation, at this fixed nuclear geometry, results in one or more electronic energies which constitute the potential energy surfaces that govern the motion of the nuclei.

If only a single or the lowest electronic energy surface is utilized in a dynamical calculation, then we obtain the results for a single surface or electronically adiabatic calculation. Until now in this book these are the only type of calculations we have considered.

When we solve the electronic Schrödinger equation, we will in general obtain many solutions for the electronic wavefunctions corresponding to different electronic energies and states. Each of these energies represents a different potential energy surface. When these surfaces closely approach each other the Born–Oppenheimer approximation is no longer valid[16] and we should consider both surfaces in the dynamical calculation. There are also other situations in which more than a single surface needs to be taken into account: when spin–orbit interactions, which were neglected in the initial electronic structure calculations, are important;[17,18] when electronic states become degenerate due to a particular symmetry arrangement such as passing through a linear geometry;[19] or when a conical intersection is present.[20–22]

Initially the potentials and potential energy surfaces used were analytic model surfaces. For reactive scattering the London–Eyring–Polanyi–Sato

(LEPS)[23,24] surface and then the Porter–Karplus[25] surface served as excellent models for the $H + H_2 \rightarrow H_2 + H$ reaction. If we wish to have any hope of performing dynamical calculations whose results can be compared with experiment, we must use highly accurate potential energy surfaces. In recent times it has become possible to compute such surfaces for moderately sized systems. Typically in a chemical reaction some bonds are broken and others formed. Only a limited number of molecular electronic structure methods are suitable for performing calculations in such situations. The most accurate calculations to date have been performed using the Multi-Reference Configuration Interaction (MRCI) method and the MOLPRO computer code.[26,27]

6.2 Hyperspherical Coordinates

In this chapter we will concentrate on describing the use of hyperspherical coordinates[11,28–35] in solving the quantum reactive scattering problem. The coordinates, the Hamiltonian operator and the coupled differential equations resulting from their use are described in some detail in Appendix G and the reader is referred to this appendix for the more technical details. The coupled differential equations may be solved using the same techniques as used for inelastic scattering problems which have been described in Chapter 4.

The first step in defining hyperspherical coordinates is the definition of a system of mass-scaled Jacobi coordinates.[11,31,36] The use of this scaling (see Appendix G eqn (G-3) and (G-4)) results in a simplification of the kinetic energy operator which now involves only a single effective mass. For the reactant arrangement channel in eqn (6.1), which we label with a subscript α, the scaling takes the form

$$\begin{aligned}\tilde{\boldsymbol{R}}_\alpha &= d_\alpha \boldsymbol{R}_\alpha,\\ \tilde{\boldsymbol{r}}_\alpha &= d_\alpha^{-1} \boldsymbol{r}_\alpha\end{aligned} \tag{6.2}$$

where the dimensionless scaling factor d_α is

$$d_\alpha = \left[\frac{m_A}{\mu}\left(1 - \frac{m_A}{M}\right)\right]^{\frac{1}{2}}, \tag{6.3}$$

$M = m_A + m_B + m_C$ is the total mass of the triatomic system and the three-body reduced mass, μ, is

$$\mu = \left(\frac{m_A m_B m_C}{M}\right)^{\frac{1}{2}}. \tag{6.4}$$

In terms of this mass-scaled Jacobi coordinate system the Schrödinger equation for the radial wavefunction remains the same as before (see eqn (3.18)), except that the reduced mass is replaced by the three-body reduced mass of eqn (6.4). Another advantage of using mass-scaled Jacobi coordinates is that it becomes easy to switch between different arrangement channels. The change is achieved by a unitary coordinate transformation termed a "kinematic rotation" (see eqn (G-8)).

Thus if we identify the arrangement channel A+BC by a subscript α and the channel AB+C by a subscript γ we can write a transformation between the two sets of Jacobi coordinates as

$$\begin{pmatrix} \tilde{\boldsymbol{R}}_\gamma \\ \tilde{\boldsymbol{r}}_\gamma \end{pmatrix} = \mathbf{T}(\chi_{\gamma\alpha}) \begin{pmatrix} \tilde{\boldsymbol{R}}_\alpha \\ \tilde{\boldsymbol{r}}_\alpha \end{pmatrix} \tag{6.5}$$

where $\begin{pmatrix} \tilde{\boldsymbol{R}}_\alpha \\ \tilde{\boldsymbol{r}}_\alpha \end{pmatrix}$ is a column vector of length 6

$$\begin{pmatrix} \tilde{\boldsymbol{R}}_\alpha \\ \tilde{\boldsymbol{r}}_\alpha \end{pmatrix} = \begin{pmatrix} \tilde{X}_\alpha \\ \tilde{Y}_\alpha \\ \tilde{Z}_\alpha \\ \tilde{x}_\alpha \\ \tilde{y}_\alpha \\ \tilde{z}_\alpha \end{pmatrix} \tag{6.6}$$

and $\mathbf{T}(\chi_{\gamma\alpha})$ is a 6×6 unitary transformation matrix. The angle $\chi_{\gamma\alpha}$ is entirely determined by the atomic masses (see Appendix G for further details).

Having defined the mass-scaled Jacobi coordinates the next step is to relate them to hyperspherical coordinates. There are several variants of these coordinates.[11,28–32] The key aspect is the replacement of two spatial variables by one spatial variable and an angular one. In mass-scaled Jacobi coordinates the three internal variables are $\tilde{R}_\alpha$ and $\tilde{r}_\alpha$ and the angle γ_α (the angle between the two vectors $\tilde{\boldsymbol{R}}_\alpha$ and $\tilde{\boldsymbol{r}}_\alpha$, see eqn (G-2)). In Delves[11,28,29,31] hyperspherical coordinates the distance variables $\tilde{R}_\alpha$ and $\tilde{r}_\alpha$ are replaced by a hyperradius ρ and a new angle α_α

$$\begin{aligned} \rho &= (\tilde{R}_\alpha^2 + \tilde{r}_\alpha^2)^{\frac{1}{2}} \\ \alpha_\alpha &= \tan^{-1}\left(\frac{\tilde{r}_\alpha}{\tilde{R}_\alpha}\right). \end{aligned} \tag{6.7}$$

The hyperradius provides a measure of the overall size of the system and is independent of arrangement channel, while the hyperangle α_α reflects the shape of the triatomic system. Kuppermann's definition of hyperspherical coordinates[30,31] differs from those of Delves by using $2 \times \alpha_\alpha$ as the hyperangle. We follow the original prescription of Delves.[11,28,29,31]

The hyperspherical coordinates defined by eqn (6.7) may be referred to as "arrangement channel hyperspherical coordinates" as two of the angles, γ_α, the angle between the two Jacobi vectors $\tilde{\boldsymbol{R}}_\alpha$ and $\tilde{\boldsymbol{r}}_\alpha$, and α_α, depend on the arrangement channel. Using such coordinates would require us to solve for the wavefunctions in all three arrangement channels separately and then to join the solutions at some selected planes in a smooth manner.

Following the method of Pack and Parker[11] we can define a set of hyperspherical coordinates that are independent of the arrangement channel. This is done by changing the transformation angle $\chi_{\gamma\alpha}$ to a variable $\chi = \chi_{Q\alpha}$.

The transformation angle is then determined by the condition that $|\tilde{\boldsymbol{R}}_\mathrm{Q}|$ is maximized and $|\tilde{\boldsymbol{r}}_\mathrm{Q}|$ minimized. The resulting angle obeys the condition

$$\tan(2\chi) = \frac{2\,\tilde{\boldsymbol{R}}_\alpha \cdot \tilde{\boldsymbol{r}}_\alpha}{\tilde{R}_\alpha^2 - \tilde{r}_\alpha^2}. \tag{6.8}$$

The set of coordinates obtained when this angle is used in the kinematic transformation of eqn (6.5) are such that $\tilde{\boldsymbol{R}}_\mathrm{Q}$ lies in the molecular plane along the principal axis of inertia associated with the smallest moment of inertia and $\tilde{\boldsymbol{r}}_\mathrm{Q}$ is perpendicular to it in the molecular plane (*i.e.* $\tilde{\boldsymbol{r}}_\mathrm{Q}$ lies along the principal axis of inertia associated with the middle value of the moment of inertia).

The body-fixed axes are then defined such that the Y axis lies perpendicular to the molecular plane and is therefore in the direction of the principal axis of inertia associated with the largest moment. The Z axis lies along $\tilde{\boldsymbol{R}}_\mathrm{Q}$ and the X axis is along $\tilde{\boldsymbol{r}}_\mathrm{Q}$. The six hyperspherical coordinates are therefore the hyperradius

$$\rho = (\tilde{R}_\mathrm{Q}^2 + \tilde{r}_\mathrm{Q}^2)^{\frac{1}{2}}, \tag{6.9}$$

an angle closely related to the hyperangle $\alpha_\mathrm{Q} = \tan^{-1}(r_\mathrm{Q}/R_\mathrm{Q})$

$$\theta = \pi/2 - 2 \cdot \tan^{-1}(r_\mathrm{Q}/R_\mathrm{Q}), \tag{6.10}$$

the kinematic transformation angle χ of eqn (6.8) and the three Euler angles needed to rotate the space-fixed axes to the body-fixed ones.

This set of hyperspherical coordinates are termed adiabatically adjusting, principal axes hyperspherical (APH) coordinates.[11] They have the great advantage that they are not associated with any particular arrangement channel but rather change *adiabatically* from reactant-like coordinates to product-like ones as the reaction proceeds. Note however that even in a situation when the A–BC separation is large the coordinates are not identical to the reactant Jacobi coordinates because $\tilde{\boldsymbol{R}}_\mathrm{Q}$ is always perpendicular to $\tilde{\boldsymbol{r}}_\mathrm{Q}$. The reader is referred to Appendix G and to articles by Pack and Parker[11] and by Johnson[31] for a more detailed discussion. The APH coordinates are similar to ones initially proposed by Whitten and Smith.[31,37]

The overall strategy for solving the Schrödinger equation is therefore to solve the differential equations in the inner region using the APH coordinates. At large separations, as in the case of inelastic scattering (see Appendix D Section D.5), the **S** matrix must be obtained by analyzing the wavefunction expressed in terms of space-fixed Jacobi coordinates. To achieve this asymptotic analysis the wavefunction in the inner region must be joined smoothly to solutions of the Schrödinger equation expressed in different coordinates as the hyperradius increases. The APH wavefunction is progressively changed first to body-fixed arrangement channel hyperspherical coordinates, then to body-fixed Jacobi coordinates and finally to space-fixed Jacobi coordinates. The last of these wavefunction transformations is identical to the body-fixed → space-fixed transformation discussed in Appendix D while all of them are carefully described in the article of Pack and Parker.[11]

6.3 The Wavefunction, S Matrix and Cross Sections

The centrally important **S** matrix is obtained from an analysis of the asymptotic form of the wavefunction which results from solving the coupled differential eqn (G-19) and (G-43). All of the relevant equations, such as the asymptotic form of the wavefunction and the relationship between the **S** matrix and the cross sections are very similar to those we have previously outlined in Chapter 3 for the case of inelastic molecular collisions. Nevertheless we repeat some of the key formulae here together with the appropriate generalizations to take account of the fact that there are now different arrangement channels which must be taken into account.

In analogy to eqn (3.8), the asymptotic form of the wavefunction in the space-fixed reference frame is

$$\psi(\boldsymbol{R},\boldsymbol{r},E)\,\widetilde{_{R\to\infty}}\left(\frac{k_{vj}\mu_\alpha}{2\pi h^2}\right)^{\frac{1}{2}}\Bigg\{\mathrm{e}^{ik_{vj}Z}\chi_{vj}(r)\,Y_{j,m_j}(\theta_r,\phi_r)\delta_{\beta\alpha} + \sum_{\beta v'j'm_j'} f_{\beta v'j'm_{j'}\leftarrow\alpha vjm_j}(\theta_R,\phi_R)\,\frac{\mathrm{e}^{ik_{v'j'}R}}{R} \times \chi_{v'j'}(r)\,Y_{j',m_{j'}}(\theta_r,\phi_r)\Bigg\}. \tag{6.11}$$

This equation is similar to eqn (3.8) where we were discussing atom–diatom inelastic scattering. Now however there is the possibility of a reaction and a consequent change in arrangement channel. This has led to the introduction of an arrangement channel index: α for the initial arrangement channel and β for the final one. So as not to make the notation too cumbersome we have omitted the arrangement channel indices from all the quantum numbers and variables, but the arrangement channel, with which quantum numbers such as v and variables such as R are associated, should always be clear from the context. In eqn (6.11) and subsequent equations, θ_R, ϕ_R and $k_{v'j'}$ are associated with arrangement channel β and k_{vj} with arrangement channel α.

The differential cross sections for elastic, inelastic and reactive scattering are then given by (see eqn (10.10) of ref. 38)

$$\sigma_{\beta v'j'm_{j'}\leftarrow\alpha vjm_j}(\theta_R,\phi_R,E) = \frac{\mu_\alpha k_{v'j'}}{\mu_\beta k_{vj}}\left|f_{\beta v'j'm_{j'}\leftarrow\alpha vjm_j}(\theta_R,\phi_R)\right|^2 \tag{6.12}$$

where μ_α and μ_β are the reduced masses for the relative motion of the reactants and products respectively. These extra mass factors, as compared with eqn (3.9), arise so as to conserve the inward and outward mass flows, as the relative mass of the reactants and products differ.

In analogy with eqn (3.14) the expansion of the wavefunction in terms of the radial wavefunctions for the various arrangement channels using the space-fixed reference frame is now given by

$$\psi(\mathbf{R},\mathbf{r},E) = \left(\frac{k_{vj}\mu_\alpha}{2\pi h^2}\right)^{\frac{1}{2}} \sum_{JM\ell}\{4\pi(2\ell+1)\}^{1/2} i^\ell \times \sum_{\beta v' j' \ell'} \frac{\phi^{J\alpha vj\ell}_{\beta v' j' \ell'}(R,E)}{k_{vj}R} \chi_{v' j'}(r) \mathcal{Y}^{JM}_{j'\ell'}(\hat{\mathbf{R}},\hat{\mathbf{r}})(j\ell JM|jm_j\ell 0). \quad (6.13)$$

The boundary condition for the space-fixed radial wavefunction now becomes

$$\begin{aligned}\phi^{J\alpha vj\ell}_{\beta v' j' \ell'}(R,E) &\underset{R\to 0}{\widetilde{\quad}} 0,\\ \phi^{J\alpha vj\ell}_{\beta v' j' \ell'}(R,E) &\underset{R\to\infty}{\widetilde{\quad}} \hat{j}_{\ell'}(k_{v'j'}R)\delta_{\alpha,\beta}\delta_{v,v'}\delta_{j,j'}\delta_{\ell,\ell'}\\ &\quad - T^J_{\beta v' j' \ell',\alpha vj\ell}\frac{1}{2i}\left(\frac{\mu_\beta k_{vj}}{\mu_\alpha k_{v'j'}}\right)^{\frac{1}{2}} e^{i\left(k_{v'j'}R-\frac{\ell'\pi}{2}\right)}. \end{aligned} \quad (6.14)$$

In analogy to eqn (3.20) this set of boundary conditions provide the means of evaluating the centrally important **T** and **S** matrices. They differ from eqn (3.20) through the presence of the arrangement channel indices α and β and the ratio of the masses $\left(\frac{\mu_\beta}{\mu_\alpha}\right)^{\frac{1}{2}}$. The **T** and **S** matrices are related by the equation (see eqn (3.22))

$$S^J_{\beta v' j' \ell',\alpha vj\ell} = \delta_{\alpha,\beta}\delta_{v,v'}\delta_{j,j'}\delta_{\ell,\ell'} - T^J_{\beta v' j' \ell',\alpha vj\ell}. \quad (6.15)$$

The body-fixed **T** matrix is given by eqn (3.47) and eqn (D-41):

$$T^J_{\beta v' j' \Omega',\alpha vj(\Omega=m_j)} = \sum_{\ell''\ell'} B_{\ell',\Omega'}\, T^J_{\beta v' j' \ell',\alpha vj\ell''} B_{\ell'',m_j}. \quad (6.16)$$

The detailed differential cross section is then obtained from the equation (see eqn (D-43))

$$\sigma_{\beta v' j' \Omega' \leftarrow \alpha vjm_j}(\theta_R,\phi_R,E) = \frac{1}{4k^2_{vj}}\left|\sum_J (2J+1) T^J_{\beta v' j' \Omega',\alpha vjm_j} d^J{}_{m_j,\Omega'}(\theta_R)\right|^2. \quad (6.17)$$

This cross section is often averaged over the initial m_j states and summed over the final Ω' states to give the averaged state-to-state differential cross section

$$\sigma_{\beta v' j' \leftarrow \alpha v j}(\theta_R, \phi_R, E) = \frac{1}{4 k_{vj}^2} \frac{1}{2j+1} \sum_{\Omega'=-|\min(J,j')|}^{\Omega'=|\min(J,j')|} \sum_{m_j=-|\min(J,j)|}^{m_j=|\min(J,j)|} \times \left| \sum_J (2J+1) T^J_{\beta v' j' \Omega', \alpha v j m_j} d^J{}_{m_j,\Omega'}(\theta_R) \right|^2 . \quad (6.18)$$

The state-to-state integral cross section is given by (see Appendix D eqn (D-45) and eqn (3.49))

$$\sigma_{\beta v' j' \Omega' \leftarrow \alpha v j m_j}(E) = \frac{\pi}{k_{vj}^2} \sum_J (2J+1) \left| T^J_{\beta v' j' \Omega', \alpha v j m_j j} \right|^2 . \quad (6.19)$$

The integral cross section averaged over the initial m_j and summed over the final Ω' quantum numbers is

$$\sigma_{\beta v' j' \leftarrow \alpha v j}(E) = \frac{\pi}{k_{vj}^2} \frac{1}{2j+1} \sum_J \sum_{\Omega'=-|\min(J,j')|}^{\Omega'=|\min(J,j')|} \sum_{m_j=-|\min(J,j)|}^{m_j=|\min(J,j)|} (2J+1) \times \left| T^J_{\beta v' j' \Omega', \alpha v j m_j} \right|^2 . \quad (6.20)$$

All the above expressions for the various cross sections (eqn (6.17)–(6.20)) apply to all processes: elastic scattering ($\beta v' j' = \alpha v j$), inelastic scattering ($\beta = \alpha$ but $v' j' \neq v j$) and reactive scattering ($\beta \neq \alpha$).

We can now define a reaction probability for each total J such that

$$\sigma_{\beta v' j' \leftarrow \alpha v j}(E) = \sum_J P_{\beta v' j' \leftarrow \alpha v j}(J, E) \quad (6.21)$$

where $\beta \neq \alpha$ and

$$P_{\beta v' j' \leftarrow \alpha v j}(J, E) = \frac{\pi}{k_{vj}^2} \frac{1}{2j+1} \sum_{\Omega'=-|\min(J,j')|}^{\Omega'=|\min(J,j')|} \sum_{m_j=-|\min(J,j)|}^{m_j=|\min(J,j)|} (2J+1) \times \left| T^J_{\beta v' j' \Omega', \alpha v j m_j} \right|^2 . \quad (6.22)$$

6.4 Exact Reactive Scattering Calculations for Atom–Diatom Reactions

6.4.1 H + H_2 Reactive Scattering

Because it is the simplest reaction, and because reasonable potential energy surfaces were available from a very early stage, the H + $H_2 \rightarrow H_2$ + H reaction has received an immense amount of attention and has served as the model system used in the development of all reactive scattering methods and computer codes. Scattering resonances were first identified in collinear model calculations[39,40] and also later in full three-dimensional ones for total angular momentum $J = 0$.[41] Miller and Zhang[42] gave the first in-depth discussion of a consequence of a reactive scattering resonance which could be experimentally observed. They found that although clear signs of resonances were apparent in the energy dependence of the reaction probabilities for fixed values of the total angular momentum J, when these were summed to yield the integral cross section, no sign of the resonances survived. This arises because the resonance features for each J value occur at different energies and therefore mask each other when the summation is performed.

Figure 6.1 shows calculated integral cross sections for the reaction $H + H_2 \rightarrow H_2 + H$[43] in which the indistinguishability of the three H atoms has not been taken into account. The main point of showing the figure is to demonstrate that there is no sign of any resonance feature.

Figure 6.2 shows the calculated reaction probability (eqn (6.22)) for the reaction $H + H_2(\nu = 0, j = 0) \rightarrow H_2(\nu' = 0, j' = 1)$ for several fixed values of the total angular momentum J as a function of the collision energy.[42] The three maxima seen in these plots arise from reactive scattering resonances. If we focus on the maximum at around $E = 0.7$ eV, we see that there is no indication of this in the integral cross section shown in Figure 6.1. If however we look at the differential cross section, shown in Figure 6.3, plotted as a function of both scattering angle and energy, we see that the reactive scattering resonances show up as ridges in the three-dimensional landscape. The presence of reactive scattering resonances in this reaction was

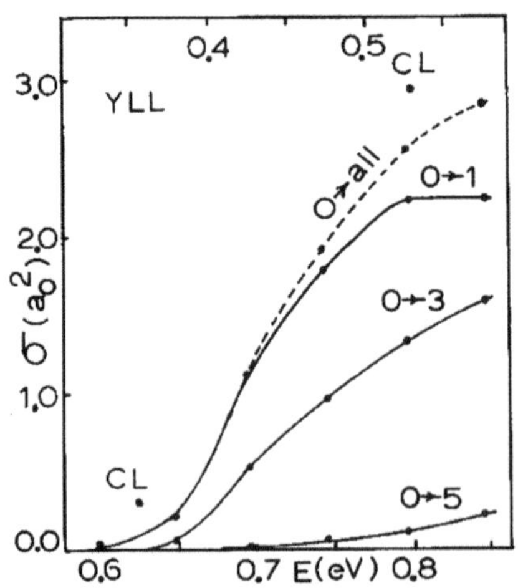

Figure 6.1 Computed integral cross section for the reaction $H + H_2(\nu = 0, j = 0) \rightarrow H_2(\nu' = 0, j')$. Figure reproduced with permission from A. B. Elkowitz and R.E. Wyatt, *J. Chem. Phys.* **62**, 2504. Copyright 1975, AIP Publishing LLC.[43] DOI: 10.1063/1.430734.

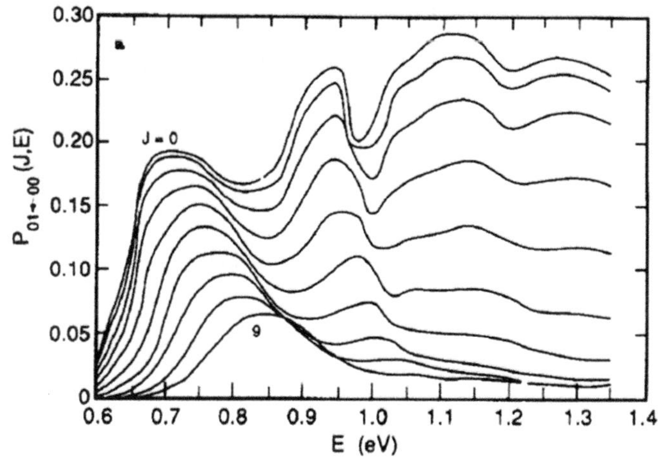

Figure 6.2 Computed reaction probabilities for the reaction $H+H_2(\nu=0, j=0)\rightarrow H_2(\nu'=0, j'=1)$ for fixed values of the total angular momentum J as a function of the collision energy E. Figure reproduced with permission from W. H. Miller and J. Z. H. Zhang, *J. Phys. Chem.*, 1991, **95**, 12.[42]

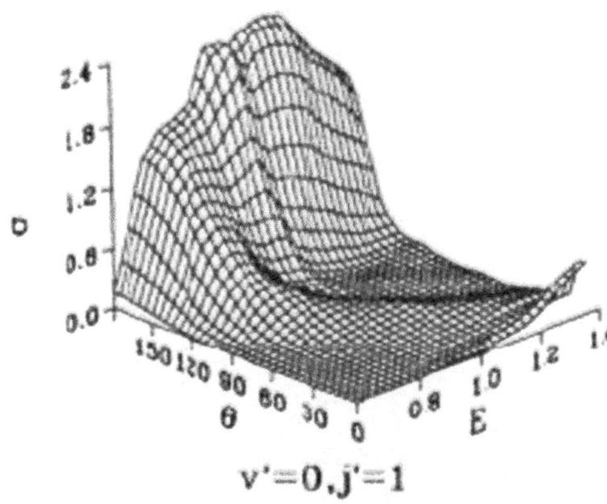

Figure 6.3 Differential cross section averaged over m_j states for the reaction $H+H_2(\nu=0, j=0)\rightarrow H_2(\nu'=0, j'=1)$ as a function of both scattering angle and collision energy. Figure reproduced with permission from W. H. Miller and J. Z. H. Zhang, *J. Phys. Chem.*, 1991, **95**, 12.[42]

established at an early stage by Hipes and Kuppermann[41] using a lifetime analysis (see Chapter 2 Section 2.6).

More recently, time-dependent quantum scattering calculations and comparisons with experimental results[44] have confirmed that for the isotopically related reaction $H+D_2(\nu=0, j=0)\rightarrow DH(\nu'=3, j'=0)+D$ there is a resonance mechanism which is associated with a time delay in the reaction dynamics and manifests itself in a ridge in the three-dimensional angle–energy map of the differential cross section as well as in a forward scattering peak at some energies.

6.4.2 F + HD → HF + D Reactive Scattering

After $H + H_2$, the most studied reaction at the detailed levels we are considering here has been the $F + H_2$ reaction. Experimentally it was first studied in infrared chemiluminescence experiments by Polanyi and co-workers.[45] Extensive quantum mechanical theoretical calculations, carried out on a collinear model of the reaction,[46–48] strongly indicated the participation of

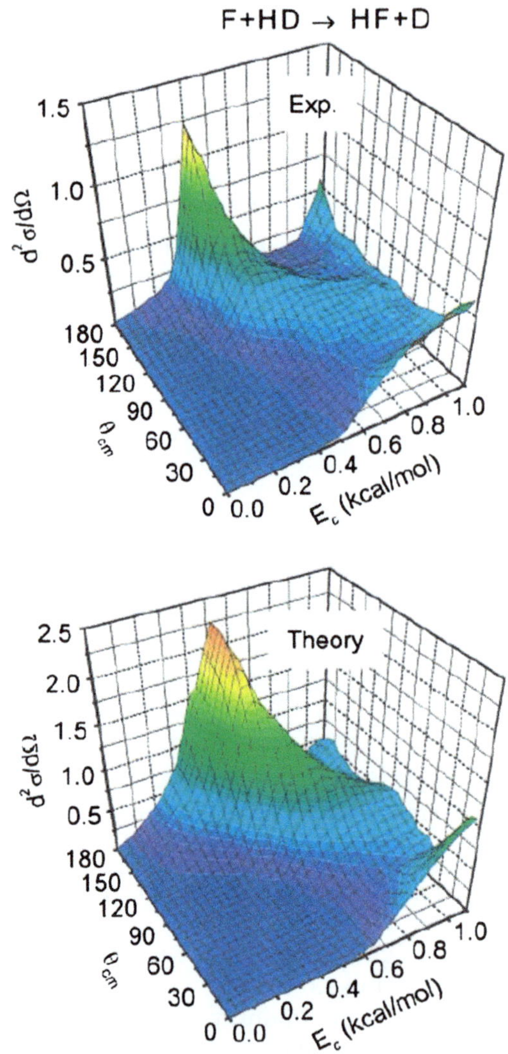

Figure 6.4 Differential cross section averaged over reactant m_j states and summed over product vibrational–rotational states for the reaction $F + HD(v = 0, j = 0) \rightarrow HF + D$ as a function of both scattering angle and collision energy. Figure reproduced with permission from R. T. Skodje, D. Skouteris, D. E. Manolopoulos, S.-H. Lee, F. Dong and K. Liu, *Phys. Rev. Lett.*, 2000, **85**, 1206. Copyright 2000, American Physical Society.[51]

reactive scattering (Feschbach) resonances (see Chapter 2 Section 2.6). Even in the early collinear model calculations, when different isotopic variants were examined, the largest resonance features were identified as occurring in the F+HD→HF+D reaction.[48] The F+H_2 system was the subject of extensive experimental[49] and theoretical[50] study, with great attention

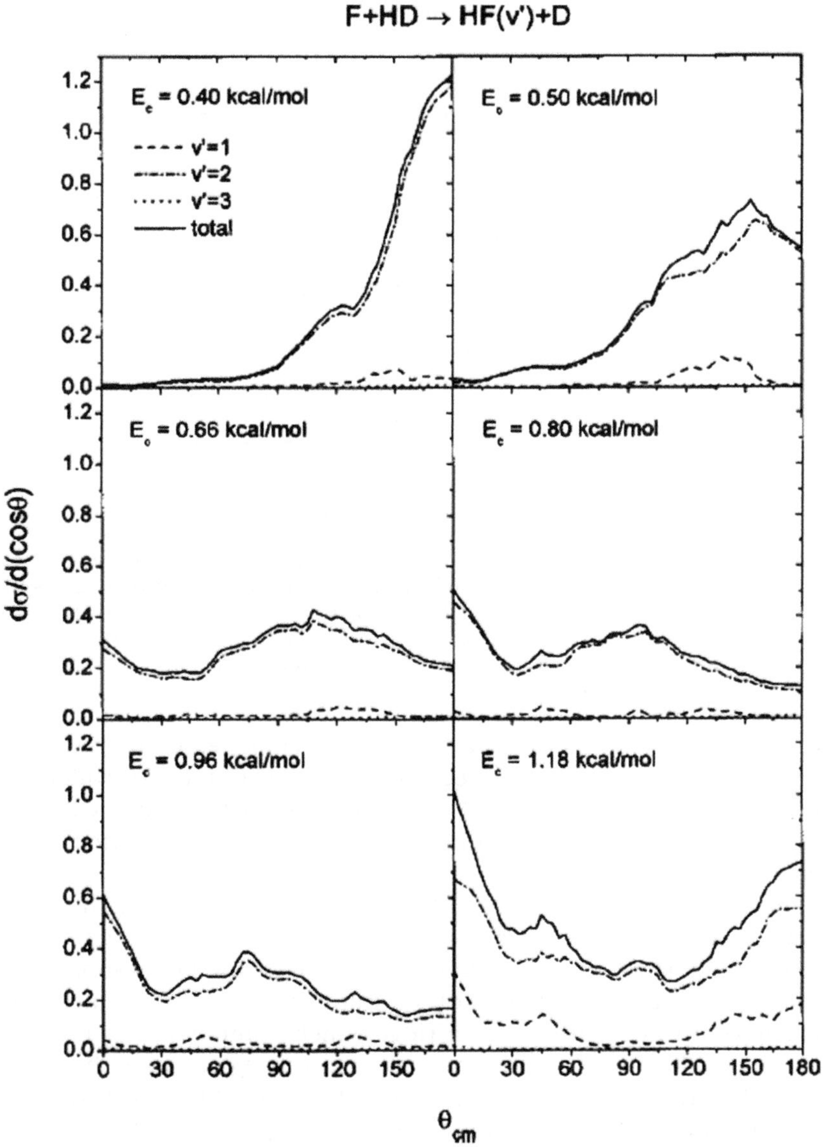

Figure 6.5 Experimentally determined, vibrational state resolved, differential cross section for the reaction F+HD($v = 0, j = 0$)→HF(v')+D. Reproduced with permission from S.-H. Lee, F. Dong and K. Liu, *J. Chem. Phys.*, 2002, **116**, 7839. Copyright 2002, AIP Publishing LLC.[52] DOI: 10.1063/1.430734.

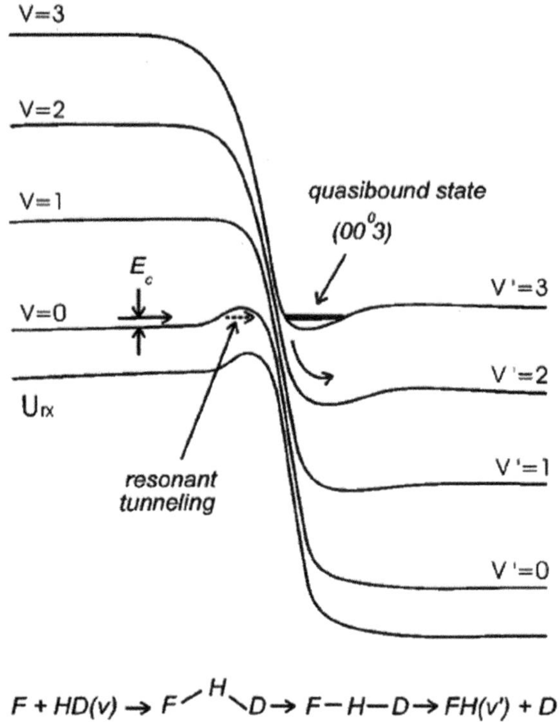

Figure 6.6 Schematic diagram of vibrationally adiabatic potential energy curves showing the correlation of reactant to product vibrational levels for the reaction F + HD(ν) $\rightarrow$ HF(ν') + D. Reproduced with permission from S.-H. Lee, F. Dong and K. Liu, *J. Chem. Phys.*, 2002, **116**, 7839. Copyright 2002, AIP Publishing LLC.[52] DOI: 10.1063/1.1468880.

always being paid to the observation of resonance features. In 2000–2002 Liu and co-workers first observed conclusive experimental evidence of a resonance feature in the F + HD $\rightarrow$ HF + D reaction[51,52] cross section. The work was executed in collaboration with extensive exact *ab initio* theoretical calculations.[51]

Figure 6.4 shows the reactive differential cross section for the reaction. The cross section is shown as a function of both angle and energy. As in the case of the H + H_2 reaction an important signature of a resonance is the sharply changing angular nature of the scattering over a small energy range. This shows up in Figure 6.4 as a ridge which commences as a backward scattering peak and progresses to smaller scattering angles with increasing energy. Near the resonance energy at 1.18 kcal mol^{-1} (*i.e.* close to the high energy edge of the figure) the cross section is roughly symmetrically forward-backward peaked, this being the signature of a long-lived intermediate. Both experimental and theoretical results, computed using the techniques outlined earlier in the chapter together with the best available potential energy surfaces, are shown in the figure. The two pictures agree reasonably well. The small differences are attributed mainly to small residual errors in the *ab initio* potential energy surface. Note also that the fluorine atom has two possible spin–orbit states. These have not been properly accounted for in the theoretical calculations. However when proper account is taken of these states[17,53] it makes only a small difference to the outcome. The differential cross sections for the production of the FD product do not show any analogous behavior. They are backward peaked and show no sudden change in character with small variations in energy.

Figure 6.5 shows experimentally determined vibrational state resolved differential cross sections for the reaction $F+HD(\nu=0, j=0)\rightarrow HF(\nu')+D$ at six closely spaced low collision energies. We see clearly how the cross sections change from being entirely backward peaked to being nearly symmetric in the backward-forward directions. The highest energy shown corresponds to the resonance energy of the system, just 0.02 kcal mol^{-1} above the threshold energy for producing the $HF(\nu'=3)$ product. The $\nu'=2$ state dominates the cross section throughout this energy range.

The origin of the resonance state is shown in Figure 6.6 as being a vibrationally excited compound state which correlates asymptotically with the $HF(\nu'=3)$ level. The molecular system becomes temporarily trapped in a quasi-bound state which does not have sufficient energy in the dissociative coordinate to break up. After a time the energy is redistributed and the system breaks up to yield $HF(\nu'=2)+D$. This corresponds to a classical Feschbach resonance.

References

1. R. A. Marcus, On the analytical mechanics of chemical reactions: quantum mechanics of linear collisions, *J. Chem. Phys.*, 1966, **45**, 4493.
2. R. A. Marcus, Analytical mechanics of chemical reactions. III. Natural collision coordinates, *J. Chem. Phys.*, 1968, **49**, 2610.
3. J. N. L. Connor and M. S. Child, Differential cross sections for chemically reactive systems, *Mol. Phys.*, 1970, **18**, 653.
4. R. B. Walker, J. C. Light and A. Altenberger-Siczek, Chemical reaction theory for asymmetric atom–molecule collisions, *J. Chem. Phys.*, 1976, **64**, 1166.
5. A. B. Elkowitz and R. E. Wyatt, Three-dimensional natural coordinate asymmetric top theory of reactions: application to $H+H_2$, *J. Chem. Phys.*, 1975, **63**, 702.
6. G. C. Schatz and A. Kuppermann, Quantum mechanical reactive scattering for three-dimensional atom plus diatom systems. I. Theory, *J. Chem. Phys.*, 1976, **65**, 4642.
7. G. C. Schatz and A. Kuppermann, Quantum mechanical reactive scattering for three-dimensional atom plus diatom systems. II. Accurate cross sections for $H + H_2$, *J. Chem. Phys.*, 1976, **65**, 4668.
8. J. Z. H. Zhang and W. H. Miller, Quantum reactive scattering *via* the S-matrix version of the Kohn variational principle: differential and

integral cross sections for D + $H_2 \rightarrow$ HD + H, *J. Chem. Phys.*, 1989, **91**, 1528.
9. F. Webster and J. C. Light, Atom–diatom reactive scattering. I. Quantum theory, *J. Chem. Phys.*, 1989, **90**, 265.
10. F. Webster and J. C. Light, Atom–diatom reactive scattering. II. H + H_2 and its isotopomers, $J = 0$, *J. Chem. Phys.*, 1989, **90**, 300.
11. R. T. Pack and G. A. Parker, Quantum reactive scattering in three dimensions using hyperspherical (APH) coordinates. Theory, *J. Chem. Phys.*, 1987, **87**, 3888.
12. L. Pauling and E. B. Wilson, *Introduction to Quantum Mechanics*, McGraw-Hill Book Co., 1935.
13. H. Eyring, J. Walter and G. E. Kimball, *Quantum Chemistry*, John Wiley and Sons Inc., 1st edn, 1944.
14. L. D. Landau and E. M. Lifshitz, *Quantum Mechanics: Non-Relativistic Theory*, Butterworth-Heinemann, Oxford, 3rd edn, 1977 (Reprinted with corrections 1991).
15. P. Atkins and J. de Paula, *Atkins' Phys. Chem.*, Oxford University Press, 9th edn, 2009.
16. M. Born and J. R. Oppenheimer, Zur Quantentheorie der Molekeln, *Ann. Phys.*, 1927, **85**, 457.
17. M. H. Alexander, H.-J. Werner and D. E. Manolopoulos, Spin–orbit effects in the reaction of F(^{2}P) with H_2, *J. Chem. Phys.*, 1998, **109**, 5710.
18. M. H. Alexander, G. Capecchi and H.-J. Werner, Details and consequences of the nonadiabatic coupling in the Cl(^{2}P) + H_2 reaction, *Faraday Discuss.*, 2004, **127**, 59.
19. P. Gamallo, P. Defazio, M. Gonzlez and C. Petrongolo, Renner–Teller coupled-channel dynamics of the N(^{2}D) + H_2 reaction and the role of the NH_2 $\widetilde{A}^2A_1$ electronic state, *J. Chem. Phys.*, 2008, **129**, 244307.
20. S. Mahapatra, H. Köppel and L. S. Cederbaum, Reactive scattering dynamics on conically intersecting potential energy surfaces: the H + H_2 exchange reaction, *J. Phys. Chem. A*, 2001, **105**, 2321.
21. J. C. Juanes-Marcos and S. C. Althorpe, On the role of the conical intersection in H + H_2 reactive scattering, *Chem. Phys. Lett.*, 2003, **381**, 743.
22. S. K. Gray, G. G. Balint-Kurti, G. C. Schatz, J. J. Lin, X. Liu, S. Harich and X. Yang, Probing the effect of the H_2 rotational state in O(^{1}D) + H_2 $\rightarrow$ OH + H: theoretical dynamics including nonadiabatic effects and a crossed molecular beam study, *J. Chem. Phys.*, 2000, **113**, 7330.
23. S. Sato, On a new method of drawing the potential energy surface, *J. Chem. Phys.*, 1955, **23**, 592.
24. S. Sato, Potential energy surface of the system of three atoms, *J. Chem. Phys.*, 1955, **23**, 2465.
25. R. N. Porter and M. Karplus, Potential energy surface for H_3, *J. Chem. Phys.*, 1964, **40**, 1105.
26. H. J. Werner, P. J. Knowles, G. Knizia, F. R. Manby and M. Schütz, Molpro: a general-purpose quantum chemistry program package, *WIREs Comput. Mol. Sci.*, 2012, **2**, 242.

27. H.-J. Werner, P. J. Knowles, G. Knizia, F. R. Manby, M. Schütz, *et al.*, Molpro, version 2012.1, a package of ab initio programs, See http://www.molpro.net, Cardiff, UK, 2012.
28. L. M. Delves, Tertiary and general-order collisions, *Nucl. Phys.*, 1958–1959, **9**, 391.
29. L.M. Delves, Tertiary and general-order collisions (II), *Nucl. Phys.*, 1960, **20**, 275.
30. A. Kuppermann, A useful mapping of triatomic potential energy surfaces, *Chem. Phys. Lett.*, 1975, **32**, 374.
31. B. R. Johnson, On hyperspherical coordinates and mapping the internal configurations of a three body system, *J. Chem. Phys.*, 1980, **73**, 5051.
32. V. Aquilanti, S. Cavalli and D. De Fazio, Hyperquantization algorithm. I. Theory for triatomic systems, *J. Chem. Phys.*, 1998, **109**, 3792.
33. D. Skouteris, J. F. Castillo and D. E. Manolopoulos, ABC: a quantum reactive scattering program, *Comput. Phys. Commun.*, 2000, **133**, 128.
34. P. Honvault and J.-M. Launay, A quantum-mechanical study of the dynamics of the $O(^1D) + H_2 \rightarrow OH + H$ insertion reaction, *J. Chem. Phys.*, 2001, **114**, 1057.
35. J. Crawford and G. A. Parker, State-to-state three-atom time-dependent reactive scattering in hyperspherical coordinates, *J. Chem. Phys.*, 2013, **138**, 054313.
36. J. O. Hirschfelder and J. S. Dahler, The kinetic energy of relative motion, *Proc. Natl. Acad. Sci. U. S. A.*, 1956, **42**, 363.
37. R. C. Whitten and F. T. Smith, Symmetric representation for three-body problems. II. Motion in space, *J. Math. Phys.*, 1968, **9**, 1103.
38. M. S. Child, *Molecular Collision Theory*, Academic Press, 1974. Corrected Dove Publication 1996.
39. R.D. Levine and S.-F. Wu, Resonances in reactive collisions: computational study of the $H + H_2$ collision, *Chem. Phys. Lett.*, 1971, **11**, 557.
40. G. C. Schatz and A. Kuppermann, Role of direct and resonant (compound state) processes and of their interferences in the quantum dynamics of the collinear $H + H_2$ exchange reaction, *J. Chem. Phys.*, 1973, **59**, 964.
41. P. G Hipes and A. Kuppermann, Lifetime analysis of high-energy resonances in three-dimensional reactive scattering, *Chem. Phys. Lett.*, 1987, **133**, 1.
42. W. H. Miller and J. Z. H. Zhang, How to observe the elusive resonances in H or $D + H_2 \rightarrow H_2$ or HD + H reactive scattering, *J. Phys. Chem.*, 1991, **95**, 12.
43. A. B. Elkowitz and R. E. Wyatt, Quantum mechanical reaction cross sections for the three-dimensional hydrogen exchange reaction, *J. Chem. Phys.*, 1975, **62**, 2504.
44. S. C. Althorpe, F. Ferńandez-Alonso, B. D. Beanand J. D. Ayers, A. E. Pomerantz, R. N. Zare and E. Wrede, Observation and interpretation of a time-delayed mechanism in the hydrogen exchange reaction, *Nature*, 2002, **416**, 67.

45. J. C. Polanyi and K. B. Woodall, Energy distribution among reaction products. VI. $F + H_2$, D2, *J. Chem. Phys.*, 1972, **57**, 1574.
46. G. C. Schatz, J. M. Bowman and A. Kuppermann, Large quantum effects in the collinear $F + H_2 \rightarrow FH + H$ reaction, *J. Chem. Phys.*, 1973, **58**, 4023.
47. G. C. Schatz, J. M. Bowman and A. Kuppermann, Exact quantum, quasiclassical, and semiclassical reaction probabilities for the collinear $F + H_2 \rightarrow FH + H$ reaction, *J. Chem. Phys.*, 1975, **63**, 674.
48. A. Kuppermann and J. A. Kaye, Collision-lifetime matrix analysis of the first resonance in the collinear fluorine + hydrogen reaction and its isotopically substituted analogs, *J. Phys. Chem.*, 1981, **85**, 1969.
49. D. M. Neumark, A. M. Wodtke, G. N. Robinson, C. C. Hayden and Y. T. Lee, Molecular beam studies of the $F + H_2$ reaction, *J. Chem. Phys.*, 1985, **82**, 3045.
50. J. F. Castillo, D. E. Manolopoulos, K. Stark and H.-J. Werner, Quantum mechanical angular distributions for the $F + H_2$ reaction, *J. Chem. Phys.*, 1996, **104**, 6531.
51. R. T. Skodje, D. Skouteris, D. E. Manolopoulos, S.-H. Lee, F. Dong and K. Liu, Resonance-Mediated Chemical Reaction: $F + HD \rightarrow HF + D$, *Phys. Rev. Lett.*, Aug. 2000, **85**, 1206.
52. S.-H. Lee, F. Dong and K. Liu, Reaction dynamics of $F + HD \rightarrow HF + D$ at low energies: resonant tunneling mechanism, *J. Chem. Phys.*, 2002, **116**, 7839.
53. Y.-R. Tzeng and M. H. Alexander, Angular distributions for the $F + H_2 \rightarrow HF + H$ reaction: the role of the F spin–orbit excited state and comparison with molecular beam experiments, *J. Chem. Phys.*, 2004, **121**, 5812.

CHAPTER 7

Wavepackets and Time-Dependent Quantum Theory of Reactive Scattering

In the present book we concentrate on the quantum mechanical calculation of the cross section as defined for instance in eqn (3.9). These cross sections do not involve the variable time. So what is the motivation for involving time in the calculation process? It turns out that the use of wavepackets and of the time-dependent Schrödinger equation simplifies the process and changes the mathematical nature of the computations. The calculation becomes an "initial value problem" rather than one in which the wavefunction must satisfy particular boundary conditions at small and large R values. Consequently each computation relates to just a single initial quantum state of the colliding molecules and only a single column of the **S** matrix is evaluated. On the other hand a single time-dependent calculation provides results over a large range of energies, not just a single energy as in the case of a time-independent calculation. In the time-dependent approach we do not need to invest computational effort calculating cross sections for highly excited initial quantum states which are of no real interest to us.

Time and energy are conjugate variables and as such a function of time is related to a function of energy by a Fourier transform. This is the key to the use of the time-dependent Schrödinger equation to compute an **S** matrix or a scattering cross section. The time-dependent quantities we will calculate are transformed into energy dependent quantities through the use of Fourier transforms.

RSC Theoretical and Computational Chemistry Series No. 7
Theory of Molecular Collisions
By Gabriel G. Balint-Kurti and Alexander P. Palov

Published by the Royal Society of Chemistry, www.rsc.org

The critical formula connecting the time-independent and time-dependent formulations is the representation of the Dirac delta function[1–3] in terms of a Fourier transform over time

$$\delta(\hat{\boldsymbol{H}} - E) = \frac{1}{2\pi\hbar}\int_{-\infty}^{\infty} \mathrm{d}t \exp\left[\frac{-i(\hat{\boldsymbol{H}} - E)t}{\hbar}\right]. \tag{7.1}$$

This delta function acts as a projection operator onto the specified energy. If it acts on a wavepacket, $\Psi(\boldsymbol{R},\boldsymbol{r})$, which because of its localized nature will encompass a range of energies, it will project out solutions of just the desired energy

$$\begin{aligned}\psi(\boldsymbol{R},\boldsymbol{r},E) &= \delta(\hat{\boldsymbol{H}} - E)\Psi(\boldsymbol{R},\boldsymbol{r}) \\ &= \frac{1}{2\pi\hbar}\int_{-\infty}^{\infty} \mathrm{d}t \exp\left[\frac{-i(\hat{\boldsymbol{H}} - E)t}{\hbar}\right]\Psi(\boldsymbol{R},\boldsymbol{r}) \\ &= \frac{1}{2\pi\hbar}\int_{-\infty}^{\infty} \mathrm{d}t\, \mathrm{e}^{\frac{iEt}{\hbar}}\left\{\exp\left[\frac{-i\hat{\boldsymbol{H}}t}{\hbar}\right]\Psi(\boldsymbol{R},\boldsymbol{r})\right\}. \end{aligned} \tag{7.2}$$

We will see below that the term in curly brackets in eqn (7.2) is just the time-dependent wavepacket and we may therefore write

$$\psi(\boldsymbol{R},\boldsymbol{r},E) = \frac{1}{2\pi\hbar}\int_{-\infty}^{\infty} \mathrm{d}t\, \mathrm{e}^{\frac{iEt}{\hbar}}\Psi(\boldsymbol{R},\boldsymbol{r},t). \tag{7.3}$$

This equation then relates the time-dependent wavepacket, $\Psi(\boldsymbol{R},\boldsymbol{r},t)$, to the energy dependent wavefunction, $\psi(\boldsymbol{R},\boldsymbol{r},E)$.

Some of the mathematically complicated aspects of time-dependent atom–diatom scattering theory are presented in Appendix H while Appendix I discusses in detail the more technical aspects of the theory. We will frequently refer to these appendices. J. Z. H. Zhang has written a valuable book[4] covering many of the points discussed in this chapter.

Time-dependent wavepacket theory has now been largely superseded by other closely related methods, but all of the details discussed below are also relevant to these newer approaches. In Chapter 8 one of these new methods is discussed and some examples of wavepacket calculations are presented.

The wavepackets in time-dependent methods are represented on a coordinate grid. As they are propagated forward in time they must be prevented from reaching the edge of the grid and the techniques used for doing this are discussed later in this chapter. At low collision energies severe problems arise due both to the long wavelengths associated with low energies and to the difficulty of absorbing these components of the wavepacket when we try to prevent them reaching the edge of the grid. For these reasons neither time-dependent nor time-independent wavepacket techniques are recommended for investigating processes involving low collision energies.

7.1 Solving the Time-Dependent Schrödinger Equation

The time-dependent Schrödinger equation is

$$i\hbar\frac{\mathrm{d}}{\mathrm{d}t}\Psi(t)=\hat{\boldsymbol{H}}\Psi(t). \tag{7.4}$$

If the Hamiltonian operator does not depend on time, the formal solution is

$$\begin{aligned}\Psi(t)&=\mathrm{e}^{-i\hat{\boldsymbol{H}}t/\hbar}\,\Psi(t=0)\\&=\left\{1+\left[-i\frac{\hat{\boldsymbol{H}}t}{\hbar}\right]+\frac{1}{2!}\left[-i\frac{\hat{\boldsymbol{H}}t}{\hbar}\right]^2+\frac{1}{3!}\left[-i\frac{\hat{\boldsymbol{H}}t}{\hbar}\right]^3\cdots\right\}\Psi(t_0)\end{aligned} \tag{7.5}$$

where $\Psi(t_0)$ is the initial wavefunction at $t=0$.

$\mathrm{e}^{-i\hat{\boldsymbol{H}}t/\hbar}$ is called the time evolution operator. Many different methods have been proposed for solving eqn (7.4).[5,6] As may be seen from the formal solution in eqn (7.5) an essential aspect of all these methods will be the repeated operation of the Hamiltonian operator on a wavefunction or wavepacket.

The choice of coordinates remains a problem in time-dependent reactive scattering theory. To a large extent this problem has been solved in time-independent theory through the use of hyperspherical coordinates. Their use, however, is very much in its infancy in time-dependent theory. So far there have been only a very few publications which used them and then only for $J=0$.[7,8] The initial wavepacket must always be defined in reactant coordinates which we denoted by the subscript α. In this chapter, as in the previous ones, we concentrate on atom–diatom reactive scattering. For the reactants we will use the Jacobi coordinates R_α, r_α and γ_α. The analysis of the product quantum states must be carried out in product Jacobi coordinates. We will use the subscript β to label these. Although methods for computing reactive scattering **S** matrices using time propagation either exclusively in terms of reactant coordinates[9,10] or using both reactant and product coordinate systems[11,12] have been formulated, we will set up the initial wavepacket in reactant coordinates and then transform immediately to product Jacobi coordinates. The subsequent propagation and analysis is then carried out in product Jacobi coordinates.

7.1.1 Chebyshev Expansion of the Time Evolution Operator

The most efficient method of solving the time-dependent Schrödinger equation is to expand the exponential in eqn (7.5) in terms of a Chebyshev polynomial series,[13] rather than as a straightforward power series as done in the second line of eqn (7.5). This approach was first recommended by Tal-Ezer and Kosloff[14] and found to be the best when compared to other time propagation methods.[5,15] The argument of the Chebyshev polynomials, $T_n(x)$, is required to be in the range $-1<x<1$. For our purposes this means that we need scale and shift the "spectrum" of the Hamiltonian operator so that it is also contained within this range. Let us suppose that the lowest energy of our system is $E_{\min}$ (*i.e.* the lowest possible eigenvalue of $\hat{\boldsymbol{H}}$) and that the highest possible energy is $E_{\max}$. The spectral range of the Hamiltonian is therefore $\Delta E = E_{\max} - E_{\min}$. The new scaled and shifted Hamiltonian $\hat{\boldsymbol{H}}_{\mathrm{s}}$ is defined as

$$\hat{\boldsymbol{H}}_{\mathrm{s}} = \frac{\hat{\boldsymbol{H}} - \left(E_{\min} + \frac{1}{2}\Delta E\right)}{\frac{1}{2}\Delta E}. \tag{7.6}$$

The time evolution operator is now expressed as

$$\mathrm{e}^{-i\hat{\boldsymbol{H}}t/\hbar} = \mathrm{e}^{-i\left(E_{\min} + \frac{1}{2}\Delta E\right)t/\hbar}\mathrm{e}^{-i\Delta E\,\hat{\boldsymbol{H}}_{\mathrm{s}}\,t/2\hbar}. \tag{7.7}$$

Following the work of Tal-Ezer and Kosloff[14,16,17] we can expand the second of the exponentials above in terms of complex Chebyshev polynomials. Using this form of the time evolution operator, the wavefunction at time t evolved from that at time $t = 0$ may be written as (see Appendix A of ref. 18)

$$\begin{aligned}\Psi(t) &= \mathrm{e}^{-i\hat{\boldsymbol{H}}t/\hbar}\Psi(t=0)\\ &= \exp\left[\frac{-i\left(E_{\min} + \frac{1}{2}\Delta E\right)t}{\hbar}\right] \times \sum_{n=0}^{N}(2 - \delta_{no})J_n\left(\frac{\Delta E\,t}{2\,\hbar}\right)\Phi_n.\end{aligned} \tag{7.8}$$

The wavefunction iterates, Φ_n, are defined as

$$\Phi_n = Q_n(-i\hat{\boldsymbol{H}}_{\mathrm{s}})\,\Psi(t=0) \tag{7.9}$$

and successive iterates are can be generated using the recursion relationship

$$\Phi_{n+1} = -2i\hat{\boldsymbol{H}}_{\mathrm{s}}\Phi_n + \Phi_{n-1}. \tag{7.10}$$

The quantities Q_n are Chebyshev polynomials of complex argument and are related to the standard Chebyshev polynomials,[19] T_n, by the relationship

$$Q_n(-iz) = (i)^n\,T_n(-z). \tag{7.11}$$

The $J_n\left(\frac{\Delta E\,t}{2\,\hbar}\right)$ appearing in eqn (7.8) are regular Bessel functions.

The first two wavefunction iterates, Φ_n, are

$$\begin{aligned}\Phi_0 &= \Psi(t=0)\\ \Phi_1 &= -i\hat{\boldsymbol{H}}_{\mathrm{s}}\Psi(t=0).\end{aligned} \tag{7.12}$$

The Bessel functions $J_n\left(\frac{\Delta E\,t}{2\,\hbar}\right)$ in the summation of eqn (7.8) have the property that for values of n greater than the argument they decrease exponentially in magnitude. This property can be used to determine the length of the summation whose upper limit will therefore be slightly greater than $N = \frac{\Delta E\,t}{2\,\hbar}$. It is important that the quantity $E_{\max}$ be set as small as possible, compatible, of course, with including all physically accessible energies contributing to the physics of the molecular dynamics. Some aspects of this problem, particularly for the case $J > 0$, are a little technical and are discussed in Appendix I (see Section I.4).

Up to this point we have generally expanded the wavefunction in a basis set of vibrational–rotational functions and then taken matrix elements of the Hamiltonian with respect to these functions. This led to sets of coupled differential equations (see Chapter 3, eqn (3.18)). For the present application (*i.e.* the time-dependent wavepacket approach) we do not utilize any basis, but rather specify the wavefunction by giving its values on a grid of coordinate points. The angular grid is chosen to be Gauss–Legendre quadrature points[20] and evenly spaced grids are used for the radial variables R and r. It is clear from eqn (7.5) and (7.8) that the most important numerical step in the solution of the time-dependent Schrödinger equation is the repeated action of the Hamiltonian on the wavepacket. Appendix I describes this centrally important step in detail while the form of the Hamiltonian operator is derived in Appendix H. The key concept which permits the operation to be executed efficiently is the use of pseudo-spectral methods. Such methods involve transforming the grid-based representation of the wavefunction to one expressed as a linear combination of eigenfunctions of the operator of interest, acting with the operator (which is now trivial) and then transforming back to the grid-based representation. For the angular variables the Legendre or associated Legendre polynomials, which are eigenfunctions of the angular momentum operators, are used. Hence the choice of Gauss–Legendre quadrature points for the angular variables. The transformation between the grid representation and the so-called "Finite Basis Set" representation[21] is achieved using simple matrix multiplications. For the radial coordinates the transformations are accomplished using Fourier transforms. This approach was introduced by the two Kosloff brothers.[17,22]

7.2 Setting Up the Initial Wavepacket

The initial wavepacket for an atom–diatom collision consists of a vibrational–rotational wavefunction for the diatomic multiplied by a localized wavepacket, maybe a Gaussian in R, and a wavefunction of the form e^{-ikR} for the relative motion of the two. The wavenumber, k, approximately determines the relative kinetic energy of the collision process, $E_{\text{collision}} \approx k^2\hbar^2/(2\mu_R)$ (Note that momentum $= k\hbar$). We say that this collision energy is approximate because the wavepacket is localized in space and by the Heisenberg uncertainty principle it will have an uncertainty in its momentum, $\Delta x \Delta p = \hbar \Delta x \Delta k = \hbar/2$. If we use a Gaussian of the form $e^{-\alpha(R-R_0)^2}$ for the localized wavepacket the R dependence of our initial wavepacket becomes

$$\psi_0(R) = e^{-\alpha(R-R_0)^2} e^{-ik_0(R-R_0)}. \tag{7.13}$$

The momentum space representation of the wavepacket, $g(k)$, is obtained by taking the Fourier transform over R

$$\begin{aligned} g(k) &= \frac{1}{2\pi}\int_{-\infty}^{\infty} \psi_0(R)\mathrm{e}^{-ikR}\,\mathrm{d}R \\ &= \left(\frac{1}{4\pi\alpha}\right)^{\frac{1}{2}} \mathrm{e}^{-ikR_0}\mathrm{e}^{-(k_0+k)^2/(4\alpha)}. \end{aligned} \tag{7.14}$$

Thus we see that if the initial wavepacket is taken to be a Gaussian, then the wavepacket has a Gaussian distribution also in momentum space with its maximum at $k = -k_0$. In the analysis of the time dependent wavepacket propagation to find the **S** matrix we will need to use $g(-k)$, which is the amplitude of the component of the initial wavepacket with momentum $-k\hbar$.

An alternative "sinc" form of the initial wavepacket has been discussed.[23] This form has the advantage that its momentum space representation is more square and does not have the long tail possessed by the Gaussian. When analyzing the wavepacket the energy components in the tail of the Gaussian are associated with a small wavepacket amplitude. Consequently they are liable to suffer from greater numerical inaccuracy and must be discarded. The use of the "sinc" wavepacket minimizes the portion of the wavepacket which needs to be discarded in this manner.

The angular dependence of the wavepacket is given by a normalized associated Legendre polynomial ($\Theta_{j,\Omega}(\gamma)$, see ref. 24), corresponding to the desired initial rotational state (j) and its body-fixed z component ($\Omega = m_j$).†

The initial vibrational wavefunction, $\chi_{\nu j}(r_\alpha)$, is found by solving the vibrational rotational Schrödinger equation, eqn (3.6). In general the position R_0 for the center of the initial wavepacket in eqn (7.13) is near but not entirely in the asymptotic region where the atom–diatom interaction has decreased to zero. We can take account of this by using the expectation value of $V(R_0, r, \gamma)$ over the rotational wavefunction as the vibrational potential ($V_{\mathrm{diatom}}(r)$). Two simple algorithms have been developed to evaluate such vibrational wavefunction with the potential being specified on an evenly spaced grid of points. One of these, the Fourier Grid Hamiltonian method[25–28] was developed by Marston and Balint-Kurti while the other is due to Colbert and Miller.[29] The former method is outlined in Section I.5 of Appendix I.

The initial wavepacket is specified in terms of reactant Jacobi coordinates and may be written in the form

$$\Phi^{J\nu j\Omega}(R_\alpha, r_\alpha, \gamma_\alpha, t=0) = \psi_0(R_\alpha)\chi_{\nu j}(r_\alpha)\Theta_{j,\Omega}(\gamma_\alpha) \tag{7.15}$$

†Note that as we choose the initial relative velocity to lie along the space-fixed Z axis the initial value of the Z component of the rotational angular momentum is the same in both the space-fixed and body-fixed reference frames.

where the subscripts α indicate the use of reactant Jacobi coordinates and all the functions on the right hand side of the equation are known analytically except for the vibrational wavefunctions, which have been calculated numerically and are evaluated on a grid of points in r_α.

We need the initial wavepacket to be evaluated on a grid of coordinate points defined in product Jacobi coordinates. We therefore define this grid using evenly spaced grids in R_β and r_β and Gauss–Legendre grid points for the angle γ_β. For the radial coordinates (R and r), the spacing of the grid points determines the maximum kinetic energy that can be represented on the grid. This is limited by the fact that the spacing between grid points is half of a de Broglie wavelength corresponding to the maximum kinetic energy that can be represented on the grid (*i.e.* $E_{\max} = \hbar^2\pi^2/2\mu\Delta x^2$ where x is either one of the radial Jacobi coordinates). For each point on our product coordinate grid we find the equivalent set of reactant coordinates and also the angle $\chi^{\alpha\beta}$ between the Jacobi R vectors in the two coordinate systems. This angle is different for each product coordinate grid point. The initial wavepacket at the product coordinate grid point is then evaluated as

$$\begin{aligned}\Phi_{\Omega'}^{Jvj\Omega}(R_\beta, r_\beta, \gamma_\beta, t=0) &= \frac{R_\beta\, r_\beta}{R_\alpha\, r_\alpha}\Phi^{Jvj\Omega}(R_\alpha, r_\alpha, \gamma_\alpha, t=0)D^J{}_{\Omega,\Omega'}(0, \chi^{\alpha\beta}, 0)\\ &= \frac{R_\beta\, r_\beta}{R_\alpha\, r_\alpha}\Phi^{Jvj\Omega}(R_\alpha, r_\alpha, \gamma_\alpha, t=0)d^J{}_{\Omega,\Omega'}(\chi^{\alpha\beta}) \qquad (7.16)\end{aligned}$$

where $D^J{}_{\Omega,\Omega'}$ is a Wigner **D** matrix and $d^J{}_{\Omega,\Omega'}$ is a small Wigner **d** matrix. The Jacobian factor $(R_\beta\, r_\beta)/(R_\alpha\, r_\alpha)$ is included to account for the factor of $1/(R\,r)$ in the expansion of the original wavefunction (see Appendix H, eqn (H-2)).

For each calculation there is only one value of the initial Ω $(= m_j)$, but if $J > 0$, we will generally wish to perform the calculation separately for all $2J + 1$ allowed values of Ω (see discussion of parity below). The initial Z component of the total angular momentum, Ω, is defined in the reactant reference frame. When we transform to the product reference frame each initial Ω value maps onto all $(2J + 1)$ possible Ω' values in the product reference frame. This means that our initial wavepacket expressed in product coordinates possesses a non-zero component for each of the $(2J+1)$ possible values of Ω'.

Parity is conserved in all collisions and enables us to halve the number of mutually coupled Ω states we need to treat at any one time. The treatment of parity is discussed in Appendix D Section D.4. It also applies to the time-dependent case and will not be discussed further.

7.3 Absorbing the Wavepacket at the Edge of the Grid

The mathematical grids that we use to represent the wavepacket and the relative motion of the collision partners are of necessity finite in length. Problems arise if, or when, the wavepacket nears the grid edge. This is particularly true when we use Fourier transform techniques which assume that the functions being treated are cyclic, that is that the value of the function at the last and first grid points are identical. If, for instance, the wavepacket goes off the far or large R end of the grid it will reappear with the same amplitude at the beginning or small R end of the grid. This is in the classically forbidden region of the potential and consequently the associated energy will be very high. This general problem is known as "aliasing" in Fourier transform theory. In order to avoid it we need to prevent the wavepacket reaching the end of the grid. This can be done by introducing a negative imaginary potential which damps out the wavepacket at the edge of the grid.

Such potentials have long been used,[30–32] and are known as optical potentials, but within the context of time-dependent scattering theory they were introduced by Neuhauser and Baer.[33,34] One of the present authors has also studied the optimal parameters and forms of the negative imaginary potential to be used for such purposes.[35–37] All absorbing potentials break down at low energies when the wavelength of the relative motion becomes longer than or comparable to the width assigned to the absorbing potential. Manolopoulos[38,39] has developed what is probably the best absorbing potential. It has been developed so as not to allow any transmission through the potential and in its exact form it is quite complicated. An approximate easily usable form has, however, also been proposed[39]

$$-iV_{\mathrm{abs}}(R) = -i\frac{\hbar^2}{2\mu}\left(\frac{2\pi}{L}\right)^2\left\{\frac{4}{(2.62-x)^2}+\frac{4}{(2.62+x)^2}-1.16543325\right\} \tag{7.17}$$

where

$$x = \frac{2.62(R-R_{\mathrm{abs}})}{L_{\mathrm{abs}}}. \tag{7.18}$$

R_{abs} is the start of the absorbing region and L is the length of the absorbing region. The end of the absorbing region will generally be at the large R or r end of the grid.

7.4 The Scattering Equations and Calculation of the Cross Sections

We see from eqn (7.8) that in principle we could propagate the initial wavepacket from time $t = 0$ to some final time such that the collision had completed in a single time step. Such a propagation would involve a very large number of wavepacket iterates, Φ_n and a correspondingly long summation. The extraction of the **S** matrix from the time-dependent propagation relies on a Fourier transform over time. In practice this is accomplished as a summation over discrete time steps. The length of time steps or intervals (δt) is related to the maximum range of energies that the results encompass (ΔE) through Heisenberg's uncertainty relationship, $\Delta E = \hbar/(2\delta t)$. This relationship is the major factor in determining the length of the time steps taken in the calculation.

In order to propagate the wavepacket from time t to time $t + \delta t$ we need to apply eqn (7.9) and compute all the wavepacket iterates Φ_n using eqn (7.9), (7.10) and (7.12). This involves acting on the set of wavepackets with the Hamiltonian operator (see Appendix H eqn (H-16)), the set being composed of $(2J + 1)$ wavepackets with all the allowed values of Ω'

$$\begin{aligned}\{\hat{\boldsymbol{H}}\Phi^{J\Omega}(R,r,\gamma)\}_{\Omega'} = \Bigg\{ &-\frac{\hbar^2}{2\mu_R}\frac{\partial^2}{\partial R^2} - \frac{\hbar^2}{2\mu_r}\frac{\partial^2}{\partial r^2} + \frac{\hat{\boldsymbol{j}}^2}{2}\left\{\frac{1}{\mu_r r^2} + \frac{1}{\mu_R R^2}\right\} \\ &+ \frac{\hbar^2\left(J(J+1) - 2\Omega'^2\right)}{2\mu_R R^2} + V(R,r,\gamma)\Bigg\}\Phi^{J\Omega}_{\Omega'}(R,r,\gamma) \\ &- \frac{\hbar}{2\mu_R R^2}\Big\{\lambda_+(J,\Omega')\,\hat{\boldsymbol{j}}_-\,\Phi^{J\Omega}_{\Omega'+1}(R,r,\gamma) \\ &+ \lambda_-(J,\Omega')\,\hat{\boldsymbol{j}}_+\,\Phi^{J\Omega}_{\Omega'-1}(R,r,\gamma)\Big\} \qquad (7.19)\end{aligned}$$

where the notation $\{\hat{\boldsymbol{H}}\Phi^{J\Omega}(R,r,\gamma)\}_{\Omega'}$ indicates the new Ω' component of the set of wavepackets after acting on them with the Hamiltonian operator and

$$\lambda_\pm(J,\Omega) = [J(J+1) - \Omega(\Omega \pm 1)]^{\frac{1}{2}}. \qquad (7.20)$$

Wavepackets with different (adjacent) values of Ω' are coupled to each other by the operators $\lambda_+(J,\Omega')\,\hat{\boldsymbol{j}}_-$ and $\lambda_-(J,\Omega')\,\hat{\boldsymbol{j}}_+$ in the last curly bracket of eqn (7.19). The wavepackets are defined by their values on a grid of coordinate points and the details of how the operators act on these wavepackets is described in Appendix I, eqn (I-16), (I-18), (I-20) and (I-24).

Our technique for evaluating the **S** matrix is to draw a line, called the analysis line, at a fixed value of the scattering coordinate $R_\beta = R_{\beta\infty}$ across the exit channel of the reaction and to watch the wavepacket as it crosses this line. A schematic diagram of such a line is shown in Figure 7.1. The line is superimposed on a potential energy surface and an initial wavepacket is also depicted.

A cut is then taken through the wavepacket along the analysis line at each time step. This yields a two-dimensional function $\Phi^{J\nu j\Omega}_{\Omega'}(R_{\beta\infty}, r_\beta, \gamma_\beta, t)$ which we then project onto the different product vibrational–rotational states (see Appendix H, eqn (H-17) and (H-18))

$$C^J_{\nu j\Omega\to\nu' j'\Omega'}(t) = \int \chi^*_{\nu' j'}(r_\beta)\,\Theta^*_{j'\Omega'}(\gamma_\beta)\,\Phi^{J\nu j\Omega}_{\Omega'}(R_{\beta\infty}, r_\beta, \gamma_\beta, t)\mathrm{d}r_\beta \sin\gamma_\beta \mathrm{d}\gamma_\beta \qquad (7.21)$$

yielding a time-dependent quantity $C^J_{\nu j\Omega\to\nu' j'\Omega'}(t)$. We then take the Fourier transform of this time-dependent quantity to obtain an energy dependent one (see Appendix H, eqn (H-19))

$$A^J_{\nu j\Omega\to\nu' j'\Omega'}(E) = \frac{1}{2\pi}\int_0^\infty \mathrm{e}^{iEt/\hbar} C^J_{\nu j\Omega\to\nu' j'\Omega'}(t)\,\mathrm{d}t. \qquad (7.22)$$

The helicity coupling terms connecting the different values of Ω' are unfortunately very long ranged, falling off only as $1/R^2$.[41] As these terms are

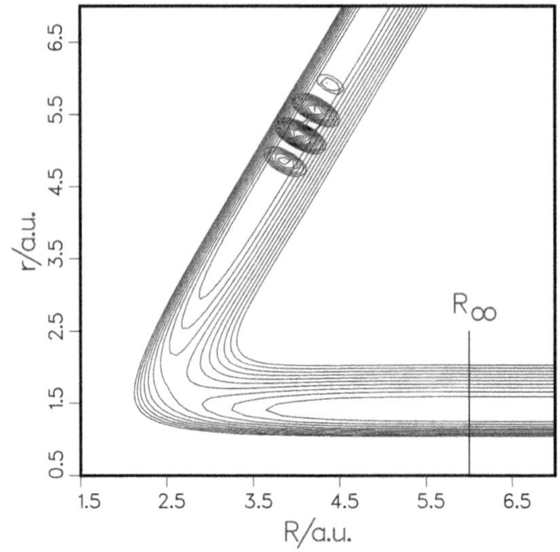

Figure 7.1 Schematic diagram of an initial wavepacket superimposed on a contour map of a potential energy surface. The analysis line in the asymptotic region of the product channel is labeled by R_∞. Reproduced with permission from S. K. Gray and G. G. Balint-Kurti, *J. Chem. Phys.*, **108**, 950. Copyright 1998, AIP Publishing LLC.[40] DOI: 10.1063/1.430453.

absent in the space-fixed reference frame we choose to perform the **S** matrix analysis in a space-fixed reference frame (see Appendix D Section D.5 and Chapter 6 Section 6.2). For this reason we first transform the **A** matrix of eqn (7.22) to the space-fixed reference frame (see Appendix H Section H.3 eqn (H-47)) to obtain a matrix of the form $A^J_{\nu j\ell\to\nu' j'\ell'}(E)$ labeled by space-fixed ℓ indices rather than the body-fixed Ω ones. We then compute the space-fixed **S** matrix (see Appendix H eqn (H-53))

$$S^{J,\beta,\alpha}_{\nu' j'\ell',\nu j\ell} = -\hbar\left(\frac{k^\alpha_{\nu j}k^\beta_{\nu' j'}}{\mu^\alpha\mu^\beta}\right)^{\frac{1}{2}}\frac{A^J_{\nu j\ell\to\nu' j'\ell'}(E)}{g(-k^\alpha_{\nu j})}\mathrm{e}^{-i(k^\beta_{\nu' j'}R_{\beta\infty}+\delta\eta_{\nu' j'\ell'}+\delta\eta_{\nu j\ell})} \tag{7.23}$$

where some phase correction factors ($\delta\eta_{\nu' j'\ell'}$ and $\delta\eta_{\nu j\ell}$) have been applied to account for the fact that the initial wavepacket and the analysis line were in regions where the centrifugal potential was still not zero. The **S** matrix computed in this manner is then transformed back to the body-fixed representation (see Appendix H eqn (H-54)) as the formulae for computing the cross sections are simpler in this reference frame (see Chapter 3 eqn (3.48)–(3.50)).

We have also developed an alternative way of computing the **S** matrix from the time evolution of the wavepacket. This method is again based on drawing a line (or, more correctly stated, a surface at a fixed value of $R_\beta = R_{\beta\infty}$) across the exit channel of the reaction, but instead of simply monitoring the wavepacket as it passes by, we now explicitly compute the flux in the different quantum product states passing through the surface.[42] It is sometimes useful to apply both methods to the same calculation to check that the results are converged.

7.5 An Example: HD + OH $\rightarrow$ H_2O + D Reactive Differential Cross Section

Figure 7.2 shows a comparison between an experimental differential state-to-state reactive cross section for the reaction HD + OH $\rightarrow$ H_2O + D and a time-dependent quantum theoretical calculation of the same quantity.[43] The calculations were performed by Dong H. Zhang and his research group. The methods used differ somewhat from those described above in the following respects: (1) they used an expansion in terms of a basis set of sine functions for the radial variables instead of Fourier transforms and (2) they used a technique called reactant–product decoupling[11] to transform the wavepacket to the product coordinate system beyond a fixed plane dividing the reactant and product coordinate spaces. We see that the theoretical results provide a remarkably reliable picture of the experimental results.

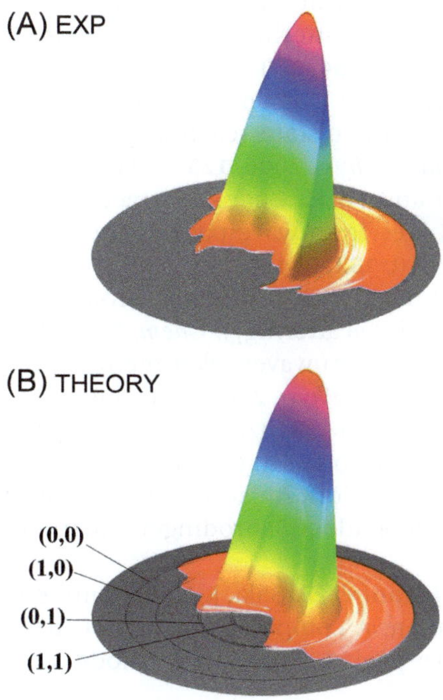

Figure 7.2 Comparison of experimental and theoretical differential reactive cross sections for the reaction HD + OH $\rightarrow$ H_2O + D. Reproduced with permission from C. Xiao *et al.* Science, **333**, 440 (2011). Reprinted with permission from AAAS.[43]

References

1. A. Messiah, *Quantum Mechanics*, North Holland Publishing Co., Amsterdam, 1965.
2. P. A. M. Dirac, *The Principles of Quantum Mechanics*, Oxford University Press, Oxford, 4th edn, 1958.
3. C. Cohen-Tannoudji, B. Diu and F. Laloë, *Quantum mechanics, Vol. 2, Appendix II*, Wiley, New York, 1977.
4. J. Z. H. Zhang, *Theory and Application of Quantum Molecular Dynamics*, World Scientific, Singapore, 1998.
5. C. Leforestier, R. H. Bisseling, C. Cerjan, M. D. Feit, R. Friesner, A. Guldberg, A. Hammerich, G. Jolicard, W. Karrlein, H.-D. Meyer, N. Lipkin, O. Roncero and R. Kosloff, A comparison of different propagation schemes for the time dependent Schrödinger equation, *J. Comput. Phys.*, 1991, **94**, 59.
6. N. Balakrishnan, C. Kalyanaraman and N. Sathyamurthy, Time-dependent quantum mechanical approach to reactive scattering and related processes, *Phys. Rep.*, 1997, **280**, 79.
7. J. Crawford and G. A. Parker, State-to-state three-atom time-dependent reactive scattering in hyperspherical coordinates, *J. Chem. Phys.*, 2013, **138**, 054313.
8. S. Adhikari and A. J. C. Varandas, The coupled 3D wave packet approach for triatomic reactive scattering in hyperspherical coordinates, *Comput. Phys. Commun.*, 2013, **184**, 270.
9. S. Gómez-Carrasco and O. Roncero, Coordinate transformation methods to calculate state-to-state reaction probabilities with wave packet treatments, *J. Chem. Phys.*, 2006, **125**, 054102.
10. Z. Sun, H. Guo and D. H. Zhang, Extraction of state-to-state reactive scattering attributes from wave packet in reactant Jacobi coordinates, *J. Chem. Phys.*, 2010, **132**, 084112.
11. T. Peng and J. Z. H. Zhang, A reactant–product decoupling method for state-to-state reactive scattering, *J. Chem. Phys.*, 1996, **105**, 6072.
12. S. C. Althorpe, Quantum wavepacket method for state-to-state reactive cross sections, *J. Chem. Phys.*, 2001, **114**, 1601.
13. I. S. Gradshteyn and I. M. Ryhzik, *Table of Integrals, Series and Products*, Academic Press, New York, 4th edn, 1965.
14. H. Tal-Ezer and R. Kosloff, An accurate and efficient scheme for propagating the time dependent Schrödinger equation, *J. Chem. Phys.*, 1984, **81**, 3967.
15. T. N. Truong, J. J. Tanner, P. Bala, J. A. McCammon, D. J. Kouri, B. Lesyng and D. K. Hoffman, A comparative study of time dependent quantum mechanical wave packet evolution methods, *J. Chem. Phys.*, 1992, **96**, 2077.
16. H. Tal-Ezer, Spectral methods in time for hyperbolic equations, *SIAM J. Num. Anal.*, 1986, **23**, 11.
17. R. Kosloff, Time-dependent quantum-mechanical methods for molecular dynamics, *J. Phys. Chem.*, 1988, **92**, 2087.

18. G. G. Balint-Kurti, Time-dependent and time-independent wavepacket approaches to reactive scattering and photodissociation dynamics, *Int. Rev. Phys. Chem.*, 2008, **27**, 507.
19. M. Abromowitz and I. A. Stegun, *Handbook on Mathematical Functions, Applied Mathematics Series 55*, National Bureau of Standards, 10th edn, 1972.
20. W. A. Vetterling W. H. Press, S. A. Teukolsky and B. P. Flannery, *Numerical Recipes in Fortran*, Cambridge University Press, Cambridge, 2nd edn, 1992.
21. J. C. Light, I. P. Hamilton and V. J. Lill, Generalized discrete variable approximation in quantum mechanics, *J. Chem. Phys.*, 1985, **82**, 1400.
22. D. Kosloff and R. Kosloff, A Fourier method solution for the time dependent Schrodinger equation as a tool in molecular dynamics, *J. Comput. Phys.*, 1983, **52**, 35.
23. M. Hankel, G. G. Balint-Kurti and S. K. Gray, Sinc wave packets: New form of wave packet for time-dependent quantum mechanical reactive scattering calculations, *Int. J. Quantum Chem.*, 2003, **92**, 205.
24. R. N. Zare. *Angular Momentum*. John Wiley and Sons, New York, 1988. See also list of corrections at web address: http://web.stanford.edu/group/Zarelab/Errata.pdf.
25. C. C. Marston and G. G. Balint-Kurti, The Fourier grid Hamiltonian method for bound state eigenvalues and eigenfunctions, *J. Chem. Phys.*, 1989, **91**, 3571.
26. G. G. Balint-Kurti, C. L. Ward and C. C. Marston, Two computer programs for solving the Schrödinger equation for bound-state eigenvalues and eigenfunctions using the Fourier grid Hamiltonian method, *Comput. Phys. Commun.*, 1991, **67**, 285.
27. G. G. Balint-Kurti, R. N. Dixon and C. C. Marston, Grid methods for solving the Schrodinger equation and time dependent quantum dynamics of molecular photofragmentation and reactive scattering processes, *Int. Rev. Phys. Chem.*, 1992, **11**, 317.
28. F. Gögtas, G. G. Balint-Kurti and C. C. Marston, The Fourier grid Hamiltonian method based on even grid numbers, *Turk. J. of Phys.*, 1997, **21**, 923.
29. D. T. Colbert and W. H. Miller, A novel discrete variable representation for quantum mechanical reactive scattering *via* the S-matrix Kohn method, *J. Chem. Phys.*, 1992, **96**, 1982.
30. J. L. J. Rosenfeld and J. Ross, Calculation of chemical reaction probabilities from elastic scattering data, *J. Chem. Phys.*, 1966, **44**, 188.
31. C. Nyeland and J. Ross, Interpretation of elastic scattering in reactive systems, *J. Chem. Phys.*, 1968, **49**, 843.
32. D. A. Micha, Optical potentials in molecular collisions, *J. Chem. Phys.*, 1969, **50**, 722.
33. D. Neuhauser and M. Baer, The time-dependent Schrödinger equation: application of absorbing boundary conditions, *J. Chem. Phys.*, 1989, **90**, 4351.

34. D. Neuhauser, M. Baer and D. J. Kouri, The application of optical potentials for reactive scattering: a case study, *J. Chem. Phys.*, 1990, **93**, 2499.
35. Á. Vibók and G. G. Balint-Kurti, Reflection and transmission of waves by a complex potential—a semiclassical Jeffreys–Wentzel–Kramers–Brillouin treatment, *J. Chem. Phys.*, 1992, **96**, 7615.
36. Á. Vibók and G. G. Balint-Kurti, Parametrization of complex absorbing potentials for time-dependent quantum dynamics, *J. Chem. Phys.*, 1992, **96**, 8712.
37. G. G. Balint-Kurti and Á. Vibók, Complex absorbing potentials in time dependent quantum dynamics, in *Numerical Grid Methods and Their Application to Schrödinger's Equation*, ed. C. Cerjan, volume 412 of *NATO ASI Series*, Springer, Netherlands, 1993, pp. 195.
38. D. E. Manolopoulos, Derivation and reflection properties of a transmission-free absorbing potential, *J. Chem. Phys.*, 2002, **117**, 9552.
39. T. Gonzalez-Lezana, E. J. Rackham and D. E. Manolopoulos, Quantum reactive scattering with a transmission-free absorbing potential, *J. Chem. Phys.*, 2004, **120**, 2247.
40. S. K. Gray and G. G. Balint-Kurti, Quantum dynamics with real wave packets, including application to three-dimensional ($J=0$) $D+H_2 \rightarrow HD+H$ reactive scattering, *J. Chem. Phys.*, 1998, **108**, 950.
41. G. G. Balint-Kurti and M. Shapiro, Photofragmentation of triatomic molecules. Theory of angular and state distribution of product fragments, *J. Chem. Phys.*, 1981, **61**, 137.
42. A. J. H. M. Meijer, E. M. Goldfield, S. K. Gray and G. G. Balint-Kurti, Flux analysis for calculating reaction probabilities with real wavepackets, *Chem. Phys. Lett.*, 1998, **293**, 270.
43. C. Xiao, X. Xu, S. Liu, T. Wang, W. Dong, T. Yang, Z. Sun, D. Dai, X. Xu, D. H. Zhang and X. Yang, Experimental and theoretical differential cross sections for a four-atom reaction: $HD + OH \rightarrow H_2O + D$, *Science*, 2011, **333**, 440.

CHAPTER 8

The Real Wavepacket Method and Time-Independent Wavepackets

In this chapter we describe the Real Wavepacket method.[1] This method is a development of the time-dependent wavepacket method outlined in the previous chapter and shares many aspects with it. It turns out that it is entirely equivalent to two other methods developed independently of it: the time-independent wavepacket approach[2–10] and Guo's order domain propagation method.[11,12] Its advantages lie in the fact that the propagation is carried out using entirely real algebra and that fewer evaluations of the action of the Hamiltonian operator on the real part of the wavepacket are required to achieve the same accuracy. The Real Wavepacket method was introduced in a paper by Gray and Balint-Kurti[1] and a computer code, DIFFREALWAVE, which implements the method for atom–diatom reactive scattering is available.[13–15] A review article comparing the Real Wavepacket method to the time-independent wavepacket approach has also been published.[16]

8.1 Iterative Time Propagation

If we assume that the Hamiltonian does not depend on time we can use the time evolution operator of eqn (7.5) and write the equation for propagating a wavepacket forward from time t to time $t+\tau$ as

$$\begin{aligned}\Psi(t+\tau) &= \mathrm{e}^{-i\hat{\boldsymbol{H}}\tau/\hbar}\,\Psi(t)\\ &= \{\cos(\hat{\boldsymbol{H}}\tau/\hbar) - i\sin(\hat{\boldsymbol{H}}\tau/\hbar)\}\Psi(t).\end{aligned} \tag{8.1}$$

RSC Theoretical and Computational Chemistry Series No. 7
Theory of Molecular Collisions
By Gabriel G. Balint-Kurti and Alexander P. Palov

Published by the Royal Society of Chemistry, www.rsc.org

The formula for backward propagation is then given by

$$\Psi(t-\tau) = \{\cos(\hat{\boldsymbol{H}}\tau/\hbar) + i\sin(\hat{\boldsymbol{H}}\tau/\hbar)\}\Psi(t). \tag{8.2}$$

By adding together these two equations we obtain an iterative equation for $\Psi(t+\tau)$ in terms of $\Psi(t)$ and $\Psi(t-\tau)$

$$\Psi(t+\tau) = 2\cos(\hat{\boldsymbol{H}}\tau/\hbar)\Psi(t) - \Psi(t-\tau). \tag{8.3}$$

This equation may be referred to as an iterative equation equivalent of the Schrödinger equation and was first pointed out and used by S. K. Gray.[17] Note that the equation does not contain the imaginary number $i=\sqrt{-1}$ and can therefore be used to propagate the real part of a wavepacket without reference to any imaginary component. We denote the wavepacket as the sum of a real and imaginary part: $\Psi(t) = q(t) + ip(t)$. The initial wavepacket is then set up as in eqn (7.13) and can now be written as

$$\Psi(t=0) = q(t=0) + ip(t=0) \tag{8.4}$$

and the real part q can be propagated using only real algebra:

$$q(t+\tau) = 2\cos(\hat{\boldsymbol{H}}\tau/\hbar)q(t) - q(t-\tau). \tag{8.5}$$

In order to start the iterations we need the real part of the wavepacket evaluated at $t=0$ and at $t=\tau$. The initial wavepacket provides $q(t=0)$, but $q(t=\tau)$ must be evaluated using eqn (8.1)

$$q(\tau) = \cos(\hat{\boldsymbol{H}}\tau/\hbar)q(0) + \sin(\hat{\boldsymbol{H}}\tau/\hbar)p(0)S. \tag{8.6}$$

This equation involves the use of $p(0)$, the imaginary part of the initial wavepacket.

8.2 Scaling and Mapping the Hamiltonian Operator

The main problem with using eqn (8.5) to propagate the wavepacket forward in time is the fact that we need to take the cosine of the Hamiltonian operator. This is just as difficult as evaluating the exponential of the operator. We can circumvent this difficulty by mapping the Hamiltonian onto a function of itself, that is, using a function of the Hamiltonian, $f(\hat{\boldsymbol{H}})$, instead of the Hamiltonian itself. The mapping we choose is

$$f(\hat{\boldsymbol{H}}_{\mathrm{s}}) = -\frac{\hbar}{\tau}\cos^{-1}\hat{\boldsymbol{H}}_{\mathrm{s}}. \tag{8.7}$$

In eqn (8.7) we have used a shifted and scaled Hamiltonian such that its spectrum or range of possible eigenvalues spans the range $-1 \leq E \leq 1$. We use the same shifting and scaling as in eqn (7.6) and we need to do this because the value of the cosine is limited to lie in this range. It is important to note that this is a one-to-one mapping and that for every possible eigenvalue E_{s} of $\hat{\boldsymbol{H}}_{\mathrm{s}}$ there is a unique value of the function $f(E_{\mathrm{s}})$ and *vice versa*. Making the mapping $\hat{\boldsymbol{H}}_{\mathrm{s}} \to f(\hat{\boldsymbol{H}}_{\mathrm{s}})$ in eqn (8.5) we obtain

$$q(t+\tau) = 2\hat{\boldsymbol{H}}_{\mathrm{s}}q(t) - q(t-\tau). \tag{8.8}$$

This is clearly a much simpler iterative propagation procedure than that of eqn (8.5). A consequence of making this mapping is that the propagation procedure is no longer solving the time-dependent Schrödinger equation but is rather solving a "mapped" equation

$$i\hbar\frac{\mathrm{d}}{\mathrm{d}t}\Psi(t) = f(\hat{\boldsymbol{H}}_{\mathrm{s}})\Psi(t). \tag{8.9}$$

As time plays no role in the cross sections and other quantities we seek to compute, the fact that we are no longer solving the correct time-dependent Schrödinger equation does not affect us. Applying the mapping now to eqn (8.6) we obtain

$$q(\tau) = \hat{\boldsymbol{H}}_{\mathrm{s}}\, q(0) - \sqrt{1 - \hat{\boldsymbol{H}}_{\mathrm{s}}^2}\, p(0). \tag{8.10}$$

This equation must be solved to obtain the real part of the wavepacket at "time" $t = \tau$ as this is needed to start the iteration process (see eqn (8.8)). The operator $\sqrt{1 - \hat{\boldsymbol{H}}_{\mathrm{s}}^2}$ can be expanded in terms of Chebyshev polynomials[18]

$$\sqrt{1 - \hat{\boldsymbol{H}}_{\mathrm{s}}^2} = \frac{2}{\pi}\left[1 - 2\sum_{n=1}^{\infty}(4n^2 - 1)^{-1}T_{2n}(\hat{\boldsymbol{H}}_{\mathrm{s}})\right] \tag{8.11}$$

where $T_n(x)$ is the Chebyshev polynomial of the first kind[19] and terms of the form $T_{2n}(\hat{\boldsymbol{H}}_{\mathrm{s}})\, p(0)$ are evaluated using the recursion equations of the Chebyshev polynomials

$$T_{n+1}(\hat{\boldsymbol{H}}_{\mathrm{s}})\, p(0) = 2\,\hat{\boldsymbol{H}}_{\mathrm{s}}[T_n(\hat{\boldsymbol{H}}_{\mathrm{s}})\, p(0)] - [T_{n-1}(\hat{\boldsymbol{H}}_{\mathrm{s}})\, p(0)]. \tag{8.12}$$

8.3 Absorbing the Wavepacket

The wavepacket is represented by its value on a coordinate grid. As in the time-dependent case the wavepacket needs to be damped out or absorbed near the edges of the grid to avoid unphysical reflection from the edge and other complications (see also Section 7.3). Gray[17] has shown that the correct method of doing this is to define some absorbing operator $\hat{\boldsymbol{A}}$ which could take a form similar to

$$\begin{aligned} \hat{\boldsymbol{A}} &= 1.0 && \text{for } R \le R_{\mathrm{a}} \\ \hat{\boldsymbol{A}} &= \mathrm{e}^{-\alpha(R-R_{\mathrm{a}})^2} && \text{for } R > R_{\mathrm{a}} \end{aligned} \tag{8.13}$$

and then to use a modified recursion equation which includes a damping operation at every step

$$q(t + \tau) = \hat{\boldsymbol{A}}[2\hat{\boldsymbol{H}}_{\mathrm{s}}q(t) - \hat{\boldsymbol{A}}q(t - \tau)]. \tag{8.14}$$

This equation then replaces eqn (8.8) as the iterative propagation recursion formula.

8.4 Calculating the S Matrix

As in the case of time-dependent wavepacket theory, the key quantity we compute is the value of the wavepacket at the analysis line at each time step (see eqn (7.21))

$$C^{J}_{vj\Omega\to v'j'\Omega'}(t) = \int \chi^{*}_{v'j'}(r_\beta)\,\Theta^{*}_{j'\Omega'}(\gamma_\beta)\,q^{Jvj\Omega}_{\Omega'}(R_{\beta\infty}, r_\beta, \gamma_\beta, t)\,\mathrm{d}r_\beta \sin\gamma_\beta\,\mathrm{d}\gamma_\beta \tag{8.15}$$

where $q^{Jvj\Omega}_{\Omega'}(R_{\beta\infty}, r_\beta, \gamma_\beta, t)$ is the real part of the wavepacket and as each iteration step progresses time by τ we can write $t = n\tau$, where n is an iteration counter.

The energy has been scaled and shifted in the same way as the Hamiltonian operator. The relationship between the scaled-shifted Hamiltonian $\hat{\boldsymbol{H}}_{\mathrm{s}}$ and the original unaltered Hamiltonian is

$$\hat{\boldsymbol{H}}_{\mathrm{s}} = a_{\mathrm{s}}\hat{\boldsymbol{H}} + b_{\mathrm{s}}. \tag{8.16}$$

where $a_{\mathrm{s}} = 2/\Delta E$ and $b_{\mathrm{s}} = -1 - a_{\mathrm{s}}E_{\mathrm{min}}$ (see eqn (7.6)).

The scaled-shifted energy and the unaltered energy are related in exactly the same way. There is now also the further complication of the mapping of the Hamiltonian we have introduced, $\hat{\boldsymbol{H}}_{\mathrm{s}} \to f$, see eqn (8.7). This means that when we Fourier transform C^J in eqn (8.15) over time we obtain a function of f rather than a function of E_{s} (the scaled-shifted energy). As time now progresses through an iterative process, we only evaluate the time-dependent quantities at intervals of τ. This means that we need to replace the continuous integral over time in the time-energy Fourier transform (see eqn (7.22)) by a discretized summation

$$A^{J}_{vj\Omega\to v'j'\Omega'}(f) = \frac{\tau}{2\pi}\sum_{n=0}^{\infty}\left(1 - \frac{\delta_{n,0}}{2}\right)\mathrm{e}^{ifn\tau/\hbar}C^{J}_{vj\Omega\to v'j'\Omega'}(n\tau). \tag{8.17}$$

The "time step" τ in eqn (8.14) and (8.17) is completely arbitrary, so we choose it to be $\tau = 1$, thus eliminating it and replacing time by the Chebyshev iteration number

$$A^{J}_{vj\Omega\to v'j'\Omega'}(f) = \frac{1}{2\pi}\sum_{n=0}^{\infty}\left(1 - \frac{\delta_{n,0}}{2}\right)\mathrm{e}^{ifn/\hbar}C^{J}_{vj\Omega\to v'j'\Omega'}(n). \tag{8.18}$$

In order to compute quantities such as cross sections that are functions of energy we now need to reverse the mapping and the scaling and shifting of eqn (8.7) and (8.16). The continuum wavefunctions that we have used to develop our theories are consistently normalized on the energy scale (see Appendix A), that is

$$\int \psi^{*}(x,E)\psi(x,E')\mathrm{d}x = \delta(E - E') \tag{8.19}$$

where x stands for all the spatial coordinates.

We have made the mapping $\hat{H}_s \rightarrow f$ and consequently we are solving eqn (8.9) instead of the Schrödinger equation (eqn (7.4)). Our new mapped wavefunctions are therefore normalized on the f scale

$$\int \psi^*(x, f)\psi(x, f')\mathrm{d}x = \delta(f - f'). \tag{8.20}$$

The Dirac delta functions $\delta(E-E')$ and $\delta(f-f')$ are related through the following equation[18,20]

$$\delta(f - f') \left| \frac{\mathrm{d}f}{\mathrm{d}E} \right| = \delta(E - E'). \tag{8.21}$$

Wavefunctions normalized on the f scale are therefore related to those normalized on the energy scale by

$$\psi(x, E) = \left| \frac{\mathrm{d}f}{\mathrm{d}E} \right|^{\frac{1}{2}} \psi(x, f). \tag{8.22}$$

From eqn (8.7) and (8.16) we see that

$$\begin{aligned} \left| \frac{\mathrm{d}f}{\mathrm{d}E} \right| &= \frac{\hbar}{\tau} \frac{a_s}{\sqrt{1 - E_s^2}} \\ &= \frac{2\,\hbar}{\Delta E \sqrt{1 - E_s^2}} \end{aligned} \tag{8.23}$$

where we have set $\tau = 1$ in the last line of the above equation.

The $\mathbf{C}^\mathrm{J}$ matrix, and therefore also the $\mathbf{A}^\mathrm{J}$ matrix, are proportional to the square of the continuum wavefunction (see Appendix H eqn (H-27)) and $A^J_{vj\Omega \rightarrow v'j'\Omega'}(f)$ must therefore be multiplied by $\left| \frac{\mathrm{df}}{\mathrm{d}E} \right|$ to convert it to the un-mapped form

$$A^J_{vj\Omega \rightarrow v'j'\Omega'}(E) = \frac{2\,\hbar}{\Delta E \sqrt{1 - E_s^2}} A^J_{vj\Omega \rightarrow v'j'\Omega'}(f). \tag{8.24}$$

As in time-independent wavepacket theory, we now need to transform the $\mathbf{A}^\mathrm{J}$ matrix from the body-fixed to the space-fixed representation so as to avoid the long range coupling of the different Ω channels which are present in the body-fixed representation (see discussions below eqn (7.22) and above eqn (H-47)). This transformation gives us the matrix elements $A^J_{vj\ell \rightarrow v'j'\ell'}(f)$ instead of those in eqn (8.17) and (8.24) above.

The space-fixed $\mathbf{A}^\mathrm{J}$ is then used to generate a space-fixed $\mathbf{S}^\mathrm{J}$ matrix as detailed in Appendix J and Appendix H eqn (H-47) to (H-53)

$$\begin{aligned} S^{J,\beta,\alpha}_{v'j'\ell',vj\ell} = &-2\,\hbar \left(\frac{k^\alpha_{vj} k^\beta_{v'j'}}{\mu^\alpha \mu^\beta} \right)^{\frac{1}{2}} \frac{A^J_{vj\ell \rightarrow v'j'\ell'}(E)}{g(-k^\alpha_{vj})} \\ &\times e^{-i(k^\beta_{v'j'} R_{\beta\infty} + \delta\eta_{v'j'\ell'} + \delta\eta_{vj\ell})}. \end{aligned} \tag{8.25}$$

where, following the reasoning outlined in Appendix J, we have multiplied the expression for the $\mathbf{S}^J$ matrix (eqn (H-53)) by a factor of 2 as only the real part of the complex wavepacket has been analyzed. The other quantities in the equation are explained in Appendix H.

The space-fixed $\mathbf{S}^J$ matrix is then transformed to the body-fixed reference frame using eqn (D-41) of Appendix D, and the differential and integral cross sections may then be evaluated using eqn (D-43) and (D-45). Note that the equations in Appendix D are expressed in terms of the **T** matrix rather than the **S** matrix, but for reactive or inelastic scattering processes we have from eqn (3.22) that $\mathbf{T} = -\mathbf{S}$.

8.5 Applications

8.5.1 $O(^1D) + H_2 \rightarrow OH + H$: Conical Intersections and Electronically Non-Adiabatic Transitions

The ^1D excited electronic state of the oxygen atom has a total electronic angular momentum quantum number $J = 2$ and possesses five M_J sub-states. Thus when considering collisions of $O(^1D)$ with H_2 the process involves five separate electronic states and the corresponding number of potential energy surfaces.[21] Two of the electronic states have $^1\Delta$ symmetry in a collinear O–H–H arrangement and do not connect $O+H_2$ to $OH+H$ products. Collisions on any of the three remaining potential energy surfaces can all lead to reaction.[22]

Figure 8.1 displays a cut along the reaction path through the two lowest $^1A'$ surfaces in both collinear and slightly bent O–H–H geometries.[23–25] (A' surfaces are even with respect to reflection of the electronic wavefunction in the molecular plane, while A'' surfaces are odd.) In the linear geometry the $^1\Sigma$ and $^1\Pi$ potential energy curves cross at around 35 on the arbitrary scale measuring progress along the reaction coordinate. This is perfectly allowed as they correspond to electronic wavefunctions possessing different symmetries. The Π state surface has a very small barrier to reaction. The lower panel shows the same two surfaces in a slightly bent geometry. When the system is bent the component of the Π state whose electronic wavefunction is symmetric with respect to reflection in the molecular plane has the same symmetry as that of the Σ state. There is therefore a coupling between the Π and Σ states in a bent configuration and the resulting surfaces display an avoided crossing.[20] This situation is known as a symmetry allowed conical intersection.

In the bent geometry the lower $1^1A'$ adiabatic potential energy surface does not have any barrier to reaction. We will see that this has a marked effect on the reaction cross section. From Figure 8.1 it would at first sight seem that a collision initiated on the excited $2^1A'$ surface could not lead to reaction. This is however not correct. The two $^1A'$ surfaces approach close to each other and non-adiabatic transitions can occur between the two surfaces. Therefore

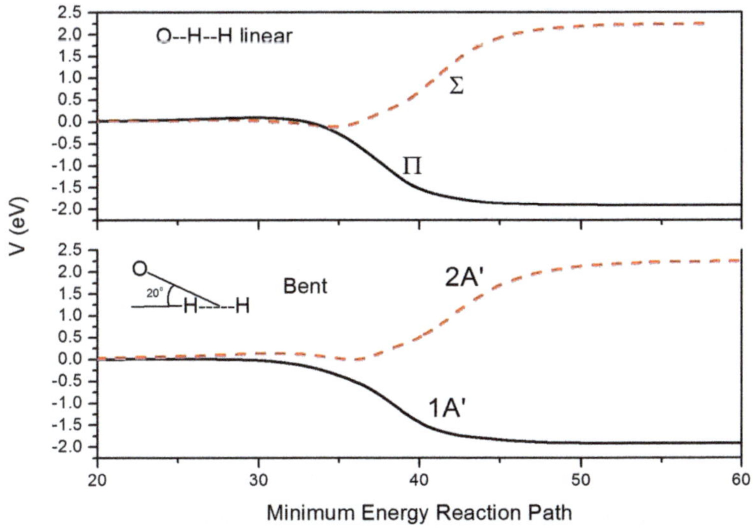

Figure 8.1 Plot of the two lowest (adiabatic) $^1A'$ potential energy surface along a minimum energy path connecting reactants and products for the reaction $O(^1D)+H_2 \rightarrow OH+H$. The upper panel shows the surfaces in an O–H–H collinear geometry while the lower panel shows the same two surfaces in a slightly bent geometry. The surfaces are from the work of Dobbyn and Knowles.[24,25] Reproduced (in modified form) with permission from S. K. Gray *et al.*, *J. Chem. Phys.* **113**, 7330. Copyright 2000. DOI: 10.1063/1.1313785.

collisions which start on the $2^1A'$ can result in reaction by making a non-adiabatic transition to the lower $1^1A'$.

Figure 8.2 shows computed cross sections for the reaction $O(^1D)+H_2$ $(j=0) \rightarrow OH+H$.[23] The cross sections were computed using the Real Wavepacket method as described in the first sections of this chapter. The Helicity Decoupling approximation was used in these calculations so as to simplify the computations (see Section 4.2.1). When the collision partners start on either of the symmetric A' surfaces the cross sections become infinite as the collision energy tends towards zero. This arises from the fact that collision on these two surfaces encounter no barrier and reaction can occur for even the slowest collisions. In such cases the factor of $\frac{1}{k_{vj}^2}$ in eqn (6.20) results in a sharp increase of magnitude of the cross sections at low energies.

Experimental measurements were made of the ratio of the integral reactive cross sections for reactions with H_2 in rotational states $j=0$ and $j=1$. Figure 8.3 shows a comparison of the calculated and the experimental ratios of these cross sections.[23] The single measured experimental measurement agrees exactly, probably somewhat fortuitously, with the calculated value.

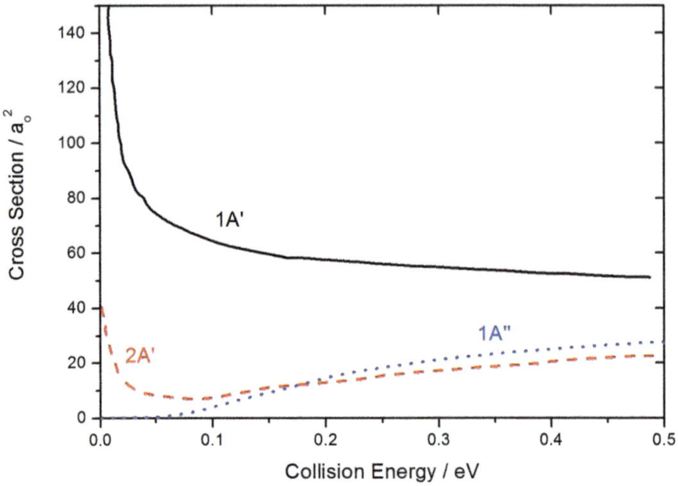

Figure 8.2 Computed integral reactive scattering cross sections for $O(^1D)+H_2$ $(j=0)\rightarrow OH+H$. The different lines show the results of the collision being started on the three different electronic state potential energy surfaces. See S. K. Gray *et al.*,[23] DOI: 10.1063/1.1313785.

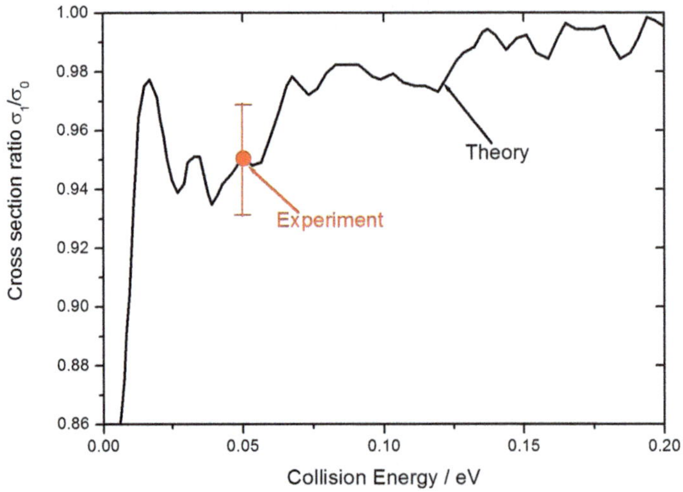

Figure 8.3 Ratio of the integral reactive cross sections for the reactions $O(^1D)+H_2$ $(j=1)\rightarrow OH+H$ and $O(^1D)+H_2(j=0)\rightarrow OH+H$. The solid line shows the computed ratio while the circle with its accompanying error bars shows the experimentally measured value. Reproduced (in modified form) with permission from S. K. Gray *et al.*, *J. Chem. Phys.*, **113**, 7330. Copyright 2000.[23] DOI: 10.1063/1.1313785.

8.5.2 $N(^4S) + CH(X^2\Pi) \rightarrow CN + H$: Reaction Proceeding *via* a Deep Well

Reactions which proceed through a deep well often involve an intermediate long lived complex. The fingerprint of such dynamics is a differential reactive cross section which is symmetrically forward-backward peaked. The

$O(^1D) + H_2 \rightarrow OH(^2\Pi) + H$ is a prototype of such behavior.[26–28] Here we outline some recent results from the reaction $N(^4S) + CH(X^2\Pi) \rightarrow CN + H$.[29] The reaction can take place on either of the two lowest lying potential energy surfaces. The lowest surface ($^3A'$) is associated with an electronic wavefunction which is symmetric with respect to reflection in the molecular plane and mediates the reaction $N(^4S) + CH(X^2\Pi) \rightarrow CN(X^2\Sigma^+) + H(^2S)$, while the reaction $N(^4S) + CH(X^2\Pi) \rightarrow CN(A^2\Pi) + H(^2S)$ takes place on the first excited $^3A''$ surface.

Figure 8.4 shows the differential cross sections, summed over all final quantum states of the fragments, for the reaction $N(^4S) + CH(X^2\Pi)(v = 0, j = 0)$ on the two surfaces. Both cross sections show backwards-forwards peaking to some degree, but the cross section on the odd or excited state surface ($^3A''$) is much more symmetric and we could therefore conclude that this reaction proceeds predominantly *via* a long-lived intermediate complex.

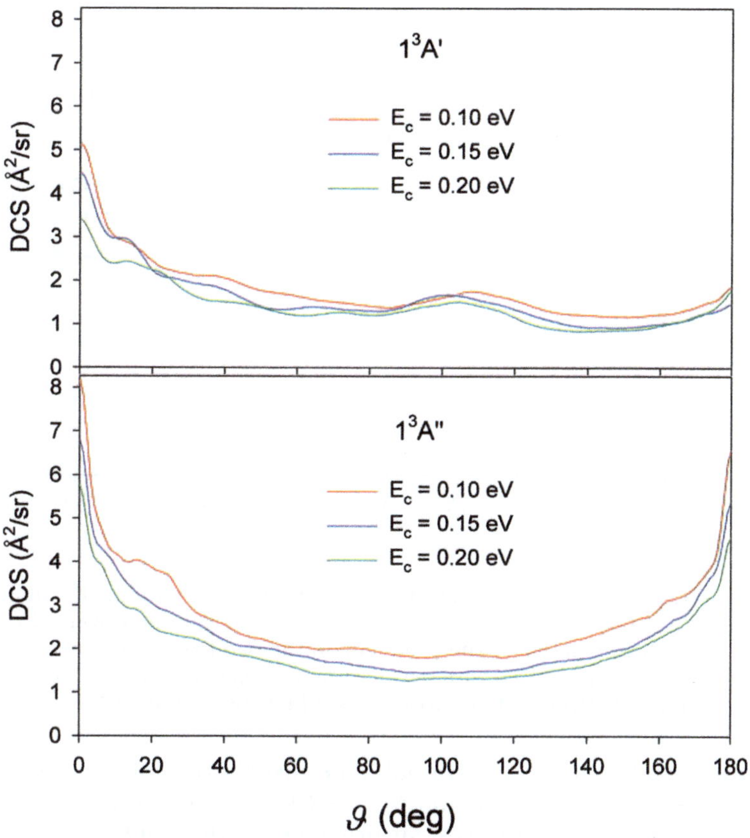

Figure 8.4 Computed differential cross sections for the reaction $N(^4S) + CH(X^2\Pi)(v = 0, j = 0)$ summed over all quantum states of the products on the two lowest potential energy surfaces. Reproduced with permission from X. Hu, C Xie, D. Xie and H. Guo, *J. Chem. Phys.*, **139**, 124313. Copyright 2013.[29] DOI: 10.1063/1.4822003.

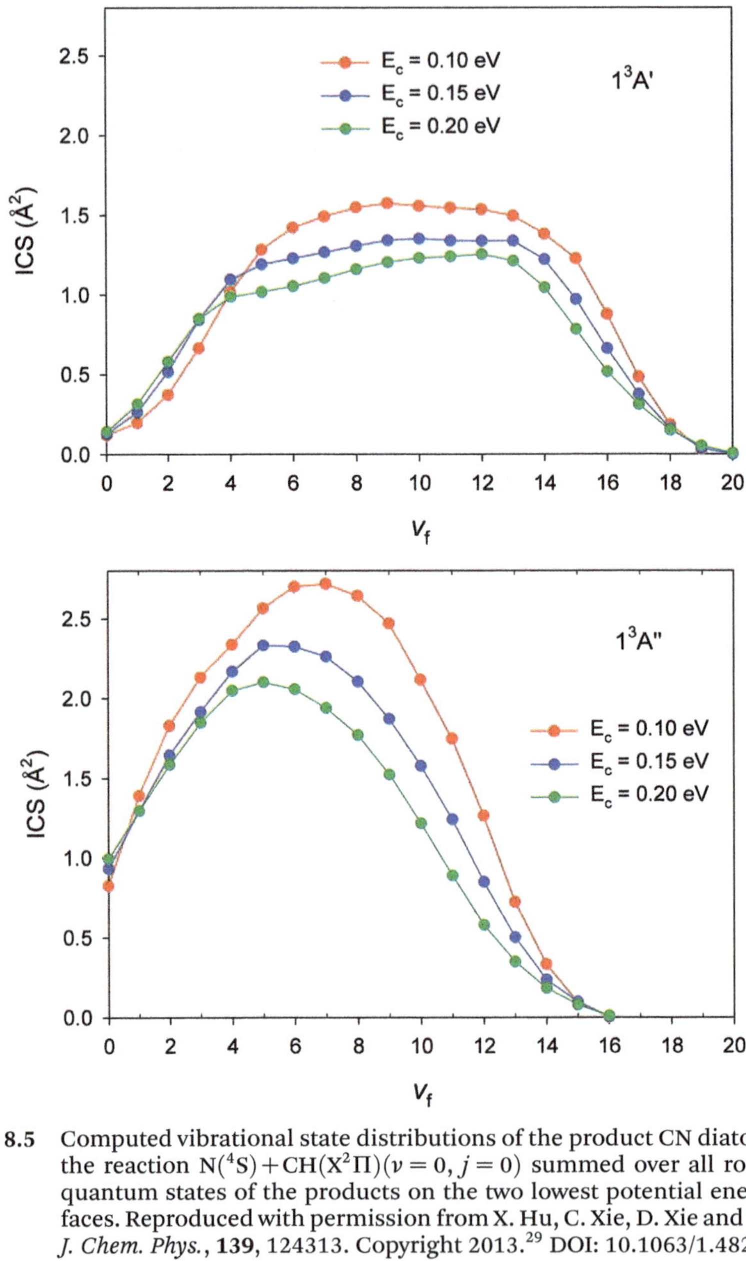

Figure 8.5 Computed vibrational state distributions of the product CN diatomic for the reaction $N(^4S)+CH(X^2\Pi)(\nu=0, j=0)$ summed over all rotational quantum states of the products on the two lowest potential energy surfaces. Reproduced with permission from X. Hu, C. Xie, D. Xie and H. Guo, *J. Chem. Phys.*, **139**, 124313. Copyright 2013.[29] DOI: 10.1063/1.4822003.

Figure 8.5 depicts the vibrational state distribution of the CN diatomic product arising from the reaction $N(^4S)+CH(X^2\Pi)(\nu=0, j=0)$ on the two lowest potential energy surfaces. Surprisingly the distribution shows a peak at non-zero values of the vibrational quantum number. This is in contrast to other similar reactions which proceed through complex formation[29] and does not agree with a phase-space statistical model of the product distribution arising from the disintegration of a complex.[30]

References

1. S. K. Gray and G. G. Balint-Kurti, Quantum dynamics with real wave packets, including application to three-dimensional ($J = 0$) $D + H_2 \rightarrow HD + H$ reactive scattering, *J. Chem. Phys.*, 1998, **108**, 950.
2. S. C. Althorpe, Quantum wavepacket method for state-to-state reactive cross sections, *J. Chem. Phys.*, 2001, **114**, 1601.
3. Y. Huang, W. Zhu, D. J. Kouri and D. K. Hoffman, A general time-to-energy transform of wavepackets. Time-independent wavepacket-Schrödinger and wavepacket-Lippmann–Schwinger equations, *Chem. Phys. Lett.*, 1993, **206**, 96.
4. Y. Huang, D. J. Kouri and D. K. Hoffman, A general, energy-separable polynomial representation of the time-independent full Green operator with application to time-independent wavepacket forms of Schrödinger and Lippmann–Schwinger equations, *Chem. Phys. Lett.*, 1994, **225**, 37.
5. V. A. Mandelshtam and H. S. Taylor, Spectral projection approach to the quantum scattering calculations, *J. Chem. Phys.*, 1995, **102**, 7390.
6. V. A. Mandelshtam and H. S. Taylor, A simple recursion polynomial expansion of the Green's function with absorbing boundary conditions. Application to the reactive scattering, *J. Chem. Phys.*, 1995, **103**, 2903.
7. H. W. Jang and J. C. Light, Artificial boundary inhomogeneity method for quantum scattering solutions in an ℓ^2 basis, *J. Chem. Phys.*, 1995, **102**, 3262.
8. Y. Huang, S. S. Iyengar, D. J. Kouri and D. K. Hoffman, Further analysis of solutions to the time-independent wave packet equations of quantum dynamics. II. Scattering as a continuous function of energy using finite, discrete approximate Hamiltonians, *J. Chem. Phys.*, 1996, **105**, 927.
9. G.-J. Kroes and D. Neuhauser, Performance of a time-independent scattering wave packet technique using real operators and wave functions, *J. Chem. Phys.*, 1996, **105**, 8690.
10. W. H. Miller, Quantum and semiclassical Green's functions in chemical reaction dynamics, *J. Chem. Soc., Faraday Trans.*, 1997, **93**, 685.
11. R. Chen and H. Guo, Extraction of resonances via wave packet propagation in Chebyshev order domain: collinear $H + H_2$, scattering, *Chem. Phys. Lett.*, 1996, **261**, 605.
12. R. Chen and H. Guo, Evolution of quantum system in order domain of Chebyshev operator, *J. Chem. Phys.*, 1996, **105**, 3569.
13. M. Hankel, S. C. Smith, R. J. Allan, S. K. Gray and G. G. Balint-Kurti, State-to-state reactive differential cross sections for the $H + H_2 \rightarrow H_2 + H$ reaction on five different potential energy surfaces employing a new quantum wavepacket computer code: DIFFREALWAVE, *J. Chem. Phys.*, 2006, **125**, 164303.
14. M. Hankel, S. K. Gray and G. G. Balint-Kurti. The DiffRealWave code is available on request from M. Hankel. E-mail: m.hankel@uq.edu.au
15. M. Hankel, S. C. Smith, S. K. Gray and G. G. Balint-Kurti, DIFFREALWAVE: A parallel real wavepacket code for the quantum mechanical calculation of reactive state-to-state differential cross sections in atom plus diatom collisions, *Comput. Phys. Commun.*, 2008, **179**, 569.

16. G. G. Balint-Kurti, Time-dependent and time-independent wavepacket approaches to reactive scattering and photodissociation dynamics, *Int. Rev. Phys. Chem.*, 2008, **27**, 507.
17. S. K. Gray, Wave packet dynamics of resonance decay: an iterative equation approach with application to HCO $\rightarrow$ H + CO, *J. Chem. Phys.*, 1992, **96**, 6543.
18. G. B. Arfken and H. J. Weber, *Mathematical Methods for Physicists*, Academic Press, 7th edn, 2013.
19. M. Abromowitz and I. A. Stegun, *Handbook on Mathematical Functions, Applied Mathematics Series 55*, National Bureau of Standards, 10th edn, 1972.
20. L. D. Landau and E. M. Lifshitz, *Quantum Mechanics: Non-Relativistic Theory*, Butterworth-Heinemann, Oxford, 3rd edn, 1977 (Reprinted with corrections 1991).
21. P. J. Kuntz, B. I. Niefer and J. J. Sloan, A multisurface DIM trajectory study of the reaction: $O(^1D_g) + H_2(X^1\Sigma_g^+) \rightarrow OH(X^2\Pi) + H(^2S)$, *J. Chem. Phys.*, 1988, **88**, 3629.
22. G. C. Schatz, L. A. Pederson and P. J. Kuntz, Adiabatic and non-adiabatic dynamics studies of $O(^1D) + H_2 \rightarrow OH + H$, *Faraday Discuss.*, 1997, **108**, 357.
23. S. K. Gray, G. G. Balint-Kurti, G. C. Schatz, J. J. Lin, X. Liu, S. Harich and X. Yang, Probing the effect of the H_2 rotational state in $O(^1D) + H_2 \rightarrow OH + H$: theoretical dynamics including nonadiabatic effects and a crossed molecular beam study, *J. Chem. Phys.*, 2000, **113**, 7330.
24. A. J. Dobbyn and P. J. Knowles, A comparative study of methods for describing non-adiabatic coupling: diabatic representation of the $^1\sigma^+/^1\pi$ HOH and HHO conical intersections, *Mol. Phys.*, 1997, **91**, 1107.
25. A. J. Dobbyn and P. J. Knowles, General discussion, *Farday Discuss.*, 1998, **110**, 247.
26. P. Honvault and J.-M. Launay, A quantum-mechanical study of the dynamics of the $O(^1D) + H_2 \rightarrow OH + H$ insertion reaction, *J. Chem. Phys.*, 2001, **114**, 1057.
27. E. J. Rackham, T. Gonzalez-Lezana and D. E. Manolopoulos, A rigorous test of the statistical model for atom–diatom insertion reactions, *J. Chem. Phys.*, 2003, **119**, 12895.
28. M. H. Alexander, E. J. Rackham and D. E. Manolopoulos, Product multiplet branching in the $O(^1D) + H_2 \rightarrow OH(^2\Pi) + H$ reaction, *J. Chem. Phys.*, 2004, **121**, 5221.
29. X. Hu, C. Xie, D. Xie and H. Guo, State-to-state quantum dynamics of the $N(^4S) + CH(X^2\Pi) \rightarrow CN(X^2\Sigma^+, A^2\Pi) + H(^2S)$ reactions, *J. Chem. Phys.*, 2013, **139**, 124313.
30. J. P. Cole and G. G. Balint-Kurti, A statistical, ab initio, quantum mechanical study of the photolysis and final state distributions of singlet ketene, *J. Chem. Phys.*, 2003, **119**, 6003.

CHAPTER 9

Lasers and the Photoloc Method

Lasers have truly revolutionized the study of chemical reactions. Their power arises from the sharp definition of the frequency of the laser light as well as from the fact that it is polarized. In this chapter we first discuss the "photoloc" method which permits the measurement of a differential reactive cross section without the need to use crossed molecular beams. We then describe some other uses of lasers in the analysis of the products of a chemical reaction. Chapter 10 will address issues such as the polarization and alignment of the rotational and other angular momenta of the products of a chemical reaction. Such measurements also rely entirely on the use of lasers.

In a typical experiment the initial reactant atoms or small radicals such as OH, CH, NH, CN are prepared by laser photolysis of suitable precursors. Subsequently a tunable laser may be used to excite the molecule to some desired electronic and ro-vibrational state.[1] The products of the reaction are then probed using various laser and spectroscopic techniques, which we describe in Section 9.2.

9.1 PHOTOLOC—Photo Initiated Reaction Analyzed *via* the Law of Cosines

PHOTOLOC is abbreviation of Photo Initiated Reaction Analyzed *via* the Law of Cosines.[2–5] The amazing aspect of the method is that it permits the measurement of a differential reactive scattering cross section without the use of a crossed molecular beam experiment. Indeed the main use of a beam in such experiments is to lower the rotational temperature of the reactants. In

RSC Theoretical and Computational Chemistry Series No. 7
Theory of Molecular Collisions
By Gabriel G. Balint-Kurti and Alexander P. Palov

Published by the Royal Society of Chemistry, www.rsc.org

this section we otline the underlying theory of the PHOTOLOC method and in doing so we follow the derivations given in ref. 2 and 5.

In the PHOTOLOC process the reactant atom, A, is created by the use of linearly polarized laser light, normally in the UV region of the spectrum, to photodissociate a molecule

$$AX + h\nu \rightarrow A + X. \tag{9.1}$$

The A atom created in this way then reacts with another molecule, BC, to yield products AB

$$A + BC(\nu, j) \rightarrow C + AB(\nu', j'). \tag{9.2}$$

For the purposes of this discussion, we will assume that the AX and BC molecules are at rest. In practice they will be expanded together through a nozzle into the reaction chamber. So their velocity relative to each other will be very small. The photodissociation process injects a well defined quantity of energy, $h\nu$, into the AX molecule. As the molecule is assumed to be at rest, we can use energy balance to write the equation

$$h\nu - E_{D_{\bar{\nu}\bar{j}}} = \frac{m_A v_A^2}{2} + \frac{m_X v_X^2}{2} \tag{9.3}$$

where $E_{D_{\bar{\nu}\bar{j}}}$ is the dissociation energy from the $\bar{\nu}\bar{j}$ level of the AX molecule.

Momentum conservation, again using the fact that the AX molecule is originally at rest, gives us the relationship

$$\mathbf{v}_X = -\frac{m_A}{m_X}\mathbf{v}_A. \tag{9.4}$$

Substituting this into eqn (9.3) we obtain an equation for the magnitude of the velocity of atom A

$$v_A^2 = \frac{2m_X}{m_A m_{AX}}\left[h\nu - E_{D_{\bar{\nu}\bar{j}}}\right], \tag{9.5}$$

and therefore also for the kinetic energy, E_A of atom A

$$\begin{aligned} E_A &= \frac{m_A}{2} v_A^2 \\ &= \frac{m_X}{m_{AX}}\left[h\nu - E_{D_{\bar{\nu}\bar{j}}}\right]. \end{aligned} \tag{9.6}$$

The momentum conservation equation for the reaction (eqn (9.2)) is

$$\begin{aligned} m_A \mathbf{v}_A + m_{BC}\mathbf{v}_{BC} &= M\boldsymbol{u}_{CM} \\ &= m_{AB}\mathbf{v}_{AB} + m_C \mathbf{v}_C \end{aligned} \tag{9.7}$$

where $\boldsymbol{u}_{CM}$ is the velocity of the center-of-mass for the reaction and $M = m_A + m_B + m_C$.

But taking account of the fact that we are assuming that the BC molecule is at rest (*i.e.* $\mathbf{v}_{BC} = 0$), eqn (9.7) becomes

$$\begin{aligned} m_A \mathbf{v}_A &= M\boldsymbol{u}_{CM} \\ &= m_{AB}\mathbf{v}_{AB} + m_C \mathbf{v}_C, \end{aligned} \tag{9.8}$$

thus providing us with an equation for the velocity of the center-of-mass vector

$$\boldsymbol{u}_{CM} = \frac{m_A}{M}\mathbf{v}_A. \tag{9.9}$$

The energy conservation law for the reaction eqn (9.2) is

$$\frac{\mu_r v_r^2}{2} = \frac{\mu_p v_p^2}{2} + \Delta E \tag{9.10}$$

where μ_r and μ_p are the reduced masses of the reactants and products respectively and $\mathbf{v}_r = \mathbf{v}_A - \mathbf{v}_{BC}$ and $\mathbf{v}_p = \mathbf{v}_{AB} - \mathbf{v}_C$ are the relative velocities in the laboratory reference frame. ΔE is the reaction endoergicity. As $\mathbf{v}_{BC} = 0$, we can write $\mathbf{v}_r = \mathbf{v}_A$.

The velocity of the AB product in the laboratory frame, $\mathbf{v}_{AB}$ may be written as

$$\mathbf{v}_{AB} = \boldsymbol{u}_{CM} + \boldsymbol{u}_{AB} \tag{9.11}$$

where $\boldsymbol{u}_{AB}$ is the velocity of AB in the center-of-mass reference frame.

Using the properties of the center-of-mass (see eqn (1.35)), we can write

$$\begin{aligned} u_{AB} &= \frac{m_C}{M} v_p \\ &= \frac{m_C}{M}\sqrt{\frac{\mu_r v_r^2 - 2\Delta E}{\mu_p}}. \end{aligned} \tag{9.12}$$

Using the fact that $\mathbf{v}_r = v_A$, we can therefore write

$$u_{AB} = \frac{m_C}{M}\sqrt{\frac{\mu_r v_A^2 - 2\Delta E}{\mu_p}} \tag{9.13}$$

where all the quantities on the right hand side of the equation are known (see eqn (9.5)).

Figure 9.1 shows a Newton diagram of the reaction. From the square of eqn (9.11) we obtain

$$v_{AB}^2 = u_{CM}^2 + u_{AB}^2 - 2u_{CM}u_{AB}\cos(180 - \theta_r) \tag{9.14}$$

or

$$v_{AB}^2 = u_{CM}^2 + u_{AB}^2 + 2u_{CM}u_{AB}\cos(\theta_r). \tag{9.15}$$

The quantities u_{CM} and u_{AB} are known (see eqn (9.9) and (9.13)) so a measurement of the magnitude of the product speed, v_{AB}, in the laboratory frame will yield the center-of-mass scattering angle.

The photodissociation process, eqn (9.1), yields photofragments A and X which are traveling in different directions. The distribution of the directions in which the atomic fragments are emitted ($f(\theta_A)$) has been well studied[6,7] and the angular distribution for a single photon dissociation process is characterized by an anisotropy parameter, β.

$$f(\theta_A) = \frac{1 + \beta P_2(\cos\theta_A)}{4\pi} \tag{9.16}$$

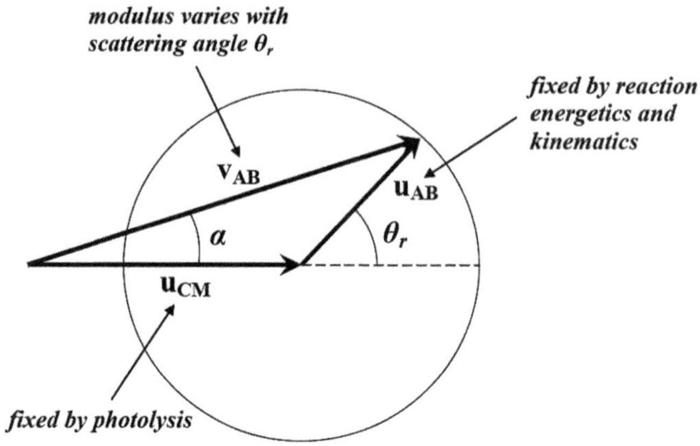

Figure 9.1 Newton diagram pertinent to the photoloc technique. For detection of a particular AB(v', j') state the only vector in this figure with a variable modulus is $\mathbf{v}_{AB}$, the laboratory product speed. This change in modulus can be mapped directly into a change in center-of-mass scattering angle θ_r. Reproduced with permission from F. Fernandez-Alonso, B. D. Bean and R. N. Zare, *J. Chem. Phys.*, **111**, 1022. Copyright 1999, AIP Publishing LLC.[5] DOI: 10.1063/1.479294.

where θ_A is the angle between velocity of the A atom photofragment and the electric vector of the linearly polarized photolysis laser light, $\hat{\varepsilon}$, and $P_2(\cos\theta_A)$ is the second Legendre polynomial. If eqn (9.16) gives the spatial distribution of the A atom photofragment velocities then the spatial distribution of AB product laboratory velocities can be shown to be[2]

$$f(\theta_{AB}) = \frac{1}{\sigma}\frac{d\sigma(\Omega)}{d(\Omega)}\frac{[1 + \beta P_2(\cos\alpha)P_2(\cos\theta_{AB})]}{2u_{CM}u_{AB}v_{AB}} \tag{9.17}$$

where θ_{AB} is the angle between the velocity of the AB product and the photolysis light polarization vector $\hat{\varepsilon}$, σ is the integral reactive cross section and $\frac{d\sigma(\Omega)}{d(\Omega)}$ is the state-to-state differential reactive cross section in the center-of-mass reference frame. u_{CM} is known from eqn (9.5) and (9.9), u_{AB} is known from eqn (9.5) and (9.13), v_{AB} and θ_{AB} must be measured and the angle α can be calculated from knowledge of u_{AB}, v_{AB} and u_{CM} (see Figure 9.1).

Thus a measurement of the scattering directions of the AB product and the magnitude of their laboratory velocity can be used to determine the state-to-state differential reactive cross section, $\frac{d\sigma(\Omega)}{d(\Omega)}$, in the center-of-mass reference frame.

Figure 9.2 shows two differential cross sections for the reaction $Cl(^2P_{3/2}) + CH_4(v_3 = 1, J = 1) \rightarrow HCl(v' = 1, j') + CH_3$ measured using the PHOTOLOC method. The results show that the rotationally excited $j' = 3$ HCl product (lower panel) leads to more backward scattering than the $j' = 1$ product.

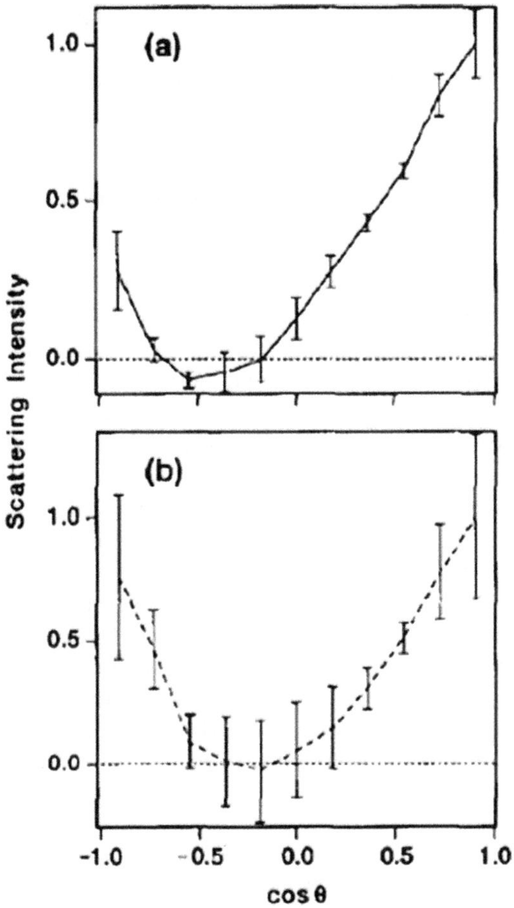

Figure 9.2 Center-of-mass differential cross sections (arbitrary units) for the reaction $Cl(^2P_{3/2}) + CH_4(\nu_3 = 1, J = 1) \rightarrow HCl(\nu' = 1, j') + CH_3$ measured by the PHOTOLOC method at a collision energy of 0.159 eV. (a) $HCl(\nu' = 1, j' = 1)$ (b) $HCl(\nu' = 1, j' = 3)$. $\cos\theta = 1.0$ corresponds to the forward scattering direction. Rotationally excited HCl (lower panel) has a greater preponderance to backward scattering. Reprinted from W. R. Simpson, A. J. Orr-Ewing and R. N. Zare, *Chem. Phy. Lett.*, **212**, 163. Copyright 1993, reproduced with permission from Elsevier.[3]

9.2 Detection Methods

Crossed molecular beam experiments initially used surface ionization detectors[8] which could only detect alkali atoms. A major advance was achieved with the introduction of mass spectrometric detection in 1969,[9,10] allowing any reaction product to be detected. Lasers have now taken our ability to interrogate the products of a chemical reaction to a new level.[11] They enable us, through the use of spectroscopic information, to discover the quantum state distribution of the product atoms or molecules and even the spatial distribution of any rotational or electronic angular momenta. In this section we will outline some of the ways in which lasers can be used to detect the products of chemical reactions.

9.2.1 Resonantly-Enhanced Multiphoton Ionization (REMPI)

The REMPI method involves the use of one or more photons to excite a molecule (or atom) to a high lying quantum level followed by absorption of a further photon which results in ionization.[12] Normally a laser in the Ultra Violet (UV) range of the spectrum is used to achieve this purpose. REMPI(2 + 1) involves an initial two-photon excitation to a "real" intermediate state followed by the absorption of a third photon which ionizes the molecule. This process is illustrated schematically in Figure 9.3.

The REMPI process is highly state-selective. This is because the laser light used is tuned to a precise frequency, unique to the molecule and the transition of interest, so that only molecules in the targeted quantum state are first excited and then ionized. For a signal to be seen the molecule must have an absorption line in its spectrum at exactly the frequency 2ν. This requirement selects only the desired quantum state of the molecule under investigation. Once ionized, the molecules are extracted from the ionization region by a strong electric field and subsequently impinge on a position sensitive detector. The basic REMPI method was reviewed in 1978 by Letokhov.[13] It

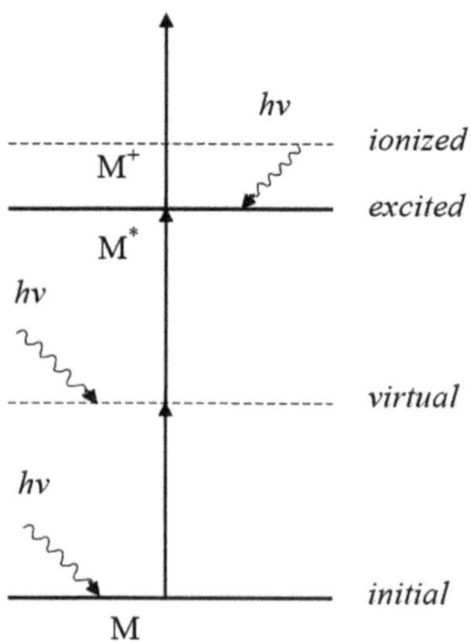

Figure 9.3 Schematic diagram illustrating the mechanism of resonant-enhanced (2 + 1) multiphoton ionization.

was used by Marinero, Rettner and Zare[14] in 1984 to examine the rotational state distribution of the HD products arising from the reaction H + D_2 → HD + D. Figure 9.4 shows the results of their experiment. The three different panels show results for different time delays between the pulsed probe REMPI laser and the photodissociation laser which photodissociated the HI to form the H atom and thus initiate the reaction. The top two panels, corresponding to shorter time delays, yield identical rotational state distributions and are therefore taken to represent the true state distribution following reaction. In the bottom panel, taken at a longer time delay, the populations have changed due to collisional processes.

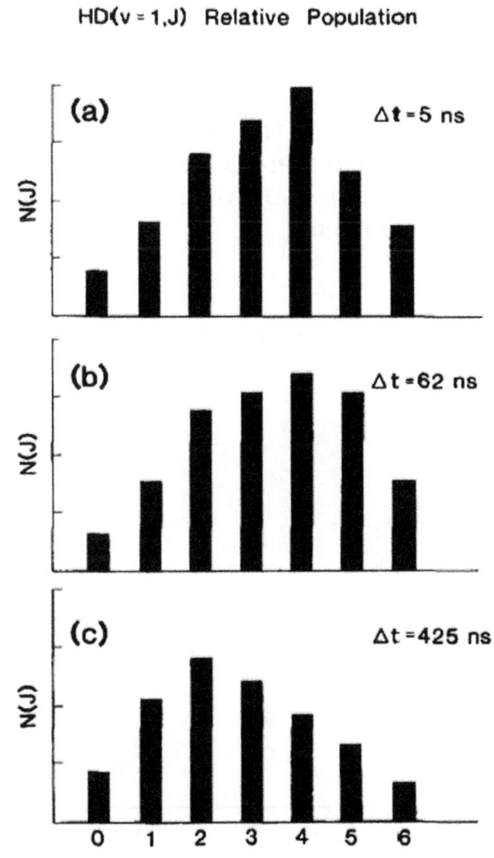

Figure 9.4 Rotational state distribution of HD(v' = 1) product resulting from the reaction H + $D_2 \rightarrow HD(v', j')$ + D analyzed using REMPI technique. The different panels show results for different time delays between the pulsed probe REMPI and the photodissociation laser. Reproduced with permission from E. E. Marinero, C. T. Rettner and R. N. Zare, *J. Chem. Phys.*, 1984, **80**, 4142. Copyright 1984, AIP Publishing LLC.[14] DOI: 10.1063/1.447242.

9.2.2 REMPI-Time-Of-Flight (REMPI-TOF)

As we have seen in the discussion of the PHOTOLOC method (Section 9.1), we need to measure the velocity of the reaction products if we wish to determine the differential cross section. Use of the REMPI method as discussed above is insufficient to enable us to do this. If a pulsed laser is used to ionize the product molecules, then the position and time of creating the ions is known. The ions are extracted from the ionization region using an electrical field and are then allowed to fly in a field free region towards the detector. If we record the time at which each ion arrives at the detector, we can calculate its velocity in the direction of the detector and this permits the determination of the state-to-state differential cross section.[5,11,15–17] This mode of detection is termed REMPI-Time-Of-Flight or REMPI-TOF mass spectrometry. Not only is the method state specific, it also has the great advantage that the detection of an ion is very efficient, while detecting a neutral molecule is far less so.

Figure 9.5 shows the differential cross section for the HD($\nu' = 2, j' = 3$) product arising from the reaction H + $D_2 \rightarrow$ HD(ν', j') + D obtained using the PHOTOLOC method and REMPI-TOF analysis of the products.[5] Six independent sets of results are shown and their spread gives an estimate of the uncertainty of the measurements.

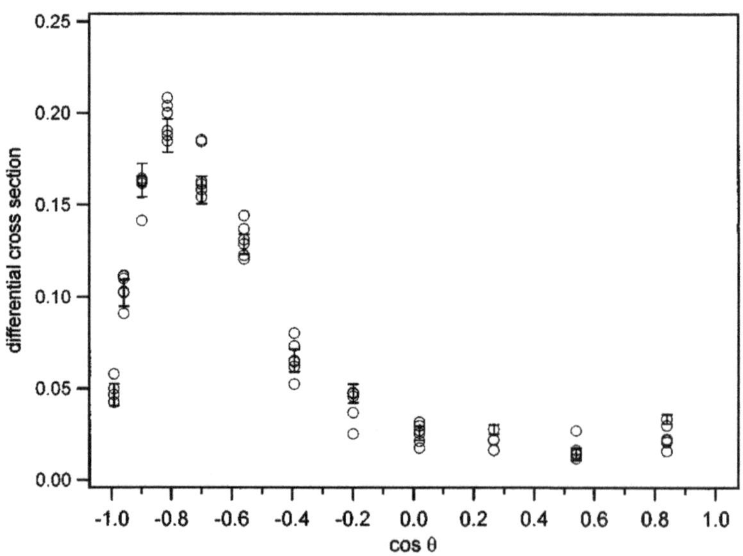

Figure 9.5 Experimental differential cross sections for the HD($\nu' = 2, j' = 3$) product arising from the reaction H + $D_2 \rightarrow$ HD(ν', j') + D obtained using the PHOTOLOC method and REMPI-TOF analysis of the products. Six independent sets of results are shown. Reproduced with permission from F. Fernandez-Alonso, B. D. Bean and R. N. Zare, *J. Chem. Phys.*, **111**, 1022. Copyright 1999, AIP Publishing LLC.[5] DOI: 10.1063/1.479294.

The lasers used for the REMPI-TOF analysis emit polarized light. If linearly polarized light is used then in principle two different measurements can be made, one with the polarization direction of the light pointed along the time-of-flight direction and the other with it polarized perpendicular to it. It can be shown[18] that these two measurements permit additional information on the rotational angular alignment of the product molecules to be extracted. This will be discussed further in the next chapter.

9.2.3 Laser-Induced Fluorescence (LIF)

When a molecule or atom is excited by absorption of radiation to a higher electronic state it will fluoresce and emit radiation in its turn. Figure 9.6 shows a schematic diagram illustrating the mechanism.

The absorption process is highly selective and the spectrum of the emitted light is unique to the particular excitation. Tunable UV lasers provide an intense source of light at a defined chosen frequency well suited for performing the initial state-selective excitation process. When a fluorescent process is initiated by laser light it is called Laser Induced Fluorescence (LIF). Collecting and measuring the intensity of the emitted light from such a process provides a means of estimating the number of molecules in a particular quantum state in the path of the laser beam. The technique has been successfully used in the 1970's by the group of R. N. Zare.[19,20] It has been reviewed by J. L. Kinsey[21] and an interesting personalized overview has recently been given by R. N. Zare.[22]

By utilizing the polarization directions of the exciting and and the emitted light the LIF method can be used, together with the Doppler profiles discussed below, to elucidate stereodynamical aspects of a reaction.

9.2.4 Doppler Spectroscopy

The frequency at which a molecule absorbs or emits radiation is affected by its motion. Thus if a molecule at rest emits radiation of frequency ν then a molecule moving towards a detector at a velocity v will be measured as emitting radiation of the higher frequency $\nu\left(1+\frac{\mathrm{v}}{c}\right)$. In an ensemble of molecules with a velocity distribution this will lead to the measurement of a

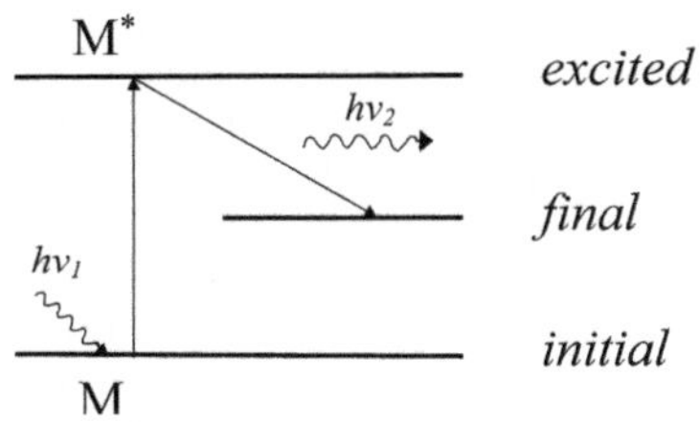

Figure 9.6 Schematic diagram showing the mechanism of laser-induced fluorescence.

typical Doppler line shape. This line shape will be influenced by the orientation of the dipole emitting the radiation and consequently by the orientation of the molecule itself relative to the line from it to the detector. The electric vector of the light is perpendicular to its direction of motion, thus if we consider a parallel transition in a diatomic molecule, the light will be emitted perpendicular to the molecular axis. So a molecule lying perpendicular to the line connecting it to the detector and moving in the direction of the detector will be measured as emitting light of frequency $\nu\left(1+\frac{\mathrm{v}}{c}\right)$ as discussed above. Conversely if it is moving away from the detector the measured frequency of the radiation will be $\nu\left(1-\frac{\mathrm{v}}{c}\right)$. For a similar molecule whose velocity does not lie directly towards the detector but rather is directed at an angle θ with respect to the line connecting it to the detector, only the component of the velocity in the direction of the detector affects the observed frequency. Thus in this case the measured frequency will be $\nu\left(1+\frac{\mathrm{v}\cos\theta}{c}\right)$.

The intensity of the light detected is also influenced by the orientation of the molecule with respect to the line connecting it to the detector. Thus for a parallel transition, if the diatomic molecule lies along the direction of the line connecting it to the detector, all the emitted light will be directed perpendicular to this line and no light at all will reach the detector.

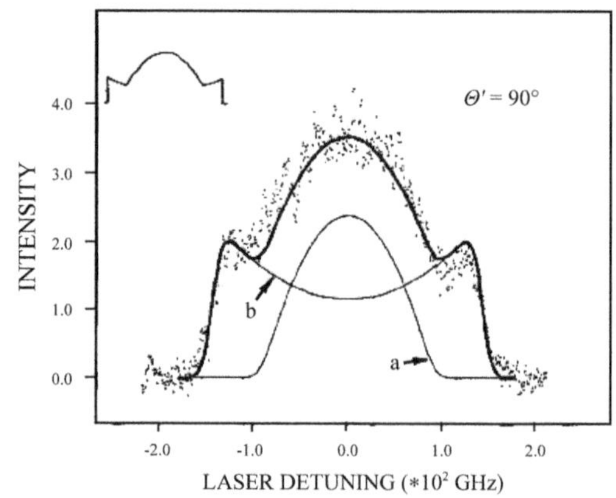

Figure 9.7 Figure showing the LIF Doppler line shape of the H atom product of the photodissociation of HI molecules. θ' is the angle between the direction of the electric field of the dissociating laser light and the probe laser beam which stimulates the LIF process. The line shape shows that there are two distinct groups of H atoms with different velocities. Lines (a) and (b) show the theoretically expected line shapes for parallel and perpendicular absorption respectively. The small dots are the experimentally measured points and the thick solid line is a composite of (a) and (b) which best fits the observations. Reproduced with permission from R. Schmiedl, H. Dugan, W. Meier and K. H. Welge, *Zeitschrift Physik A*, 1982, **304**, 137.[23]

We have talked until now of a molecule emitting light in a spontaneous transition. Exactly the same considerations apply to laser induced fluorescence (LIF). Figure 9.7 shows an LIF Doppler line shape for the H atom product arising from the photodissociation of HI molecules.[23] The Doppler line shape can be recorded with the probe laser, which gives rise to the LIF emission, situated at different angles to the direction of the electric field vector of the photodissociating light.

C. H. Greene and R. N. Zare[24] have discussed how LIF measurements may be used to determine the detailed product state distribution arising from a chemical reaction using Doppler line shapes obtained from LIF detection. Dixon has further extended the underlying theory of the method so as to permit the extraction of information on the polarization and alignment of the reaction products using LIF Doppler line shape studies.

References

1. S. Marinakis, M. L. Costen and K. G. McKendrick, Do vectors point the way to understanding energy transfer in molecular collisions?, *Chem. Soc. Rev.*, 2008, **37**, 732.
2. N. E. Shafer, A. J. Orr-Ewing, W. R. Simpson, H. Xu and R. N. Zare, State-to-state differential cross sections from photoinitiated bulb reactions, *Chem. Phys. Lett.*, 1993, **212**, 155.
3. W. R. Simpson, A. J. Orr-Ewing and R. N. Zare, State-to-state differential cross sections for the reaction $Cl + CH_4 \rightarrow HCl + CH_3$, *Chem. Phys. Lett.*, 1993, **212**, 163.
4. N. E. Shafer-Ray, A. J. Orr-Ewing and R. N. Zare, Beyond state-to-state differential cross sections: determination of product polarization in photoinitiated bimolecular reactions, *J. Phys. Chem.*, 1995, **99**, 7591.
5. F. Fernández-Alonso, B. D. Bean and R. N. Zare, Measurement of the $HD(v' = 2, J' = 3)$ product differential cross section for the $H + D_2$ exchange reaction at 1.55 ± 0.05 eV using the photoloc technique, *J. Chem. Phys.*, 1999, **111**, 1022.
6. R. N. Zare and D. R. Herschbach, Atomic and molecular fluorescence excited by photodissociation, *Appl. Opt.*, 1965, **4**, 193.
7. S.-C. Yang and R. Bersohn, Theory of the angular distribution of molecular photofragments, *J. Chem. Phys.*, 1974, **61**, 4400.
8. D. R. Herschbach, Reactive collisions in crossed molecular beams, *Discuss. Faraday Soc.*, 1962, **33**, 149.
9. Y. T. Lee, 1986. Nobel Lecture: "Molecular Beam Studies of Elementary Chemical Processes". Available from web site: http://www.nobelprize.org/nobel_prizes/chemistry/laureates/1986/lee-lecture.html.
10. Y. T. Lee, J. D. McDonald, P. R. LeBreton and D. R. Herschbach, Molecular beam reactive scattering apparatus with electron bombardment detector, *Rev. Sci. Instrum.*, 1969, **40**, 1402.
11. M. N. R. Ashfold, N. H. Nahler, A. J. Orr-Ewing, O. P. J. Vieuxmaire, R. L. Toomes, T. N. Kitsopoulos, I. A. Garcia, D. A. Chestakov, S.-M. Wu and

D. H. Parker, Imaging the dynamics of gas phase reactions, *Phys. Chem. Chem. Phys.*, 2006, **8**, 26.
12. R. V. Ambartsumian and V. S. Letokhov, Selective two-step (STS) photoionization of atoms and photodissociation of molecules by laser radiation, *Appl. Opt.*, 1972, **11**, 354.
13. V. S. Letokhov, Selective action of laser radiation on matter, *Sov. Phys.-Usp.*, 1978, **21**, 405.
14. E. E. Marinero, C. T. Rettner and R. N. Zare, H + D_2 reaction dynamics. Determination of the product state distributions at a collision energy of 1.3 eV, *J. Chem. Phys.*, 1984, **80**, 4142.
15. V. S. Antonov, V. S. Letokhov and A. N. Shibanov, Formation of molecular ions as a result of irradiation of the surface of molecular crystals by UV laser radiation, *JETP Lett.*, 1980, **31**, 441.
16. V. S. Antonov, V. S. Letokhov, Yu. A. Matveyets and A. N. Shibanov, Sputtering of neutral molecules and molecular ions from the adenine crystal surface induced by the UV picosecond laser pulse, *Laser Chem.*, 1982, **1**, 37.
17. N. E. Shafer, H. Xu, R. P. Tuckett, M. Springer and R. N. Zare, Direct measurement of the three-dimensional product velocity distribution from photoinitiated bulb reactions, *J. Phys. Chem.*, 1994, **98**, 3369.
18. A. J. Orr-Ewing, W. R. Simpson, T. P. Rakitzis, S. A. Kandel and R. N. Zare, Scattering-angle resolved product rotational alignment for the reaction of Cl with vibrationally excited methane, *J. Chem. Phys.*, 1997, **106**, 5961.
19. A. Schultz, H. W. Cruse and R. N. Zare, Laser-induced fluorescence: a method to measure the internal state distribution of reaction products, *J. Chem. Phys*, 1972, **57**, 1354.
20. H. W. Cruse, P. J. Dagdigian and R. N. Zare, Crossed-beam reactions of barium with hydrogen halides. Measurement of internal state distributions by laser-induced fluorescence, *Faraday Discuss. Chem. Soc.*, 1973, **55**, 277.
21. J. L. Kinsey, Laser-induced fluorescence, *Annu. Rev. Phys. Chem.*, 1977, **28**, 349.
22. R. N. Zare, My life with LIF: a personal account of developing laser-induced fluorescence, *Annu. Rev. of Anal. Chem.*, 2012, **5**, 1.
23. R. Schmiedl, H. Dugan, W. Meier and K. H. Welge, Laser Doppler Spectroscopy of atomic hydrogen in the photodissociation of HI, *Z. Phys. A*, 1982, **304**, 137.
24. C. H. Greene and R. N. Zare, Determination of product population and alignment using laser-induced fluorescence, *J. Chem. Phys.*, 1983, **78**, 6741.

CHAPTER 10

Polarization, Alignment and Vector Correlation

It may seem that we have discussed all possible aspects of an elementary chemical reaction in the preceding chapters of this book, this is however not the case. The area we have omitted so far is the influence of the relative orientation of the reactants and study of the so-called alignment and polarization of the products in a reaction. This topic, generally referred to as stereodynamics, will be the topic of the current chapter. It covers the relationships, or more precisely the correlations, between the various vector quantities which characterize the reactants and the products in a chemical reaction. These quantities are namely the relative velocities of the reactants and products ($\boldsymbol{k}$ and $\boldsymbol{k}'$) and their "internal" angular momenta ($\boldsymbol{j}$ and $\boldsymbol{j}'$), consisting of their rotational, spin and perhaps other component angular momenta. The problem of vector correlation in chemical reactions was pioneered by Herschbach and co-workers starting in 1975.[1–3]

10.1 Orientation and Alignment

Using the illustrative description of A. J. Orr-Ewing and R. N. Zare in ref. 4, we denote the angular momentum vectors of an ensemble of molecules by headed arrows. Figure 10.1 illustrates the concepts of alignment and orientation.

The double headed arrows on the left of the figure are "aligned", meaning that they preferentially point in the up-down direction rather than randomly in all possible directions. The single headed arrows on the right hand side of the diagram are said to be "oriented", meaning that they preferentially point in one direction, up.

RSC Theoretical and Computational Chemistry Series No. 7
Theory of Molecular Collisions
By Gabriel G. Balint-Kurti and Alexander P. Palov

Published by the Royal Society of Chemistry, www.rsc.org

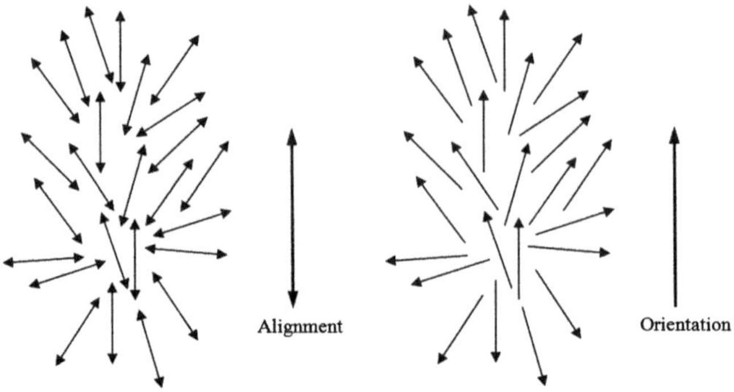

Figure 10.1 The double and single-headed arrows present aligned and oriented angular momentum of molecules, respectively. Republished with permission of Annual Reviews, Inc., from A. J. Orr-Ewing and R. N. Zare, *Annu. Rev. Phys. Chem.*, 1994, **45**, 315.[4]

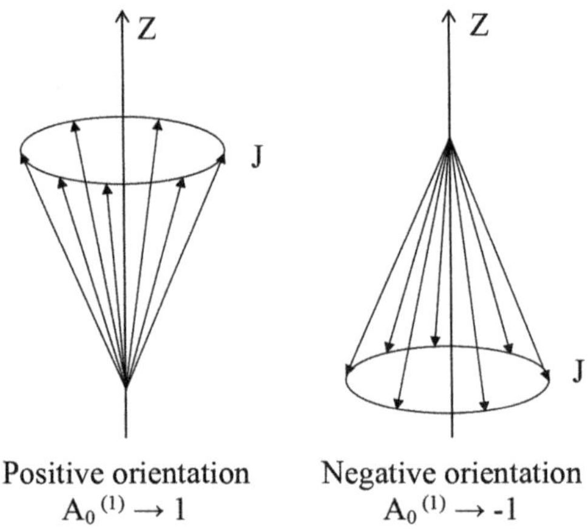

Figure 10.2 Schematic diagram of positively and negatively oriented angular momentum distributions. Republished with permission of Annual Reviews, Inc., from A. J. Orr-Ewing and R. N. Zare, *Annu. Rev. Phys. Chem.*, 1994, **45**, 315.[4]

Figure 10.2 illustrates positive and negative orientations of $\boldsymbol{J}$ vectors, while Figure 10.3 shows positive and negative alignment.

The alignment and orientation are related to the angles the vectors make with the $\boldsymbol{Z}$ axis. Thus if $\hat{\boldsymbol{J}}$ and $\hat{\boldsymbol{Z}}$ are unit vectors pointing along the $\boldsymbol{J}$ and $\boldsymbol{Z}$ directions respectively, we can denote the angle between them as θ so that

$$\hat{\boldsymbol{J}} \cdot \hat{\boldsymbol{Z}} = \cos\theta. \tag{10.1}$$

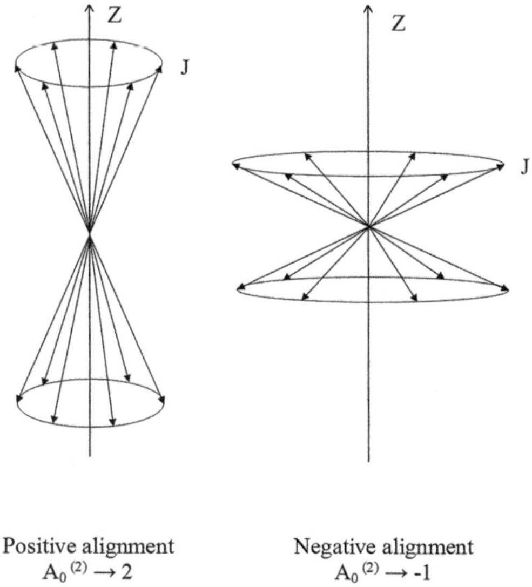

Figure 10.3 Schematic diagram of positively and negatively aligned angular momentum distributions. Republished with permission of Annual Reviews, Inc., from A. J. Orr-Ewing and R. N. Zare, *Annu. Rev. Phys. Chem.*, 1994, **45**, 315.[4]

The Legendre polynomials of $\cos\theta$ are related to the different powers of $\cos\theta$

$$P_1(\cos\theta) = \cos\theta,$$
$$P_2(\cos\theta) = \frac{1}{2}[3\cos^2\theta - 1]. \tag{10.2}$$

It has therefore become customary to define orientation and alignment parameters as the expectation values (averages) of the Legendre polynomials of $\cos\theta$[4]

$$A_0^1 = \langle P_1(\cos\theta)\rangle = \langle\cos\theta\rangle, \tag{10.3}$$
$$A_0^2 = 2\,\langle P_2(\cos\theta)\rangle = \langle 3\cos^2\theta - 1\rangle. \tag{10.4}$$

Therefore when $\boldsymbol{J}$ is parallel to $\boldsymbol{Z}$, $A_0^1 = 1$ while when $\boldsymbol{J}$ is anti-parallel to $\boldsymbol{Z}$, $A_0^1 = -1$. $A_0^2 = 2$ for both of these situations.

The polarization parameters defined in eqn (10.3) and (10.4) describe orientation and alignment with respect to a space-fixed or laboratory reference frame. Parameters of this type are better known in connection with photodissociation processes and it has been shown clearly in this case that to obtain the greatest physical insight from the measurement of such parameters they need to be related to some molecular reference frame, rather than one which is fixed in space.[5,6] In the case of reactive scattering, therefore, the polarization parameters need to be defined with reference to the relative velocity vectors of either the reactants or the products.

10.2 Polarization Dependent Differential Cross Sections

Let us consider a reaction of the type

$$A + BC(j, m_j) \rightarrow AB(j', m_{j'}) + C. \tag{10.5}$$

In such a case one might have considered that a differential cross section of the type defined in eqn (6.12) (*i.e.* $\sigma_{\beta v' j' m_{j'} \leftarrow \alpha v j m_j}(\theta_R, \phi_R, E)$) provided all the necessary information. An elegant discussion by Shafer-Ray, Orr-Ewing and Zare (see Appendix B of ref. 7) shows however that this is not sufficient. The formal expressions for the cross sections (see eqn (3.26)) are the squares of sums of complex quantities (in the mathematical sense that they involve the complex number $i = \sqrt{-1}$). This leads to interference terms arising from the relative phases of the quantities being summed. The cross sections themselves do not provide information as to the phases of the **S** matrix elements and therefore do not provide certain information concerning the correlation of different m_j and $m_{j'}$ quantum states. Shafer-Ray *et al.* show that it is insufficient to consider only the cross sections but we must rather look in detail at the wavefunctions of the ensembles involved in the reaction and use "density matrices"[8] as the basis of our discussion.

If an AB molecules exits in a pure state, we may write its wavefunction as $\psi^{AB}_{j',m_{j'}}$ or $|j', m_{j'}\rangle$. The different product j' quantum states have different energies and may be separated physically from each other as they have different relative kinetic energies. We therefore concentrate on one particular j' level with its associated $(2j' + 1)$ states each with a different $m_{j'}$ quantum number. If the initial reactants are in a pure quantum state then the products of a reactive chemical encounter will in general be in a coherent combination of all the allowed $m_{j'}$ states. We may write the j' component of the product molecule wavefunction as

$$\psi_{j'} = \sum_{m_{j'}} c_{j',m_{j'}} \, |j', m_{j'}\rangle \tag{10.6}$$

where for convenience we will impose the normalization condition $\sum_{m_{j'}} |c_{j',m_{j'}}|^2 = 1$.

Let us now inquire as to how we might calculate a property of this wavefunction, it could for instance be the average value of the angle ϕ. In any case each physical property is associated with a Hermitian operator. If we denote the operator by $\hat{\boldsymbol{A}}$, we can write the expectation value as

$$\begin{aligned}
\langle \psi_{j'} | \hat{\boldsymbol{A}} | \psi_{j'} \rangle &= \sum_{m_{j'}, m'_{j'}} c^*_{j',m_{j'}} \, c_{j',m'_{j'}} \, \langle j', m_{j'} | \hat{\boldsymbol{A}} | j', m'_{j'} \rangle \\
&= \sum_{m_{j'}, m'_{j'}} c^*_{j',m_{j'}} \, c_{j',m'_{j'}} A_{j',m_{j'};j',m'_{j'}} \\
&= \sum_{m_{j'}, m'_{j'}} \rho_{j',m_{j'};j',m'_{j'}} A_{j',m_{j'};j',m'_{j'}}
\end{aligned} \tag{10.7}$$

where $\rho_{j',m_{j'};j',m'_{j'}} = c^*_{j',m_{j'}} c_{j',m'_{j'}}$ is defined as a matrix element of the density matrix of the system.

Let us now move from the diatomic wavefunctions in eqn (10.6) and (10.7) to the wavefunction of the whole atom plus diatom system. Now the density matrix is not a constant as in a case of diatomic but a function of θ_R and ϕ_R. This density matrix is obtained by examining the wavefunction of the system at large values of the scattering coordinate R. Using eqn (6.11) we may write, for the reactive part of the density matrix in terms of the scattering amplitudes f,

$$\rho_{j',m_{j'};j',m'_{j'}} = \left\{ \frac{\mu_\alpha k_{v'j'}}{\mu_\beta k_{vj} N(E)} \right\} f^*_{\beta v'j'm_{j'} \leftarrow \alpha vjm_j}(\theta_R, \phi_R) \times f_{\beta v'j'm'_{j'} \leftarrow \alpha vjm_j}(\theta_R, \phi_R) \tag{10.8}$$

where μ_α and μ_β are the reduced masses for the relative motion of the reactants and products respectively, k_{vj} is the wavenumber for the relative motion of the reactants and $k_{v'j'}$ is that for the products. $N(E)$ and the other terms in the curly bracket on the right hand side of eqn (10.8) act as normalization factors.

By comparing this with eqn (3.9) for inelastic and (6.12) for reactive scattering, we see that the diagonal elements of the density matrix are proportional to the differential cross section

$$\rho_{j',m_{j'};j',m_{j'}}(\theta_R, \phi_R, E) = \left\{ \frac{1}{N(E)} \right\} \sigma_{\beta v'j'm_{j'} \leftarrow \alpha vjm_j}(\theta_R, \phi_R, E). \tag{10.9}$$

By imposing the normalization condition $\int_0^{2\pi} \mathrm{d}\phi_R \int_0^{\pi} \sum_{m_{j'}} \rho_{j',m_{j'};j',m_{j'}} \sin\theta_R \, \mathrm{d}\theta_R = 1$ one can show $N(E) = \sigma_{\beta v'j' \leftarrow \alpha vjm_j}(E)$, the integral reactive state-to-state cross section summed over all product $m_{j'}$ states.

Eqn (3.25) shows the relationship between the scattering amplitude and the scattering **T** matrix for inelastic scattering. For reactive scattering the analogous expression for the scattering amplitude is pre-multiplied by an extra factor of $(\mu_\beta/\mu_\alpha)^{1/2}$ (compare eqn (3.20) and (6.14)). Using this we can write the density matrix in terms of the elements of the scattering **T** matrix as

$$\begin{aligned}
\rho_{j',m_{j'};j',m'_{j'}} &= \left\{ \frac{1}{\sigma_{\beta v'j' \leftarrow \alpha vjm_j}(E)} \right\} \sum_{JM\ell} (2\ell + 1)^{\frac{1}{2}} \sum_{\ell'} i^{\ell-\ell'} T^{J*}_{v'j'\ell',vj\ell} \\
&\times \sum_{m_{\ell'}} (j'm_{j'}\ell'm_{\ell'}|j'\ell'JM)(j\ell JM|jm_j\ell 0) \\
&\times \sum_{JM\ell''} (2\ell'' + 1)^{\frac{1}{2}} \sum_{\ell'''} i^{\ell''-\ell'''} T^{J}_{v'j'\ell''',vj\ell''} \\
&\times \sum_{m_{\ell'''}} (j'm'_{j'}\ell'''m_{\ell'''}|j'\ell'''JM)(j\ell''JM|jm_j\ell''0) Y^*_{\ell'm_{\ell'}}(\hat{\boldsymbol{R}}) \\
&\times Y_{\ell'''m_{\ell'''}}(\hat{\boldsymbol{R}}).
\end{aligned} \tag{10.10}$$

The complex conjugate of a spherical harmonic is (see eqn (12), section 5.4 of ref. 9)

$$Y^*_{\ell' m_{\ell'}}(\hat{\boldsymbol{R}}) = (-1)^{m_{\ell'}} Y_{\ell' -m_{\ell'}}(\hat{\boldsymbol{R}}) \tag{10.11}$$

while the product of two spherical harmonics can be expressed as a sum over a Clebsch–Gordan coefficient and a single spherical harmonic (see eqn (9), section 5.6.2 of ref. 9)

$$\begin{aligned} &Y_{\ell' -m_{\ell'}}(\hat{\boldsymbol{R}}) Y_{\ell''' m_{\ell'''}}(\hat{\boldsymbol{R}}) \\ &= \sum_{LM} \sqrt{\frac{(2\ell' + 1)(2\ell''' + 1)}{4\pi(2L + 1)}} (\ell' 0 \ell''' 0 | \ell' \ell''' L 0) \\ &\times (\ell' - m_{\ell'} \ell''' m_{\ell'''} | \ell' \ell''' LM) Y_{LM}(\hat{\boldsymbol{R}}). \end{aligned} \tag{10.12}$$

Substituting eqn (10.11) and (10.12) into eqn (10.10) we obtain

$$\begin{aligned} &\rho_{j',m_{j'};j',m'_{j'}} \\ &= \left\{ \frac{1}{2\sqrt{\pi}\, \sigma_{\beta v' j' \leftarrow \alpha v j m_j}(E)} \right\} \sum_{JM\ell} (2\ell + 1)^{\frac{1}{2}} \sum_{\ell'} i^{\ell - \ell'} T^{J*}_{v' j' \ell', v j \ell} \\ &\times \sum_{m_{\ell'}} (j' m_{j'} \ell' m_{\ell'} | j' \ell' JM)(j \ell JM | j m_j \ell 0) \\ &\times \sum_{JM\ell''} (2\ell'' + 1)^{\frac{1}{2}} \sum_{\ell'''} i^{\ell'' - \ell'''} T^{J}_{v' j' \ell''', v j \ell''} \\ &\times \sum_{m_{\ell'''}} (j' m'_{j'} \ell''' m_{\ell'''} | j' \ell''' JM)(j \ell'' JM | j m_j \ell'' 0)(-1)^{m_{\ell'}} \\ &\times \sum_{LM} \sqrt{\frac{(2\ell' + 1)(2\ell''' + 1)}{(2L + 1)}} (\ell' 0 \ell''' 0 | \ell' \ell''' L 0) \\ &\times (\ell' - m_{\ell'} \ell''' m_{\ell'''} | \ell' \ell''' LM)\, Y_{LM}(\hat{\boldsymbol{R}}). \end{aligned} \tag{10.13}$$

This equation provides an expression for the density matrix of the product states in a reaction arising from reactants in a well defined initial state. The density matrix provides the necessary information on the correlation between different product $m_{j'}$ states arising from such a reaction, and hence the correlation between the directions of the product rotational angular momenta. The **T** matrices arise from the dynamics of the collision process and are obtained by solving the set of coupled Schrödinger equations (see eqn (3.18) and Chapter 6). Note that eqn (10.13) can be further simplified by omitting the summation over M as the Clebsch–Gordan coefficient $(j\ell JM | j m_j \ell 0)$ is zero unless $M = m_j$.

10.2.1 Spherical Multipoles

We now re-express the density matrix in terms of spherical multipoles. This permits the easy transformation of the density matrices to different reference frames. The multipoles $\rho_{KQ}^{(j')}$ are defined as (see ref. 10 page 237 eqn (62))

$$\rho_{KQ}^{(j')} = \sum_{m_{j'} m'_{j'}} (-1)^{j' - m'_{j'}} (j' m_{j'}\, j' - m'_{j'} | j' j' KQ)\, \rho_{j', m_{j'}; j', m'_{j'}} \tag{10.14}$$

where K is the "rank" of the multipole and Q is the quantum number of its projection on the Z axis of the coordinate frame.

In the present case this is the laboratory or space-fixed Z axis and is the same direction as the original relative velocity $\boldsymbol{k}$ vector. The spherical multipoles transform in the same way as the spherical harmonics under rotation of the coordinate system. Thus in order to transform the multipoles to the body-fixed reference frame we use the same transformation as in eqn (3.42)

$$\rho_{K\Omega}^{(j')} = \sum_{Q} \rho_{KQ}^{(j')} D_{Q,\Omega}^{K}(\phi_R, \theta_R, 0) \tag{10.15}$$

where Ω is now the quantum number for the projection of the angular momentum onto the body-fixed Z axis.

The inverse of eqn (10.14) is (see ref. 10 page 237 eqn 63)

$$\rho_{j', m_{j'}; j', m'_{j'}} = \sum_{KQ} (-1)^{j' - m'_{j'}} (j' m_{j'}\, j' - m'_{j'} | j' j' KQ)\, \rho_{KQ}^{(j')}. \tag{10.16}$$

The density matrices are functions of the scattering angles θ_R and ϕ_R (see eqn (10.8)) and therefore carry information about the $\boldsymbol{k}$–$\boldsymbol{k}'$ correlation. The dependence of the density matrix on the $m_{j'}$ and $m'_{j'}$, or equivalently the K and Q, quantum numbers carries information on the correlation between different orientations of the product rotational angular momentum vector. Therefore the density matrices discussed here are the key quantities needed for the study of the so-called $\boldsymbol{k} - \boldsymbol{j}' - \boldsymbol{k}'$ correlation. Analogous density matrices may be defined (and measured) that address the $\boldsymbol{k} - \boldsymbol{j} - \boldsymbol{k}'$ correlation, *i.e.* correlation involving the direction of the reactant angular momentum, $\boldsymbol{j}$.

The state multipoles $\rho_{KQ}^{(j')}(\theta_R, \phi_R)$ of eqn (10.14) are the "Polarization Dependent Differential Cross Sections" (PDDCS)†. The $Q = 0$ components of the state multipoles are related to the standard state-to-state differential cross sections.[12] This can be seen from eqn (10.14) where the Clebsch–Gordan coefficient leads to the requirement that if $Q = 0$ then $m_{j'} = m'_{j'}$. In this case we see from eqn (10.9) that the state multipoles $\rho_{K0}^{(j')}(\theta_R, \phi_R)$ are just the (re-normalized) spherical components of the standard differential cross

†In ref. 7 and 11 the symbol σ is used instead of $\sigma_{\beta v' j' \leftarrow \alpha v j m_j}(E)$ to denote the integral cross section.

section. The state multipole components with $Q > 0$ provide information about the coherence between different product $m_{j'}$ quantum states.

From eqn (10.14) and from the properties of the Clebsch–Gordan coefficients[9,13] we can show that

$$\begin{aligned}
\rho_{00}^{(j')} &= \sum_{m_{j'}m'_{j'}} (-1)^{j'-m'_{j'}} (j'm_{j'}\, j' - m'_{j'} | j'j'00)\, \rho_{j',m_{j'};j',m'_{j'}} \\
&= \sum_{m_{j'}} (2j' + 1)^{-\frac{1}{2}} \rho_{j',m_{j'};j',m_{j'}} \\
&= (2j' + 1)^{-\frac{1}{2}} \left\{ \frac{1}{\sigma_{\beta v' j' \leftarrow \alpha v j m_j}(E)} \right\} \\
&\quad \times \sum_{m_{j'}} \sigma_{\beta v' j' m_{j'} \leftarrow \alpha v j m_j}(\theta_R, \phi_R, E)
\end{aligned} \tag{10.17}$$

where we have used eqn (10.9).

We see clearly from this that $\rho_{00}^{(j')}$ is just the normalized differential cross section for the j' product state, averaged over all the $m_{j'}$ states. Its integration over all angles θ_R and ϕ_R gives a constant. The multipole moments of odd rank ($K = 1, 3, \ldots$) describe the orientation of the product rotational angular momentum vectors, $\boldsymbol{j}'$, while the multipole moments of even rank ($K = 2, 4, \ldots$) describe the alignment of these vectors. The polarization parameters can be obtained by integrating the state multipoles of eqn (10.14) over all scattering angles θ_R and ϕ_R. Thus the orientation parameter A_0^1 (see eqn (10.3)) is given by

$$A_0^1 = \int_{\theta_R=0}^{\pi} \int_{\phi_R=0}^{2\pi} \rho_{10}^{(j')}(\theta_R, \phi_R) \sin\theta_R \, d\theta_R \, d\phi_R \tag{10.18}$$

Other alignment and orientation parameters (sometimes referred to as polarization parameters) can be defined in an analogous manner.[4,10]

We have so far considered only 3-vector correlations, *i.e.* either $\boldsymbol{k} - \boldsymbol{j} - \boldsymbol{k}'$ or $\boldsymbol{k} - \boldsymbol{j}' - \boldsymbol{k}'$. Clearly it would be more general to consider 4-vector correlations involving the two directions of relative motion and the rotational angular momenta of both the reactant and product diatomics. The formalism for such 4-vector correlations has been discussed in two different publications.[12,14] The more general case where both collision partners and both products of the reaction possess internal angular momenta has also been discussed.[12]

10.2.2 Applications

In Section 10.2.1 above we considered starting with a well defined m_j state and using the density function to characterize the distribution of the product $m_{j'}$ states. This process can be reversed. We might start with some mixed initial distribution of m_j states, characterized by a density matrix, and determine the effect this has on the product differential cross section. This has been done in an application by Aldegunde *et al.*[15] where they examine the reactions $F + HD(\nu = 0, j = 1) \rightarrow HF + D$ and $F + HD(\nu = 0, j = 1) \rightarrow DF + H$. The HD diatomic was prepared in its $(j = 1, m_j = 0)$ quantum state. The difference between different calculations was that the axis of quantization for the m_j projection quantum number could be chosen to differ from the space-fixed Z axis (*i.e.* the initial $\boldsymbol{k}$ vector direction). Figure 10.4 shows some of the results of the calculations. There are three columns in the figure. The one headed $\beta = 0°$ corresponds to a "head-on" collision with the HD molecule aligned towards the oncoming F atom. The column headed $\beta = 90°$ $\alpha = 0°$ corresponds to a "side-on" collision, while the middle column corresponds to a tilted geometry. The various circles correspond to progressively smaller degrees of vibrational–rotational excitation (*i.e.* the outermost circle corresponds to $(\nu' = 0, j' = 0)$). The height of the contours gives the value of the differential reactive cross section $\sigma_{\beta\nu' j' \leftarrow \alpha\nu j}(\theta_R, E)$ summed over all final $m_{j'}$ states. The figure shows clearly that controlling the orientation of the reactants has a strong effect on the outcome of the reaction dynamics.

Figure 10.5 shows a comparison of two differential cross sections for the reaction $Li + HF\ (\nu = 0,\ j = 3) \rightarrow LiF\ (\nu' = 0,\ j' = 40) + H$ at a collision energy of 317 meV.[16] The solid line corresponds to the initial HF m_j states of the HF state being uniformly populated, *i.e.* an isotropic initial distribution, while the broken line is an initially aligned state with $m_j = 0$.

This illustration shows again that preparing the reactants in an aligned state can have a substantial effect on the cross section of a reaction.

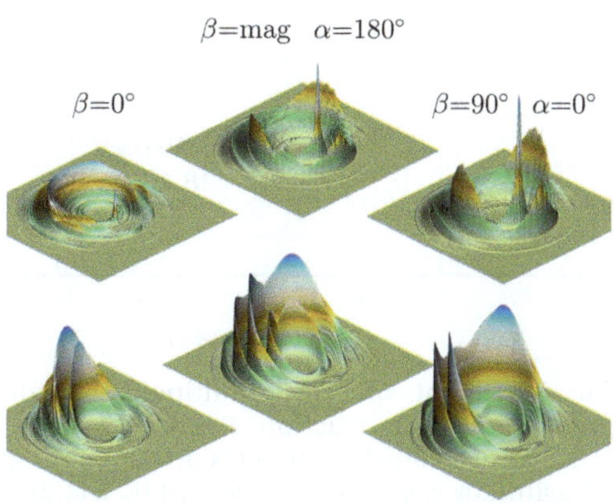

Figure 10.4 Center-of-mass scattering angle–velocity maps for the reactions $F + HD\ (\nu = 0,\ j = 1) \rightarrow HF + D$ (top row) and $F + HD\ (\nu = 0,\ j = 1) \rightarrow DF + H$ (bottom row). The angles β and α determine the orientation of the Z axis with respect to which the HD $(j = 1,\ m_j = 0)$ state is quantized. Collision energy = 156 meV. Figure reproduced with permission from ref. 15 with permission from the PCCP Owner Societies.

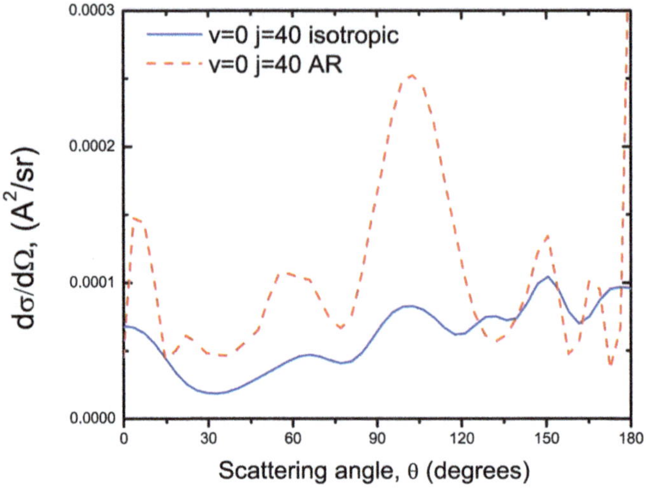

Figure 10.5 Reactive scattering cross section for the reaction Li + HF ($v = 0,\ j = 3$) $\rightarrow$ LiF ($v' = 0,\ j' = 40$) + H. Solid line: Isotropic distribution of reactant m_j states. Broken line: Reactant HF aligned in $m_j = 0$ state. Collision energy = 317 meV. Reproduced with permission from M. B. Krasilnikov *et al.*, *J. Chem. Phys.*, **138**, 244302. Copyright 2013, AIP Publishing LLC.[16] DOI: 10.1063/1.4809992.

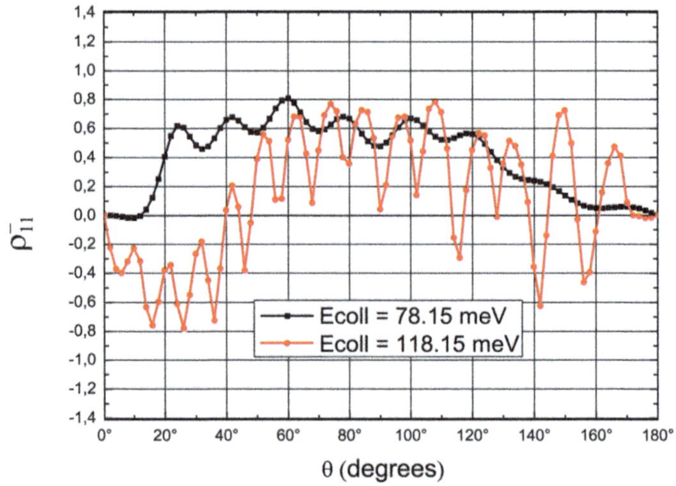

Figure 10.6 Angular dependent spherical multipole, ρ_{11}^{-} for the orientation of the products in the reaction F + HD ($v = 0,\ j = 0$) $\rightarrow$ HF ($v' = 3,\ j' = 2$) + D at two collision energies. Reproduced with permission from M. B. Krasilnikov *et al.*, *J. Chem. Phys.*, **138**, 244302. Copyright 2013, AIP Publishing LLC.[16] DOI: 10.1063/1.4809992.

Figure 10.6 shows the spherical dipole moment for the orientation of the HF products arising from the reaction F + HD ($v = 0,\ j = 0$) $\rightarrow$ HF ($v' = 3,\ j' = 2$) + D at two collision energies.[16] The spherical monopoles defined in eqn (10.14) are complex. The monopole shown in the figure corresponds to the imaginary part of the monopole, and is therefore a real quantity. In this case the components of the spherical monopoles are referred to the product scattering direction $\boldsymbol{k}'$ as the z axis. For the higher collision energy (118.15 meV) a negative value for the monopole is computed in the scattering angle range 0–40°. This indicates that the diatomic is rotating clockwise about the Z axis[16] with the rotational angular momentum pointing along the negative Z direction. There is a pronounced forward peak in the differential reactive cross section at this higher energy which coincides with the region of negative spherical monopole. The oscillations in the figure are caused by the presence of a scattering resonance associated with a van der Waals well in the exit channel of the reaction.

References

1. D. E. Case and D. R. Herschbach, Statistical theory of angular momentum polarization in chemical reactions, *Mol. Phys.*, 1975, **30**, 1537.
2. D. A. Case, G. M. McClelland and D. R. Herschbach, Angular momentum polarization in molecular collisions: classical and quantum theory for measurements using resonance fluorescence, *Mol. Phys.*, 1978, **35**, 541.
3. S. K. Kim and D. R. Herschbach, Angular momentum disposal in atom exchange reactions, *Faraday Discuss. Chem. Soc.*, 1987, **84**, 159.
4. A. J. Orr-Ewing and R. N. Zare, Orientation and alignment of reaction products, *Annu. Rev. Phys. Chem.*, 1994, **45**, 315.
5. T. P. Rakitzis and R. N. Zare, Photofragment angular momentum distributions in the molecular frame: determination and interpretation, *J. Chem. Phys*, 1999, **110**, 3341.
6. A. G. Suits and O. S. Vasyutinskii, Imaging atomic orbital polarization in photodissociation, *Chem. Rev.*, 2008, **108**, 3706.
7. N. E. Shafer-Ray, A. J. Orr-Ewing and R. N. Zare, Beyond state-to-state differential cross sections: determination of product polarization in photoinitiated bimolecular reactions, *J. Phys. Chem.*, 1995, **99**, 7591.
8. K. Blum, *Density Matrix Theory and Applications,* Plenum, New York, 2nd edn, 1996.
9. D. A. Varshalovich, A. N. Moskalev and V. K. Khersonskii, *Quantum Theory of Angular Momentum*, World Scientific, Singapore, 1988.
10. R. N. Zare, *Angular Momentum*, John Wiley and Sons, New York, 1988. See also list of corrections at web address: http://web.stanford.edu/group/Zarelab/Errata.pdf.
11. M. P. de Miranda, F. J. Aoiz, L. Bañares and V. Sáez Rábanos, A unified quantal and classical description of the stereodynamics of elementary chemical reactions: state-resolved k–k'–j' vector correlation for the $H + D_2$ ($v = 0$, $j = 0$) reaction, *J. Chem. Phys.*, 1999, **111**, 5368.
12. G. G. Balint-Kurti and O. S. Vasyutinskii, Vector correlation analysis for inelastic and reactive collisions between partners possessing spin and orbital angular momentum, *J. Phys. Chem. A*, 2009, **113**, 14281. PMID: 19642631.

13. A. R. Edmonds, *Angular Momentum in Quantum Mechanics*, Princeton University Press, Princeton, New Jersey, 1960.
14. M. P. de Miranda and D. C. Clary, Quantum dynamical stereochemistry of atom–diatom reactions, *J. Chem. Phys.*, 1997, **106**, 4509.
15. J. Aldegunde, P. G. Jambrina, M. P. de Miranda, V. S. Rabanos and F. Javier Aoiz, Stereodynamics of the F + HD ($\nu = 0,\ j = 1$) reaction: direct *vs.* resonant mechanisms, *Phys. Chem. Chem. Phys.*, 2011, **13**, 8345.
16. M. B. Krasilnikov, R. S. Popov, O. Roncero, D. De Fazio, S. Cavalli, V. Aquilanti and O. S. Vasyutinskii, Polarization of molecular angular momentum in the chemical reactions Li + HF and F + HD, *J. Chem. Phys.*, 2013, **138**, 244302.

CHAPTER 11

Collision of Larger Molecules

So far we have discussed mainly the quantum dynamics of atom plus diatom collisions. An N atom system needs $3N$ coordinates for its description. Three of these are "trivial" in that they describe the motion of the center-of-mass of the system. This motion can be entirely separated from the motion of atoms relative to each other (see Section 1.1 and ref. 1 and 2). The total angular momentum is also a conserved quantity[2,3] and this property can be used to further reduce by three the number of variables needed to describe the system (see eqn (3.19) and (D-23) which show that for a three atom system for fixed J and M we need to solve equations involving only three variables).

In order to form an idea of the problems which arise when trying to fully describe the dynamics of a system of N atoms when N becomes large, let us suppose that the wavefunction is represented on a grid of n points in each coordinate. For a three atom system we would then require n^3 or n^{3N-6} grid points to represent the potential or the wavefunction.

Using $n = 100$ as a rough estimate of the number of grid points required to represent each coordinate in the wavefunction, Table 11.1 gives the number of grid points needed to represent a wavefunction for systems with different numbers of atoms.

The number of grid points required increases exponentially with the increasing number of atoms. With the present approaches and current technologies it is clearly impossible to treat large systems in the same exact way as has been described in previous chapters of this book. The current limit for exact treatment seems to be four atoms.[4–8] There are, of course, various stratagems for reducing the number of grid points required to represent some of the variable. Foremost amongst these is the Potential Optimized Grid method,[9,10] but no such efficiencies are able to overcome the basic

RSC Theoretical and Computational Chemistry Series No. 7
Theory of Molecular Collisions
By Gabriel G. Balint-Kurti and Alexander P. Palov

Published by the Royal Society of Chemistry, www.rsc.org

Table 11.1 Estimate of number of grid points required to describe a wavefunction as a function of the number of atoms (100 grid points assumed per internal degree of freedom).

Number of atoms	Number of variables	Number of grid points
3	3	10^6
4	6	10^{12}
5	9	10^{18}
6	12	10^{24}
7	15	10^{30}
8	18	10^{36}

problem of the exponential growth in the number of grid points required to describe a wavefunction with an increasing number of atoms.

In this chapter we will describe two methods for trying to overcome this problem. The first method involves making an approximate model of the physical system under consideration so as to reduce the number of coordinates which appear explicitly as differential operators in the Schrödinger equation, while the second method is based on expanding the total wavefunction as a product of functions of a single coordinate only.

The last section of the chapter deals with methods which can be used to compute the rate constant without calculating the scattering **S** matrix as an intermediate step. While immensely simplifying the calculation, these methods do not supply a knowledge of the detailed dynamics of the collision process.

11.1 Reduced Dimensionality Approximations

There is a long history of reduced dimensionality reactive scattering calculations. These began with the early collinear quantum reactive scattering calculations whose intention was to develop and test the methodologies being developed.[11–13] As we have seen above, the calculation of a cross section becomes markedly more difficult as the number of atoms involved increases. This has motivated the systematic development of approximate methods to simplify the quantum reactive scattering calculation for systems containing more than three atoms.[14,15] The most widely used and versatile of these methods is the Rotating Bond approximation (RBA) of Clary.[15] Using the $OH + H_2 \rightarrow H_2O + H$ reaction as an example we briefly summarize the approximation below.

Figure 11.1 shows a simplified coordinate system for the $OH + H_2$ system. A four atom system requires six internal coordinates for its full description. In the RBA approximation the rotation of the H_2 are ignored. This removed two coordinates from the equations of motion. The removal of these coordinates can be realized in several different ways. One way is to fix the direction of the H_2 bond to always point towards the center-of-mass of OH.

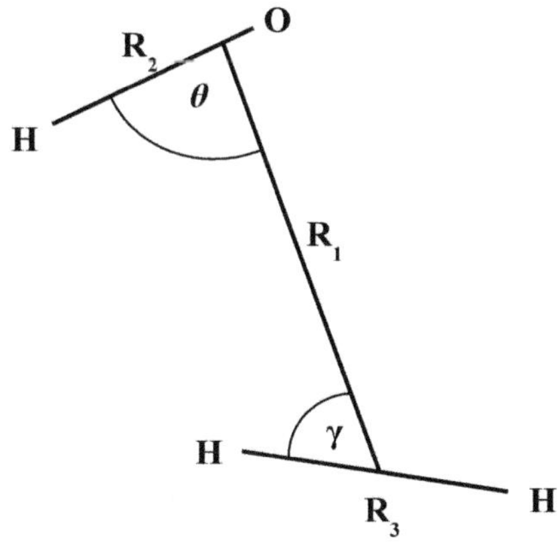

Figure 11.1 Coordinate system for the $OH + H_2 \rightarrow H_2O + H$ reaction or for the reverse $H_2O+H \rightarrow OH+H_2$ process. Reproduced with permission from D. C. Clary, *J. Chem. Phys.*, **96**, 3656. Copyright 1992, AIP Publishing LLC.[17] DOI: 10.1063/1.461919.

This approach is termed the "rotating line approximation".[16] Other approaches are: (1) to average the potential over the angle γ and the azimuthal angle, ϕ_γ, of rotation of H_2 about the OH–H_2 R_1 coordinate, or (2) the "adiabatic bend" approximation in which for all fixed values of the other four coordinates we find the minimum of the potential with respect to the γ and ϕ_γ and then add the zero point vibrational energies for the two vibrations to the potential at each coordinate grid point.

After removing the rotational motion of H_2 in one of these ways, we are left with four dynamical coordinates. Often the calculation is further simplified by using the "spectator bond" approximation which involves fixing the length of the bond which is not broken in the reaction. In the present case this is the O–H bond, which can be fixed at its value at the transition state geometry.

In contrast to many reduced dimensionality approximations the RBA approach permits the calculation of product quantum state distributions as well as differential cross sections. The method has been reviewed by Clary.[15] Figure 11.2 shows the calculated rotational state distribution of OH products arising from the reaction $H_2O + H \rightarrow H_2 + OH$ as compared with experimental results.

More recently reduced dimensional methods have been applied to 5-atom,[18] 6-atom[19–21] and 7-atom[22] systems.

11.2 Multi-Configuration Time-Dependent Hartree Method (MCTDH)

It should be clear from the preceding section that exact quantum mechanical calculations on large systems are bound to become impractical as the number of atoms in the system increases. The best solution to this problem

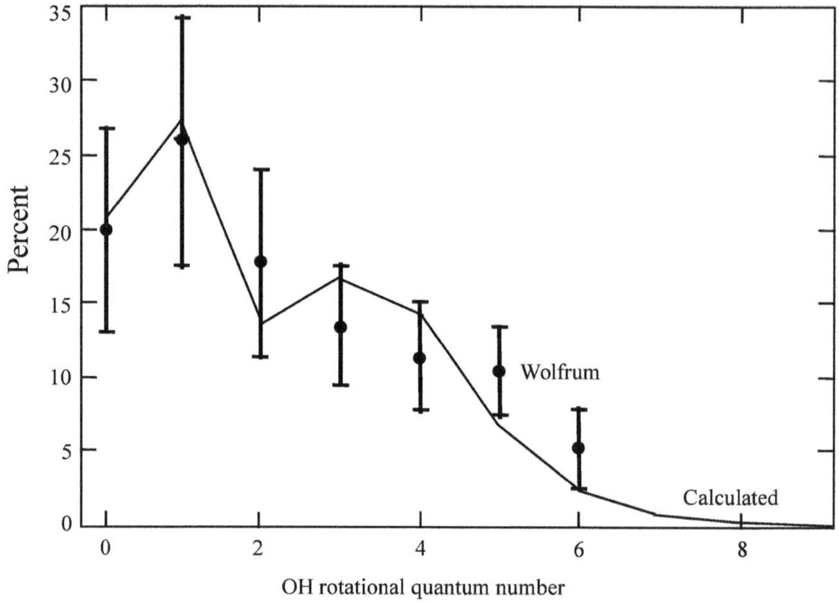

Figure 11.2 Product OH rotational state distribution arising from the reaction $H_2O + H \rightarrow H_2 + OH$ compared with experimental measurements. Redrawn version of figure reproduced with permission from G. Nyman and D. C. Clary, *J. Chem. Phys.*, 1993, **99**, 7774. Copyright 1993, AIP Publishing LLC.[16] DOI: 10.1063/1.465655.

seems to lie in a method which approximates the solution in a systematic manner and is in principal able to yield the exact results in the limit as the approximation is improved. The Multi-Configuration Time-Dependent Hartree method constitutes exactly such a method. Several reviews of the method are available, see for example.[23–25]

The simplest form of a wavefunction for a system having f coordinates is

$$\Phi_J(x_1, x_2, \ldots, x_f, t) = \phi_{j_1}^{(1)}(x_1, t)\phi_{j_2}^{(2)}(x_2, t) \cdots \phi_{j_f}^{(f)}(x_f, t) \tag{11.1}$$

where the subscript J is a combined index, $J = j_1, j_2, \ldots, j_f$.

In eqn (11.1) we have written the wavefunction as a product of "single particle functions", $\phi_{j_k}^{(k)}(x_k, t)$, each depending on a single coordinate and on the time. Such a form of wavefunction could be exact for a system in which the different coordinates are totally uncoupled from each other, *i.e.* if neither the potential nor the kinetic energy operator contained cross terms coupling any two coordinates. As this requirement is too restrictive, we must generalize the wavefunction of eqn (11.1). We therefore describe the wavefunction in eqn (11.1) as a "configuration" and form a wavefunction which is a linear combination of several such configurations

$$\begin{aligned}\Psi(x_1,x_2,\ldots,x_f,t) &= \sum_{j_1}^{n_1}\sum_{j_2}^{n_2}\cdots\sum_{j_f}^{n_f} A_{j_1\cdots j_f}(t)\phi_{j_1}^{(1)}(x_1,t)\\ &\quad\times \phi_{j_2}^{(2)}(x_2,t)\cdots\phi_{j_f}^{(f)}(x_f,t)\\ &= \sum_J A_J(t)\Phi_J(x_1,x_2,\ldots,x_f,t).\end{aligned} \tag{11.2}$$

If the single particle functions, $\phi_{j_k}^{(k)}(x_k,t)$, form a complete set then a wavefunction of this form has the ability to represent an exact solution of the multi-dimensional problem. We require the "single particle basis functions", $\phi_{j_\kappa}^{(\kappa)}(x_\kappa,t)$, to form a mutually orthogonal set[26]

$$\begin{aligned}\langle\phi_n^{(\kappa)}(x_\kappa,t)|\phi_m^{(\kappa)}(x_\kappa,t)\rangle &= \int_{-\infty}^{\infty} \mathrm{d}x_\kappa\,\phi_n^{(\kappa)}(x_\kappa,t)\,\phi_m^{(\kappa)}(x_\kappa,t)\\ &= \delta_{n,m}.\end{aligned} \tag{11.3}$$

If we now invoke the Dirac–Frenkel variational principle[26,27]

$$\langle\delta\Psi(\boldsymbol{x},t)|\left\{i\hbar\frac{\partial}{\partial t}-\hat{\boldsymbol{H}}\right\}|\Psi(\boldsymbol{x},t)\rangle = 0 \tag{11.4}$$

we obtain the set of equations of motion for both the $\boldsymbol{A}$ coefficients and for the single particle functions[26,28]

$$\begin{aligned}i\hbar\dot{A}_J &= i\hbar\frac{\partial A_J}{\partial t}\\ &= \langle\Phi_J|\,\hat{\boldsymbol{H}}\,|\Psi\rangle\\ &= \sum_L \langle\Phi_J|\,\hat{\boldsymbol{H}}\,|\Phi_L\rangle\,A_L.\end{aligned} \tag{11.5}$$

The equations of motion for the single particle functions are

$$\begin{aligned}i\hbar\sum_q^{n_k}[\boldsymbol{\rho}^{(k)}]_{s,q}\dot{\phi}_q^{(k)}(x_k,t) &= i\hbar\sum_q^{n_k}[\boldsymbol{\rho}^{(k)}]_{s,q}\frac{\partial\phi_q^{(k)}(x_k,t)}{\partial t}\\ &= \left[1-\hat{\boldsymbol{P}}^{(k)}\right]\sum_q\left[\langle\hat{\boldsymbol{H}}\rangle^{(k)}\right]_{s,q}\phi_q^{(k)}(x_k,t)\end{aligned} \tag{11.6}$$

where $\boldsymbol{\rho}^{(k)}$ is the "single particle density matrix"[26] (see Löwdin[29] for a analogous definition used in electronic structure theory)

$$\begin{aligned}&[\boldsymbol{\rho}^{(k)}]_{s,q}\\ &= \int \phi_q^{(k)*}(x_k',t)\,\left[\Psi(x_1,\ldots x_k'\ldots x_f,t)\Psi^*(x_1,\ldots x_k\ldots x_f,t)\mathrm{d}x_1\ldots\mathrm{d}x_{k-1}\mathrm{d}x_{k+1}\ldots\mathrm{d}x_f\right]\\ &\quad\times\phi_s^{(k)}(x_k,t)\,\mathrm{d}x_k'\,\mathrm{d}x_k\end{aligned} \tag{11.7}$$

and $\left[\langle\hat{\boldsymbol{H}}\rangle^{(k)}\right]_{s,q}$ is an operator in the coordinate x_k. In order to define it we must first define the "single hole functions", $\Psi_n^{(k)}$

$$\Psi_n^{(k)} = \sum_{j_1}\cdots\sum_{j_{k-1}}\sum_{j_{k+1}}\cdots\sum_{j_f} A_{j_1\cdots j_{k-1}nj_{k+1}\cdots j_f}(t)\phi_{j_1}^{(1)}\cdots\phi_{j_{k-1}}^{(k-1)}\phi_{j_{k+1}}^{(k+1)}\cdots\phi_{j_f}^{(f)}. \tag{11.8}$$

This function is identical to $\Psi(x_1, x_2, \ldots, x_f, t)$ of eqn (11.2) except for the fact that it does not contain the single particle function $\phi_n^{(k)}(x_k, t)$ and therefore does not depend on the variable x_k.

The matrix elements of the averaged or mean field operators $\left[\langle\hat{\boldsymbol{H}}\rangle^{(k)}\right]_{s,q}$ may then be defined as

$$\left[\langle\hat{\boldsymbol{H}}\rangle^{(k)}\right]_{s,q} = \int \Psi_s^{(k)*}\hat{\boldsymbol{H}}\Psi_q^{(k)}\mathrm{d}x_1\cdots\mathrm{d}x_{k-1}\mathrm{d}x_{k+1}\cdots\mathrm{d}x_f. \tag{11.9}$$

This is an operator in the variable x_k.

$\hat{\boldsymbol{P}}^{(k)}$ is a projection operator onto the space spanned by the single particle functions $\phi_{j_k}^{(k)}(x_k, t)$

$$\hat{\boldsymbol{P}}^{(k)} = \sum_{\ell} |\phi_\ell^{(k)}\rangle\langle\phi_\ell^{(k)}|. \tag{11.10}$$

Therefore the operator $[1 - \hat{\boldsymbol{P}}^{(k)}]$ in eqn (11.6) projects onto the function space which is *not* spanned by the single particle functions.

Eqn (11.5) and (11.6) are the basic equations of MCTDH theory and permit the quantum mechanical propagation of a multi-dimensional wavepacket forward in time. The method is capable of providing an "exact" time-dependent wavefunction in the limit of a increasing number of configurations $\Phi_J(x_1, x_2, \ldots, x_f, t)$ (see eqn (11.2)) and single particle functions.

The reader is referred to ref. 26 for a full derivation of eqn (11.5) and (11.6).

The MCTDH method has been successfully used to compute photodissociation cross sections in many-atom systems,[24,26] but has not been widely used to calculate reactive scattering cross sections.[30,31] In the next section we discuss its use to determine rate constants for chemical reactions.

11.3 Computation of Rate Constants: The Flux–Flux Auto-Correlation Method

The calculation of detailed state-to-state cross sections for systems containing many atoms is very difficult and this difficulty increases exponentially with increasing numbers of atoms. For many, indeed for most, applications this high degree of detail is not required. Rather, just the thermal averaged rate constant of eqn (5.8) is of interest. Eqn (5.13) shows that this rate constant may be expressed in the form

$$k(T) = \frac{1}{Q_{\mathrm{R}}(T)h}\int_0^\infty \mathrm{d}E\,\mathrm{e}^{-E/k_{\mathrm{B}}T}N(E) \tag{11.11}$$

where $N(E)$ is the "cumulative reaction probability" and is equal to the square of the **S** matrix summed over all quantum numbers

$$N(E) = \sum_{ij\Omega}\sum_{k\ell\Omega'}\sum_{J}(2J+1)\left|S^{J}_{k\ell\Omega',ij\Omega}(E)\right|^{2}. \tag{11.12}$$

W. H. Miller[32] has formulated various ways of computing $N(E)$ and the thermal rate constant $k(T)$ without the need to compute all the **S** matrix elements. The method which has become the most widely used is the flux–flux autocorrelation function. In ref. 32 Miller showed the validity of the expression for the rate constant $k(T)$

$$k(T)Q_{\mathrm{R}}(T) = \frac{1}{2}\int_{-\infty}^{\infty} \mathrm{d}t\, C_{\mathrm{f}}(t), \tag{11.13}$$

where $Q_{\mathrm{R}}(T)$ is the partition function of the reactants (see eqn (5.12)), $C_{\mathrm{f}}(t)$ is the flux–flux autocorrelation function,

$$C_{\mathrm{f}}(t) = \mathrm{tr}\left[\bar{\boldsymbol{F}}\,\mathrm{e}^{i\hat{\boldsymbol{H}}t_{\mathrm{c}}^{*}/\hbar}\,\bar{\boldsymbol{F}}\,\mathrm{e}^{-i\hat{\boldsymbol{H}}t_{\mathrm{c}}/\hbar}\right], \tag{11.14}$$

and $\bar{\boldsymbol{F}}$ is the symmetrized flux operator,

$$\begin{aligned}\bar{\boldsymbol{F}} &= \frac{1}{2}\left[\hat{\boldsymbol{F}} + \hat{\boldsymbol{F}}^{\dagger}\right]\\ &= \frac{1}{2}\left[\delta(s)\frac{\hat{\boldsymbol{p}}}{\mu} + \frac{\hat{\boldsymbol{p}}}{\mu}\delta(s)\right],\end{aligned} \tag{11.15}$$

where s is a surface chosen to divide products and reactants and $\hat{\boldsymbol{p}}$ is the momentum operator for motion perpendicular to this surface. Somewhat surprisingly, it does not matter for the accuracy of the theory which surface is chosen, though it is normal to choose a surface which includes the transition state geometry of the reaction, $t_{\mathrm{c}} = t - i\hbar\beta/2$ where $\beta = 1/(k_{\mathrm{B}}T)$. The notation $\mathrm{tr}[\hat{\boldsymbol{O}}]$ in eqn (11.14) denotes the "trace" of the operator which is the sum over all diagonal elements of the operator when represented as a matrix between a complete function basis set. If some of the "quantum numbers" of the set are continuous, for instance the energy, then an integration rather than a summation should be used.

With the above definitions we can re-express the thermal flux–flux autocorrelation function of eqn (11.14) as

$$\begin{aligned}C_{\mathrm{f}}(t) &= \mathrm{tr}\left[\bar{\boldsymbol{F}}\,\mathrm{e}^{i\hat{\boldsymbol{H}}(t/\hbar + i/(2k_{\mathrm{B}}T))}\,\bar{\boldsymbol{F}}\,\mathrm{e}^{-i\hat{\boldsymbol{H}}(t/\hbar - i/(2k_{\mathrm{B}}T))}\right]\\ &= \mathrm{tr}\left[\bar{\boldsymbol{F}}\,\mathrm{e}^{i\hat{\boldsymbol{H}}t/\hbar}\mathrm{e}^{-\hat{\boldsymbol{H}}/(2k_{\mathrm{B}}T)}\,\bar{\boldsymbol{F}}\,\mathrm{e}^{-\hat{\boldsymbol{H}}/(2k_{\mathrm{B}}T)}\mathrm{e}^{-i\hat{\boldsymbol{H}}t/\hbar}\right]\\ &= \mathrm{tr}\left[\mathrm{e}^{-\hat{\boldsymbol{H}}/(4k_{\mathrm{B}}T)}\bar{\boldsymbol{F}}\mathrm{e}^{-\hat{\boldsymbol{H}}/(4k_{\mathrm{B}}T)}\,\mathrm{e}^{i\hat{\boldsymbol{H}}t/\hbar}\mathrm{e}^{-\hat{\boldsymbol{H}}/(4k_{\mathrm{B}}T)}\,\bar{\boldsymbol{F}}\,\mathrm{e}^{-\hat{\boldsymbol{H}}/(4k_{\mathrm{B}}T)}\mathrm{e}^{-i\hat{\boldsymbol{H}}t/\hbar}\right]\\ &= \mathrm{tr}\left[\hat{\boldsymbol{F}}_{2T}\,\mathrm{e}^{i\hat{\boldsymbol{H}}t/\hbar}\,\hat{\boldsymbol{F}}_{2T}\,\mathrm{e}^{-i\hat{\boldsymbol{H}}t/\hbar}\right]\end{aligned} \tag{11.16}$$

where $\hat{\boldsymbol{F}}_{2T} = \mathrm{e}^{-\hat{\boldsymbol{H}}/(4k_{\mathrm{B}}T)}\bar{\boldsymbol{F}}\mathrm{e}^{-\hat{\boldsymbol{H}}/(4k_{\mathrm{B}}T)}$ and is called the thermal flux operator. The reader is referred to ref. 32 for a full derivation of eqn (11.13).

To complete this section we will derive an expression for the cumulative reaction probability, $N(E)$, using eqn (11.11), (11.13) and (11.16). In eqn (11.11) the thermal rate constant is related to $N(E)$ by an integration over energy, while in eqn (11.13) the rate constant is presented as an integral over time. The objective is therefore to relate eqn (11.13) to one involving an integration over energy. Our discussion follows the derivations set out in ref. 32

We use the identity

$$e^{-i\hat{\boldsymbol{H}}t/\hbar} = \int_{-\infty}^{\infty} dE\, e^{-iEt/\hbar}\delta(E - \hat{\boldsymbol{H}}) \tag{11.17}$$

in eqn (11.16) to write

$$C_f(t) = \int_{-\infty}^{\infty} dE \int_{-\infty}^{\infty} dE'\, e^{i(E-E')t/\hbar} \left\{\mathrm{tr}\left[\hat{\boldsymbol{F}}_{2T}\delta(E - \hat{\boldsymbol{H}})\,\hat{\boldsymbol{F}}_{2T}\,\delta(E' - \hat{\boldsymbol{H}})\right]\right\}. \tag{11.18}$$

Substituting this into eqn (11.13) we obtain

$$\begin{aligned} k(T)Q_R(T) = \frac{1}{2}\int_{-\infty}^{\infty} dt \int_{-\infty}^{\infty} dE \int_{-\infty}^{\infty} dE'\, e^{i(E-E')t/\hbar} \\ \times \left\{\mathrm{tr}\left[\hat{\boldsymbol{F}}_{2T}\delta(E - \hat{\boldsymbol{H}})\,\hat{\boldsymbol{F}}_{2T}\,\delta(E' - \hat{\boldsymbol{H}})\right]\right\} \end{aligned} \tag{11.19}$$

but the integral over time gives

$$\int_{-\infty}^{\infty} dt\, e^{i(E-E')t/\hbar} = 2\pi\hbar\,\delta(E - E') \tag{11.20}$$

so substituting eqn (11.20) into eqn (11.19) we obtain

$$\begin{aligned} k(T)Q_R(T) &= \frac{1}{2}\int_{-\infty}^{\infty} dE \int_{-\infty}^{\infty} dE'\, 2\pi\hbar\,\delta(E - E') \\ &\quad \times \left\{\mathrm{tr}\left[\hat{\boldsymbol{F}}_{2T}\delta(E - \hat{\boldsymbol{H}})\,\hat{\boldsymbol{F}}_{2T}\,\delta(E' - \hat{\boldsymbol{H}})\right]\right\} \\ &= \frac{h}{2}\int_{-\infty}^{\infty} dE \left\{\mathrm{tr}\left[\hat{\boldsymbol{F}}_{2T}\delta(E - \hat{\boldsymbol{H}})\,\hat{\boldsymbol{F}}_{2T}\,\delta(E - \hat{\boldsymbol{H}})\right]\right\}. \end{aligned} \tag{11.21}$$

Comparing this with eqn (11.11) we see that

$$N(E) = \frac{h^2}{2}\, e^{E/k_BT} \left\{\mathrm{tr}\left[\hat{\boldsymbol{F}}_{2T}\delta(E - \hat{\boldsymbol{H}})\,\hat{\boldsymbol{F}}_{2T}\,\delta(E - \hat{\boldsymbol{H}})\right]\right\} \tag{11.22}$$

and using eqn (11.20) twice to convert the Dirac delta functions into integrals over time we can write

$$\begin{aligned} N(E) &= \frac{h^2}{2}\, e^{E/k_BT} \left\{\mathrm{tr}\left[\hat{\boldsymbol{F}}_{2T}\frac{1}{h}\int_{-\infty}^{\infty} dt'\, e^{i(E-\hat{\boldsymbol{H}})t'/\hbar}\,\hat{\boldsymbol{F}}_{2T}\,\frac{1}{h}\int_{-\infty}^{\infty} dt\, e^{i(E-\hat{\boldsymbol{H}})t/\hbar}\right]\right\} \\ &= \frac{1}{2}\, e^{E/k_BT} \int_{-\infty}^{\infty} dt' \int_{-\infty}^{\infty} dt e^{iE(t-t')/\hbar} \left\{\mathrm{tr}\left[\hat{\boldsymbol{F}}_{2T}\, e^{i\hat{\boldsymbol{H}}t'/\hbar}\,\hat{\boldsymbol{F}}_{2T}\, e^{-i\hat{\boldsymbol{H}}t/\hbar}\right]\right\}. \end{aligned} \tag{11.23}$$

11.3.1 State Averaged MCTDH Method

In Section 11.2 we discuss the propagation of a time-dependent multi-dimensional wavepacket using the MCTDH method. Let us consider the situation at the transition state of a reactive system such as $H + CH_4$. We will be interested in the motion around the vicinity of this transition state. So we would naturally work in the normal coordinates of the transition state geometry.[33] The frequency associated with just one of these coordinates is imaginary. This is the normal coordinate associated with motion along the reaction path leading from reactants to products. We use a Gaussian wavepacket as the initial "single particle function" for this coordinate. In the other coordinates simple harmonic vibrational wavefunctions may be used for the single particle functions in eqn (11.2).

Clearly the lowest energy wavepacket corresponds to using single particle functions corresponding to the ground vibrational states in all the bound degrees of freedom. If in just one of the degrees of freedom we substitute the wavefunction of an excited vibrational level, we obtain an alternative initial wavepacket for an excited state of the complex. We see therefore that if we are interested in slightly elevated energies, such that excited states in the vicinity of the transition state geometry may play a role, we should include all the relevant initial wavepackets in the description of the system. This leads to the concept of using several initial wavepackets of the form of eqn (11.2). Each one of these wavepackets describes a different state of the system. When we use such a multi-state approach, we may also use the so-called state averaged MCTDH procedure[34] in which averages such as in eqn (11.9) are averaged over all the states as well as over all the single particle functions. We now add an additional index m to the terms in eqn (11.2) to distinguish the different states which are present

$$
\begin{aligned}
\Psi_m(x_1, x_2, \ldots, x_f, t) &= \sum_{j_1}^{n_1} \sum_{j_2}^{n_2} \cdots \sum_{j_f}^{n_f} A_{j_1 \cdots j_f, m}(t) \phi_{j_1}^{(1)}(x_1, t) \\
&\quad \times \phi_{j_2}^{(2)}(x_2, t) \cdots \phi_{j_f}^{(f)}(x_f, t) \\
m &= 1, \ldots, n_{\text{packets}}.
\end{aligned} \tag{11.24}
$$

The next step in the calculation of the rate constant is the computation of the eigenvalues and eigenfunctions of the thermal flux operator[28] (see eqn (11.16)). This is done by acting on the set of state functions Ψ_m of eqn (11.17) with the thermal flux operator to form a new set of functions $\Phi_m = \hat{\boldsymbol{F}}_{2T} \Psi_m$. The matrix of the thermal flux operator is then formed in the basis of the set of state functions

$$
\begin{aligned}
[\hat{\boldsymbol{F}}_{2T}]_{m,m'} &= \langle \Psi_m | \Phi'_m \rangle \\
&= \langle \Psi_m | \hat{\boldsymbol{F}}_{2T} | \Psi_{m'} \rangle
\end{aligned} \tag{11.25}
$$

This matrix is then diagonalized and its eigenvalues, f_j, and eigenvectors, $|f^j\rangle$, are calculated. The eigenvectors are ordered so that those corresponding to the eigenvalues with the largest absolute values are associated with the smallest indices. Note that the eigenvalues occur in pairs which differ only in sign[35] and both members of each pair are included in all the summations over the eigenvalues in the discussions below. If the matrix $U_{m,j}$ is constructed so that its columns consist of the eigenvectors we may write

$$\begin{aligned}[\hat{\boldsymbol{F}}_{2T}\mathbf{U}]_{m,j} &= \sum_{\ell}[\hat{\boldsymbol{F}}_{2T}]_{m,\ell}U_{\ell,j} \\ &= f_j U_{m,j}. \end{aligned} \tag{11.26}$$

The eigenvectors themselves, in the basis of the state functions, may be expressed as

$$|f^j\rangle = \sum_m U_{m,j}\Psi_m. \tag{11.27}$$

We now replace the set of state functions Ψ_m by the newly computed thermal flux eigenfunctions $|f^j\rangle$ and we repeat the cycle of computing the matrix elements of the thermal flux operator in the new basis and then calculating its eigenvalues and eigenfunctions. This cycle is continued until it converges.[28]

The thermal flux eigenfunctions are now used to compute the cumulative reaction probability of eqn (11.23). We first examine the calculation of the trace in this equation

$$\mathrm{tr}\left[\hat{\boldsymbol{F}}_{2T}\,\mathrm{e}^{i\hat{\boldsymbol{H}}t'/\hbar}\,\hat{\boldsymbol{F}}_{2T}\,\mathrm{e}^{-i\hat{\boldsymbol{H}}t/\hbar}\right] = \sum_j \langle f^j|\,\hat{\boldsymbol{F}}_{2T}\,\mathrm{e}^{i\hat{\boldsymbol{H}}t'/\hbar}\,\hat{\boldsymbol{F}}_{2T}\,\mathrm{e}^{-i\hat{\boldsymbol{H}}t/\hbar}\,|f^j\rangle \tag{11.28}$$

We now insert the identity operator, $\hat{\boldsymbol{I}} = \sum_j |f^j\rangle\langle f^j|$ after each of the thermal flux operators in eqn (11.28)

$$\begin{aligned}\mathrm{tr}\left[\hat{\boldsymbol{F}}_{2T}\,\mathrm{e}^{i\hat{\boldsymbol{H}}t'/\hbar}\,\hat{\boldsymbol{F}}_{2T}\,\mathrm{e}^{-i\hat{\boldsymbol{H}}t/\hbar}\right] &= \sum_j \langle f^j|\,\hat{\boldsymbol{F}}_{2T}\left\{\sum_{j'}|f^{j'}\rangle\langle f^{j'}|\right\}\mathrm{e}^{i\hat{\boldsymbol{H}}t'/\hbar}\,\hat{\boldsymbol{F}}_{2T} \\ &\quad\times\left\{\sum_{j''}|f^{j''}\rangle\langle f^{j''}|\right\}\mathrm{e}^{-i\hat{\boldsymbol{H}}t/\hbar}\,|f^j\rangle \\ &= \sum_j\sum_{j'}\sum_{j''}\langle f^j|f^{j'}\rangle\, f_{j'}\,\langle f^{j'}|\,\mathrm{e}^{i\hat{\boldsymbol{H}}t'/\hbar}\,|f^{j''}\rangle\, f_{j''} \\ &\quad\times\langle f^{j''}|\,\mathrm{e}^{-i\hat{\boldsymbol{H}}t/\hbar}\,|f^j\rangle \end{aligned} \tag{11.29}$$

but $\langle f^j|f^{j'}\rangle = \delta_{j,j'}$ and therefore

$$\begin{aligned}&\mathrm{tr}\left[\hat{\boldsymbol{F}}_{2T}\,\mathrm{e}^{i\hat{\boldsymbol{H}}t'/\hbar}\,\hat{\boldsymbol{F}}_{2T}\,\mathrm{e}^{-i\hat{\boldsymbol{H}}t/\hbar}\right] \\ &= \sum_j\sum_{j''} f_j f_{j''}\,\langle f^j|\,\mathrm{e}^{i\hat{\boldsymbol{H}}t'/\hbar}\,|f^{j''}\rangle\langle f^{j''}|\,\mathrm{e}^{-i\hat{\boldsymbol{H}}t/\hbar}\,|f^j\rangle. \end{aligned} \tag{11.30}$$

Substituting this into eqn (11.23) for the cumulative reaction probability we obtain

$$\begin{aligned} N(E) &= \frac{1}{2}\,\mathrm{e}^{E/k_{\mathrm{B}}T}\int_{-\infty}^{\infty}\mathrm{d}t'\int_{-\infty}^{\infty}\mathrm{d}t\mathrm{e}^{iE(t-t')/\hbar} \\ &\quad\times\sum_{j}\sum_{j''} f_j f_{j''}\,\langle f^j|\,\mathrm{e}^{i\hat{H}t'/\hbar}\,|f^{j''}\rangle\,\langle f^{j''}|\,\mathrm{e}^{-i\hat{H}t/\hbar}\,|f^j\rangle \\ &= \frac{1}{2}\,\mathrm{e}^{E/k_{\mathrm{B}}T}\sum_{j}\sum_{j''} f_j f_{j''}\left\{\int_{-\infty}^{\infty}\mathrm{d}t'\mathrm{e}^{-iEt'/\hbar}\,\langle f^j|\,\mathrm{e}^{i\hat{H}t'/\hbar}\,|f^{j''}\rangle\right\} \\ &\quad\times\left\{\int_{-\infty}^{\infty}\mathrm{d}t\mathrm{e}^{iEt/\hbar}\,\langle f^{j''}|\,\mathrm{e}^{-i\hat{H}t/\hbar}\,|f^j\rangle\right\} \\ &= \frac{1}{2}\,\mathrm{e}^{E/k_{\mathrm{B}}T}\sum_{j}\sum_{j''} f_j f_{j''}\left|\int_{-\infty}^{\infty}\mathrm{d}t\mathrm{e}^{iEt/\hbar}\,\langle f^{j''}|\,\mathrm{e}^{-i\hat{H}t/\hbar}\,|f^j\rangle\right|^2. \end{aligned} \quad (11.31)$$

The thermal rate constant is now obtained by substituting this into eqn (11.11)

$$k(T) = \frac{1}{2Q_{\mathrm{R}}(T)h}\int_{0}^{\infty}\mathrm{d}E\sum_{j}\sum_{j''} f_j f_{j''}\left|\int_{-\infty}^{\infty}\mathrm{d}t\mathrm{e}^{iEt/\hbar}\,\langle f^{j''}|\,\mathrm{e}^{-i\hat{H}t/\hbar}\,|f^j\rangle\right|^2. \quad (11.32)$$

Figure 11.3 shows the thermal rate constant for the reaction H + $CH_4 \rightarrow H_2$ + CH_3 computed using the methodology just discussed over a significant temperature range. The symbols on the graph show experimentally measured data points. There are in fact two theoretical curves shown. The dotted line, almost totally overlaid by the solid one, corresponds to the use of the same theory but using a previous interaction potential. The theoretical results certainly fall within the range of the experimental uncertainty, and agree very well with one of the experimental sets of data. Tunneling plays a significant role below 400 K (2.5 on the scale of the figure). The calculations confirm its importance in this reaction.

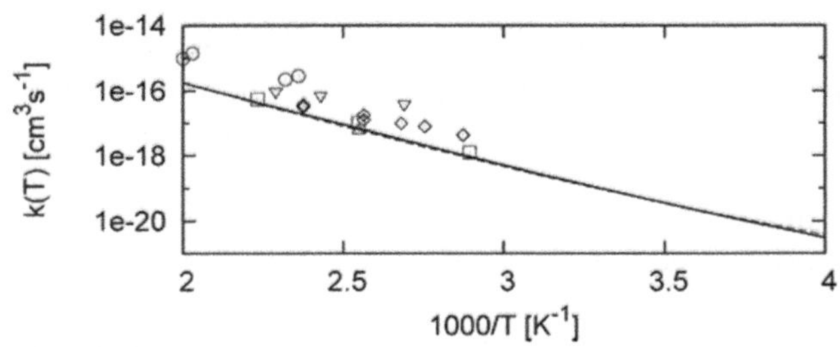

Figure 11.3 Rate constant for the reaction H + $CH_4 \rightarrow H_2$ + CH_3. The solid line shows the MCTDH calculations, while the symbols indicate experimentally measured values. Reproduced with permission from G. Schiffel, U. Manthe and G. Nyman, *J. Phys. Chem. A*, 2010, **114**, 9617.[28]

References

1. H. Goldstein, *Classical Mechanics*, Addison-Wesley, Reading, Massachusetts, USA, 1950.
2. L. D. Landau and E. M. Lifshitz, *Quantum Mechanics: Non-Relativistic Theory*, Butterworth-Heinemann, Oxford, 3rd edn, 1977. (Reprinted with corrections 1991.)
3. R. N. Zare, *Angular Momentum*, John Wiley and Sons, New York, 1988. See also list of corrections at web address: http://web.stanford.edu/group/Zarelab/Errata.pdf.
4. C. Xiao, X. Xu, S. Liu, T. Wang, W. Dong, T. Yang, Z. Sun, D. Dai, X. Xu, D. H. Zhang and X. Yang, Experimental and theoretical differential cross sections for a four-atom reaction: HD + OH $\rightarrow$ H_2O + D, *Science*, 2011, **333**, 440.
5. E. M. Goldfield and S. K. Gray, A quantum dynamics study of H_2 + OH $\rightarrow$ H_2O + H employing the Wu–Schatz–Lendvay–Fang–Harding potential function and a four-atom implementation of the real wave packet method, *J. Chem. Phys.*, 2002, **117**, 1604.
6. R. Valero, D. A. McCormack and G.-J. Kroes, New results for the OH $(\nu = 0,\ j = 0)$ + CO $(\nu = 0,\ j = 0) \rightarrow$ H + CO_2 reaction: five- and full-dimensional quantum dynamical study on several potential energy surfaces, *J. Chem. Phys.*, 2004, **120**, 4263.
7. M. T. Cvitaš and S. C. Althorpe, A Chebyshev method for state-to-state reactive scattering using reactant–product decoupling: OH + $H_2 \rightarrow H_2O$ + H, *J. Chem. Phys.*, 2013, **139**, 064307.
8. R. Otto, J. Ma, A. W. Ray, J. S. Daluz, J. Li, H. Guo and R. E. Continetti, Imaging dynamics on the F + $H_2O \rightarrow$ HF + OH potential energy surfaces from wells to barriers, *Science*, 2014, **343**, 396.
9. S. Kanfer and M. Shapiro, Inversion of the methyl iodide ($\tilde{X}^1A_1$) potential by the discrete position operator method, *J. Phys. Chem*, 1984, **88**, 3964.
10. J. Echave and D. C. Clary, Potential optimized discrete variable representation, *Chem. Phys. Lett.*, 1992, **190**, 225.
11. J. N. L. Connor and M. S. Child, Differential cross sections for chemically reactive systems, *Mol. Phys.*, 1970, **18**, 653.
12. R. D. Levine and S.-F. Wu, Resonances in reactive collisions: Computational study of the H + H_2 collision, *Chem. Phys. Lett.*, 1971, **11**, 557.
13. P. G Hipes and A. Kuppermann, Lifetime analysis of high-energy resonances in three-dimensional reactive scattering, *Chem. Phys. Lett.*, 1987, **133**, 1.
14. D. Wang and J. M. Bowman, Reduced dimensionality quantum calculations of mode specificity in OH + $H_2 \leftrightarrow H_2O$ + H, *J. Chem. Phys.*, 1992, **96**, 8906.
15. D. C. Clary, Four-atom reaction dynamics, *J. Phys. Chem*, 1994, **98**, 10678.
16. G. Nyman and D. C. Clary, Quantum scattering calculations on H_2O + H $\rightarrow H_2$ + OH and isotopes: rotational distributions and cross sections, *J. Chem. Phys.*, 1993, **99**, 7774.

17. D. C. Clary, Quantum scattering calculations on the $OH + H_2$ ($\nu = 0, 1$), $OH + D_2$, and $OD + H_2$ reactions, *J. Chem. Phys.*, 1992, **96**, 3656.
18. M. Yang and J. C. Corchado, Seven-dimensional quantum dynamics study of the $H + NH_3 \rightarrow H_2 + NH_2$ reaction, *J. Chem. Phys.*, 2007, **126**, 214312.
19. D. Wang and J. M. Bowman, A reduced dimensionality, six-degree-of-freedom, quantum calculation of the $H + CH_4 \rightarrow H_2 + CH_3$ reaction, *J. Chem. Phys.*, 2001, **115**, 2055.
20. M. Yang, D. H. Zhang and S.-Y. Lee, A seven-dimensional quantum study of the $H + CH_4$ reaction, *J. Chem. Phys.*, 2002, **117**, 9539.
21. H. F. von Horsten and D. C. Clary, Reactive resonances in the $F + CHD_3$ reaction—a quantum dynamics study, *Phys. Chem. Chem. Phys.*, 2011, **13**, 4340.
22. H. Song, J. Li, B. Jiang, M. Yang, Y. Lu and H. Guo, Effects of reactant rotation on the dynamics of the $OH + CH_4 \rightarrow H_2O + CH_3$ reaction: a six-dimensional study, *J. Chem. Phys.*, 2014, **140**, 084307.
23. M. H. Beck, A. Jäckle, G. A. Worth and H.-D. Meyer, The multiconfiguration time-dependent Hartree (MCTDH) method: a highly efficient algorithm for propagating wavepackets, *Phys. Rep.*, 2000, **324**, 1.
24. *Multidimensional Quantum Dynamics: MCTDH Theory and Applications*, ed. H. D. Meyer, F. Gatti and G. A. Worth, Wiley-VCH, Weinheim, 2009.
25. H.-D. Meyer, Studying molecular quantum dynamics with the multi-configuration time-dependent Hartree method, *Wiley Interdiscip. Rev.: Comput. Mol. Sci.*, 2012, **2**, 351.
26. U. Manthe, H. D. Meyer and L. S. Cederbaum, Wavepacket dynamics within the multiconfiguration Hartree framework: general aspects and application to NOCl, *J. Chem. Phys.*, 1992, **97**, 3199.
27. P. A. M. Dirac, Note on exchange phenomena in the Thomas atom, *Proc. Cambridge Philos. Soc.*, 1930, **26**, 376.
28. G. Schiffel, U. Manthe and G. Nyman, Full-dimensional quantum reaction rate calculations for $H + CH_4 \rightarrow H_2 + CH_3$ on a recent potential energy surface, *J. Phys. Chem A*, 2010, **114**, 9617.
29. P.-O. Löwdin, Quantum theory of many-particle systems. I. Physical interpretations by means of density matrices, natural spin–orbitals, and convergence problems in the method of configurational interaction, *Phys. Rev.*, 1955, **97**, 1474.
30. S. Bhattacharya, A. N. Panda and H.-D. Meyer, Multiconfiguration time-dependent Hartree approach to study the $OH + H_2$ reaction, *J. Chem. Phys.*, 2010, **132**, 214304.
31. S. Bhattacharya, A. N. Panda and H.-D. Meyer, Cross sections and rate constants for $OH + H_2$ reaction on three different potential energy surfaces for ro-vibrationally excited reagents, *J. Chem. Phys.*, 2011, **135**, 194302.
32. W. H. Miller, S. D. Schwartz and J. W. Tromp, Quantum mechanical rate constants for bimolecular reactions, *J. Chem. Phys.*, 1983, **79**, 4889.

33. T. Wu, H.-J. Werner and U. Manthe, Accurate potential energy surface and quantum reaction rate calculations for the $H + CH_4 \rightarrow H_2 + CH_3$ reaction, *J. Chem. Phys.*, 2006, **124**, 234311.
34. U. Manthe, The state averaged multiconfigurational time-dependent Hartree approach: vibrational state and reaction rate calculations, *J. Chem. Phys.*, 2008, **128**, 064108.
35. F. Matzkies and U. Manthe, A multi-configurational time-dependent Hartree approach to the direct calculation of thermal rate constants, *J. Chem. Phys.*, 1997, **106**, 2646.

APPENDIX A

Energy Normalization of Plane Wave

A.1 Plane Wave in One Dimension

In one dimension a plane wave traveling in the positive Z direction has the form

$$\psi(z;k) = N\mathrm{e}^{ikz} \tag{A-1}$$

where N is a normalization constant.

We now ask the question as to the normalization of this wavefunction. The normalization is obtained by multiplying the complex conjugate of the wavefunction by the wavefunction and integrating over all space

$$\int_{-\infty}^{\infty} \psi^*(z;k')\psi(z;k)\,\mathrm{d}z = N^2 \int_{-\infty}^{\infty} \mathrm{e}^{i\,(k-k')\,z}\,\mathrm{d}z. \tag{A-2}$$

But one of the representations of the Dirac delta function may be written in the form[1,2]

$$\frac{1}{2\pi}\int_{-\infty}^{\infty} \mathrm{e}^{-i(k-k')z}\,\mathrm{d}z = \delta(k-k') \tag{A-3}$$

and eqn (A-2) may therefore be rewritten as

$$\int_{-\infty}^{\infty} \psi^*(z;k')\psi(z;k)\,\mathrm{d}z = N^2 2\pi\delta(k-k'). \tag{A-4}$$

The energy, E, is related to the wave number k through the relationship

$$E = \frac{(k\hbar)^2}{2\,\mu} \tag{A-5}$$

RSC Theoretical and Computational Chemistry Series No. 7
Theory of Molecular Collisions
By Gabriel G. Balint-Kurti and Alexander P. Palov

Published by the Royal Society of Chemistry, www.rsc.org

where μ is the reduced mass associated with the motion. Differentiating eqn (A-5) we obtain

$$dE = \frac{k\hbar^2\,dk}{\mu}. \tag{A-6}$$

We wish to define a plane wave wavefunction, $\psi(z;E)$, which is normalized on the "Energy" scale, *i.e.* it has the property

$$\int_{-\infty}^{\infty} \psi^*(z;E')\psi(z;E)\,dz = \delta(E' - E). \tag{A-7}$$

We first need to relate the two Dirac delta functions $\delta(k - k')$ and $\delta(E - E')$. Let us consider the relationship

$$\int_{\infty}^{\infty} f(k)\delta(k - k')\,dk = f(k'). \tag{A-8}$$

Substituting for dk we obtain

$$\int_{\infty}^{\infty} f(k)\delta(k - k')\frac{\mu}{k\hbar^2}\,dE = f(k'). \tag{A-9}$$

We see therefore that

$$\delta(E - E') = \delta(k - k')\frac{\mu}{k\hbar^2} \tag{A-10}$$

and substituting this into eqn (A-4) we obtain

$$\int_{-\infty}^{\infty} \psi^*(z;k')\psi(z;k)\,dz = N^2 2\pi\frac{k\hbar^2}{\mu}\delta(E - E') \tag{A-11}$$

then comparing eqn (A-11) with eqn (A-7) we see that

$$\psi(z;E) = \frac{1}{N}\left\{\frac{\mu}{\hbar^2 2\pi k}\right\}^{\frac{1}{2}}\psi(z;k). \tag{A-12}$$

Now using eqn (A-1) to substitute for $\psi(z;k)$ on the right hand side of this equation, we can write

$$\psi(z;E) = \left\{\frac{\mu}{\hbar^2 2\pi k}\right\}^{\frac{1}{2}} e^{ikz}. \tag{A-13}$$

This is the energy normalized plane wave wavefunction in one dimension.

A.2 Plane Wave in Three Dimensions

In three dimensions we can write a plane wave traveling in the direction $\boldsymbol{k}$ as

$$\psi(\boldsymbol{R};\boldsymbol{k}) = \mathcal{N}e^{i\boldsymbol{k}\cdot\boldsymbol{R}} \tag{A-14}$$

where $\mathcal{N}$ is a normalization factor.

Following eqn (A.30) in Appendix 1.A of ref. 3 we can rewrite this as (see also ref. 4)

$$\psi(\boldsymbol{R};\boldsymbol{k}) = \mathcal{N}4\pi\sum_{\ell,m_\ell} i^\ell j_\ell(kR)Y^*_{\ell,m_\ell}(\hat{\boldsymbol{k}})Y_{\ell,m_\ell}(\hat{\boldsymbol{R}}) \tag{A-15}$$

where $j_\ell(kR)$ is a spherical Bessel function,[1,5] $\hat{\boldsymbol{k}}$ and $\hat{\boldsymbol{R}}$ are unit vectors pointing respectively in the directions of $\boldsymbol{k}$ and $\boldsymbol{R}$ and $Y_{\ell,m_\ell}(\hat{\boldsymbol{R}})$ is a spherical harmonic.[6,7]

The spherical harmonics form an "orthonormal set", *i.e.* they are mutually orthogonal and are normalized. The closure property of the spherical Bessel functions is[4]

$$\int_0^\infty R^2 j_\ell(kR) j_\ell(k'R)\,\mathrm{d}R = \frac{\pi}{2k^2}\delta(k-k') \tag{A-16}$$

Armed with this information we can now investigate the normalization properties of the functions $\psi(\boldsymbol{R};\boldsymbol{k})$ of eqn (A-14)

$$\begin{aligned}\langle\psi(\boldsymbol{R};\boldsymbol{k})|\psi(\boldsymbol{R};\boldsymbol{k}')\rangle &= \int_0^\infty R^2\,\mathrm{d}R \int_0^\pi \sin\theta_R\,\mathrm{d}\theta_R \int_0^{2\pi} \mathrm{d}\phi_R\,\psi^*(\boldsymbol{R};\boldsymbol{k})\,\psi(\boldsymbol{R};\boldsymbol{k}')\\ &= \frac{\pi}{2k^2}\delta(k-k')(\mathcal{N}4\pi)^2\sum_{\ell,m_\ell} Y_{\ell,m_\ell}(\hat{\boldsymbol{k}})Y^*_{\ell,m_\ell}(\hat{\boldsymbol{k}}')\\ &= \frac{\pi}{2k^2}\delta(k-k')(\mathcal{N}4\pi)^2\delta(\hat{\boldsymbol{k}}-\hat{\boldsymbol{k}}')\end{aligned} \tag{A-17}$$

where the "bra–ket" convention, $\langle|\rangle$, indicates integration over all spatial variables and we have utilized the completeness (or closure) relationship for the spherical harmonics[8]

$$\sum_{\ell,m_\ell} Y_{\ell,m_\ell}(\hat{\boldsymbol{k}})Y^*_{\ell,m_\ell}(\hat{\boldsymbol{k}}') = \delta(\hat{\boldsymbol{k}}-\hat{\boldsymbol{k}}'). \tag{A-18}$$

Now invoking eqn (A-10) we can write

$$\begin{aligned}\langle\psi(\boldsymbol{R};\boldsymbol{k})|\psi(\boldsymbol{R};\boldsymbol{k}')\rangle &= \frac{\pi}{2k^2}\delta(E-E')\frac{k\hbar^2}{\mu}(\mathcal{N}4\pi)^2\delta(\hat{\boldsymbol{k}}-\hat{\boldsymbol{k}}')\\ &= \frac{2\pi}{k\mu}(\mathcal{N}h)^2\delta(E-E')\delta(\hat{\boldsymbol{k}}-\hat{\boldsymbol{k}}').\end{aligned} \tag{A-19}$$

We wish to define a function $\psi(\boldsymbol{R};E,\hat{\boldsymbol{k}})$ representing a plane wave, normalized on the energy scale and traveling in a direction $\hat{\boldsymbol{k}}$. This wavefunction has the normalization property

$$\langle\psi(\boldsymbol{R};E,\hat{\boldsymbol{k}})|\psi(\boldsymbol{R};E',\hat{\boldsymbol{k}}')\rangle = \delta(E-E')\delta(\hat{\boldsymbol{k}}-\hat{\boldsymbol{k}}') \tag{A-20}$$

From eqn (A-14) and (A-19) we see that $\psi(\boldsymbol{R};E,\hat{\boldsymbol{k}})$ may be written as

$$\psi(\boldsymbol{R};E,\hat{\boldsymbol{k}}) = \left(\frac{k\mu}{2\pi h^2}\right)^{\frac{1}{2}} \mathrm{e}^{i\boldsymbol{k}\cdot\boldsymbol{R}}. \tag{A-21}$$

Eqn (A-21) represents a plane wave in three dimensions, traveling in a specified direction $\hat{\boldsymbol{k}}$ and normalized on the energy scale. Although we have invested some effort in normalizing the plane wave in this fashion, it can be noted that for most purposes in scattering theory this normalization plays little or no role. It does however matter when we need to use plane wavefunctions as a basis set for expanding other wavefunctions.

A.2.1 Propagation in Z Direction

As a final item in this appendix let us consider what happens when we choose $\boldsymbol{k}$ to be in the positive Z direction. We can then write

$$\begin{aligned} e^{i\boldsymbol{k}\cdot\boldsymbol{R}} &= e^{ikZ} \\ &= 4\pi\sum_{\ell} i^{\ell}\, j_{\ell}(kR) Y^{*}_{\ell,0}(\hat{\boldsymbol{k}}) Y_{\ell,0}(\hat{\boldsymbol{R}}) \end{aligned} \tag{A-22}$$

where we have used eqn (A-15) and have set $m_{\ell} = 0$ as $\boldsymbol{k}$ points along the Z direction.

We now using the facts that[6]

$$Y_{\ell,0}(\theta,\phi) = \left[\frac{(2\ell+1)}{4\pi}\right]^{\frac{1}{2}} P_{\ell}(\cos\theta) \tag{A-23}$$

and, as the polar angle θ_k corresponding to $\boldsymbol{k}$ is equal to 0,

$$P_{\ell}(\cos\theta_k) = P_{\ell}(1) = 1 \tag{A-24}$$

we can rewrite eqn (A-22) as

$$e^{ikZ} = \sum_{\ell} i^{\ell}\,(2\ell+1) j_{\ell}(kR) P_{\ell}(\cos\theta). \tag{A-25}$$

References

1. A. Messiah, *Quantum Mechanics*, North Holland Publishing Co., Amsterdam, 1965.
2. L. D. Landau and E. M. Lifshitz, *Quantum Mechanics: Non-Relativistic Theory*, Butterworth-Heinemann, Oxford, 3rd edn, 1977 (Reprinted with corrections 1991).
3. R. D. Levine, *Quantum Mechanics of Molecular Rate Processes*, Dover Publications, Oxford, 2nd edn, 1999.
4. R. Mehrem, arXiv:0909.0494v4 [math-ph].
5. P. M. Morse and H. Feschbach, *Methods of Theoretical Physics*, McGraw-Hill Book Co. Inc., New York, 1st edn, 1953. See Section 9.3.
6. A. R. Edmonds, *Angular Momentum in Quantum Mechanics*, Princeton University Press, Princeton, New Jersey, 1960.
7. R. N. Zare, *Angular Momentum*, John Wiley and Sons, New York, 1988. See also list of corrections at web address: http://web.stanford.edu/group/Zarelab/Errata.pdf.
8. G. B. Arfken and H. J. Weber, *Mathematical Methods for Physicists*, Academic Press, 7th edn, 2013.

APPENDIX B

Evaluation of the Phase Shift and the Variable Phase Approach

In this appendix we outline the Variable Phase Approach to the evaluation of the phase shift. This is one of the two approaches we have used in our model calculations for this book and permits the computation of the absolute value of the phase shift. The alternative approach, of numerically solving the radial Schrödinger equation and then comparing its form at large interatomic separations to that of the solution in the absence of a potential, yields a phase shift only modulo π, *i.e.* we need to add or subtract an arbitrary integer multiple of π to obtain the correct absolute value of the phase shift.

The radial Schrödinger equation is (see eqn (2.9))

$$\frac{\hbar^2}{2\mu}\left[-\frac{d^2}{dr^2}+\frac{\ell(\ell+1)}{r^2}\right]\chi(r)+V(r)\chi(r)=E\chi(r) \tag{B-1}$$

where $\chi(r)$ is related to the radial wavefunction $\psi(r)$ by $\chi(r)=r\,\psi(r)$.

We simplify eqn (B-1) by multiplying through by $\frac{2\mu}{\hbar^2}$ to obtain

$$\frac{d^2}{dr^2}\chi(r)+\left[-\frac{\ell(\ell+1)}{r^2}+k^2\right]\chi(r)=U(r)\chi(r) \tag{B-2}$$

where $k^2=\frac{2\mu E}{\hbar^2}$ and $U(r)=\frac{2\mu V(r)}{\hbar^2}$.

There are many derivations of the Variable Phase Approach equations.[1–3] Here we will follow that of Ronveaux.[2] We will find it advantageous now to

RSC Theoretical and Computational Chemistry Series No. 7
Theory of Molecular Collisions
By Gabriel G. Balint-Kurti and Alexander P. Palov

Published by the Royal Society of Chemistry, www.rsc.org

switch to a matrix representation of the problem. In this approach eqn (B-2) can be written in the form

$$\mathbf{X}' = \mathbf{AX} + \mathcal{V}\mathbf{X} \tag{B-3}$$

where a prime stands for differentiation and

$$\mathbf{X} = \begin{pmatrix} \chi(r) \\ \chi'(r) \end{pmatrix}, \quad \mathbf{A} = \begin{pmatrix} 0 & 1 \\ \left[\dfrac{\ell(\ell+1)}{r^2} - k^2\right] & 0 \end{pmatrix} \quad \text{and} \quad \mathcal{V} = \begin{pmatrix} 0 & 0 \\ U(r) & 0 \end{pmatrix}.$$

The homogeneous equation corresponding to eqn (B-2), *i.e.* the equation without the potential, is

$$\left[\frac{\mathrm{d}^2}{\mathrm{d}r^2} - \frac{\ell(\ell+1)}{r^2} + k^2\right] u(r) = 0. \tag{B-4}$$

This equation has two linearly independent solutions. These are spherical Riccati–Bessel functions[4] $\hat{j}_\ell(kr) = kr\, j_\ell(kr)$ and $\hat{n}_\ell(kr) = kr\, n_\ell(kr)$, where $j_\ell(kr)$ and $n_\ell(kr)$ are respectively the spherical Bessel functions of the first and second kinds. The Wronskian corresponding to these two solutions is a constant

$$W = \left[\hat{j}_\ell(kr)\frac{\mathrm{d}\hat{n}_\ell(kr)}{\mathrm{d}r} - \hat{n}_\ell(kr)\frac{\mathrm{d}\hat{j}_\ell(kr)}{\mathrm{d}r}\right] = k. \tag{B-5}$$

We now seek to solve for the solution of eqn (B-2) using the Lagrange method of variation of constants.[5] We write the solution in the form

$$\chi(r) = C_1(r)\,\hat{j}_\ell(kr) + C_2(r)\,\hat{n}_\ell(kr) \tag{B-6}$$

where the "constants" $C_1(r)$ and $C_2(r)$ are constrained to obey the relationship

$$\frac{\mathrm{d}\,C_1(r)}{\mathrm{d}r}\,\hat{j}_\ell(kr) + \frac{\mathrm{d}\,C_2(r)}{\mathrm{d}r}\,\hat{n}_\ell(kr) = 0. \tag{B-7}$$

Eqn (B-6) may therefore be rewritten in matrix form as

$$\mathbf{X} = \mathbf{WC} \tag{B-8}$$

where

$$\mathbf{W} = \begin{pmatrix} \hat{j}_\ell(kr) & \hat{n}_\ell(kr) \\ \hat{j}'_\ell(kr) & \hat{n}'_\ell(kr) \end{pmatrix} \quad \text{and} \quad \mathbf{C} = \begin{pmatrix} C_1(r) \\ C_2(r) \end{pmatrix}.$$

Note that the Wronskian (eqn (B-5)) is given by $\det\{\mathbf{W}\} = |\mathbf{W}| = k$. Differentiation of eqn (B-8) yields

$$\mathbf{X}' = \mathbf{W}'\mathbf{C} + \mathbf{WC}'. \tag{B-9}$$

Comparing this with eqn (B-3), we can equate the right hand sides of the two equations, eqn (B-3) and eqn (B-9), to obtain

$$\mathbf{WC}' = \mathcal{V}\mathbf{X} + \mathbf{AX} - \mathbf{W}'\mathbf{C}. \tag{B-10}$$

We now examine the last two terms in this equation in detail and demonstrate that they cancel

$$\mathbf{AX} - \mathbf{W}'\mathbf{C} = \begin{pmatrix} 0 & 1 \\ \left[\frac{\ell(\ell+1)}{r^2} - k^2\right] & 0 \end{pmatrix} \begin{pmatrix} \chi(r) \\ \chi'(r) \end{pmatrix} - \begin{pmatrix} \hat{j}'_\ell(kr) & \hat{n}'_\ell(kr) \\ \hat{j}''_\ell(kr) & \hat{n}''_\ell(kr) \end{pmatrix} \begin{pmatrix} C_1(r) \\ C_2(r) \end{pmatrix}$$
$$= \begin{pmatrix} \chi'(r) \\ \left[\frac{\ell(\ell+1)}{r^2} - k^2\right] \chi(r) \end{pmatrix} - \begin{pmatrix} C_1(r)\hat{j}'_\ell(kr) + C_2(r)\hat{n}'_\ell(kr) \\ C_1(r)\hat{j}''_\ell(kr) + C_2(r)\hat{n}''_\ell(kr) \end{pmatrix}. \tag{B-11}$$

We should also note the following relationships

$$\begin{aligned} \chi'(r) &= (C_1(r)\,\hat{j}_\ell(kr) + C_2(r)\,\hat{n}_\ell(kr))' \\ &= C_1(r)\,\hat{j}'_\ell(kr) + C_2(r)\,\hat{n}'_\ell(kr) + C'_1(r)\,\hat{j}_\ell(kr) + C'_2(r)\,\hat{n}_\ell(kr) \\ &= C_1(r)\,\hat{j}'_\ell(kr) + C_2(r)\,\hat{n}'_\ell(kr) \end{aligned} \tag{B-12}$$

where we have used eqn (B-7) to obtain the last line of the equation. Using eqn (B-4) and (B-6) we can now write

$$\begin{aligned} C_1(r)\hat{j}''_\ell(kr) + C_2(r)\hat{n}''_\ell(kr) &= C_1(r)\left[\frac{\ell(\ell+1)}{r^2} - k^2\right]\hat{j}_\ell(kr) \\ &\quad + C_2(r)\left[\frac{\ell(\ell+1)}{r^2} - k^2\right]\hat{n}_\ell(kr) \\ &= \left[\frac{\ell(\ell+1)}{r^2} - k^2\right]\left\{C_1(r)\hat{j}_\ell(kr) + C_2(r)\hat{n}_\ell(kr)\right\} \\ &= \left[\frac{\ell(\ell+1)}{r^2} - k^2\right]\chi(r) \end{aligned} \tag{B-13}$$

and substituting eqn (B-12) and (B-13) into eqn (B-11) we see that $\mathbf{AX} - \mathbf{W}'\mathbf{C} = 0$, and from eqn (B-10) we can therefore write

$$\mathbf{WC}' = \mathcal{V}\mathbf{X}. \tag{B-14}$$

Using eqn (B-8) and (B-14) we then obtain

$$\mathbf{C}' = \{\mathbf{W}\}^{-1}\mathcal{V}\mathbf{WC}. \tag{B-15}$$

This is a key equation in the derivation and shows that $\mathbf{C}'$ is related by a matrix multiplication to $\mathbf{C}$. We denote the matrix which performs this transformation by $\mathbf{M}$

$$\mathbf{C}' = \mathbf{MC} = \begin{pmatrix} M_{11}(r) & M_{12}(r) \\ M_{21}(r) & M_{22}(r) \end{pmatrix}\mathbf{C} \tag{B-16}$$

where $\mathbf{M}(r) = \{\mathbf{W}\}^{-1}\mathcal{V}\mathbf{W}$, and the inverse Wronskian matrix is

$$\{\mathbf{W}(r)\}^{-1} = \frac{1}{|\mathbf{W}(r)|}\begin{pmatrix} \hat{n}'_\ell(kr) & -\hat{n}_\ell(kr) \\ -\hat{j}'_\ell(kr) & \hat{j}_\ell(kr) \end{pmatrix} \tag{B-17}$$

The transformation matrix **M** is therefore

$$\mathbf{M}(r) = \frac{1}{|\mathbf{W}(r)|}\begin{pmatrix} -\hat{n}_\ell(kr)U(r)\hat{j}_\ell(kr) & -\hat{n}_\ell(kr)U(r)\hat{n}_\ell(kr) \\ \hat{j}_\ell(kr)U(r)\hat{j}_\ell(kr) & \hat{j}_\ell(kr)U(r)\hat{n}_\ell(kr) \end{pmatrix} \tag{B-18}$$

From eqn (B-6) we can write

$$\begin{aligned} \chi(r) &= C_1(r)\left[\hat{j}_\ell(kr) + \frac{C_2(r)}{C_1(r)}\hat{n}_\ell(kr)\right] \\ &= C_1(r)[\hat{j}_\ell(kr) + S(r)\,\hat{n}_\ell(kr)] \end{aligned} \tag{B-19}$$

where

$$S(r) = \frac{C_2(r)}{C_1(r)} \tag{B-20}$$

and by differentiating the $S(r)$ variable, and using eqn (B-16), we obtain the following differential equation

$$\begin{aligned} S'(r) &= \frac{C_2'(r)}{C_1(r)} - \frac{C_2(r)C_1'(r)}{C_1^2(r)} \\ &= \frac{M_{21}(r)C_1(r) + M_{22}(r)C_2(r)}{C_1(r)} - \frac{C_2(r)(M_{11}(r)C_1(r) + M_{12}(r))C_2(r)}{C_1^2(r)} \\ &= M_{21}(r) + (M_{22}(r) - M_{11}(r))S(r) - M_{12}(r)S^2(r) \\ &= \frac{1}{|\mathbf{W}(r)|}\Big\{\hat{j}_\ell(kr)U(r)\hat{j}_\ell(kr) + \left[\hat{j}_\ell(kr)U(r)\hat{n}_\ell(kr) + \hat{n}_\ell(kr)U(r)\hat{j}_\ell(kr)\right] \\ &\quad \times S(r) + \hat{n}_\ell(kr)U(r)\hat{n}_\ell(kr)S^2(r)\Big\} \\ &= \frac{1}{|\mathbf{W}(r)|}\{\hat{j}_\ell(kr) + \hat{n}_\ell(kr)S(r)\}U(r)\{\hat{j}_\ell(kr) + \hat{n}_\ell(kr)S(r)\} \\ &= \frac{1}{k}U(r)\{\hat{j}_\ell(kr) + \hat{n}_\ell(kr)S(r)\}^2. \end{aligned} \tag{B-21}$$

To understand the physical meaning of the $S(r)$ variable we need study the asymptotic behavior of the radial wavefunction eqn (B-6). At large r values the Riccati–Bessel functions behave as follows

$$\begin{aligned} \hat{j}_\ell(kr) &\xrightarrow[r\to\infty]{} \sin\left(kr - \frac{\ell\pi}{2}\right) \\ \hat{n}_\ell(kr) &\xrightarrow[r\to\infty]{} -\cos\left(kr - \frac{\ell\pi}{2}\right). \end{aligned} \tag{B-22}$$

Then $\chi(r)$ at large r values can be written

$$\begin{aligned} \chi(r) &\xrightarrow[r\to\infty]{} C_1(\infty)\sin\left(kr - \frac{\ell\pi}{2}\right) - C_2(\infty)\cos\left(kr - \frac{\ell\pi}{2}\right) \\ &\xrightarrow[r\to\infty]{} A\sin\left(kr - \frac{\ell\pi}{2} + \eta_\ell\right). \end{aligned} \tag{B-23}$$

In accordance with the definition of the phase shift η_ℓ this leads to

$$\tan(\eta_\ell) = -\frac{C_2(\infty)}{C_1(\infty)}. \tag{B-24}$$

Now we are able to introduce a function $\eta_\ell(r)$ called the phase shift function. Its tangent can be written as

$$\tan(\eta_\ell(r)) = -S(r). \tag{B-25}$$

From this equation we see the phase shift function $(\eta_\ell(r))$ becomes equal to the phase shift as $r \to \infty$.

Summarizing, therefore, we have from eqn (B-21) and (B-25) the key equation

$$\frac{\mathrm{d}\tan(\eta_\ell(r))}{\mathrm{d}r} = -\frac{1}{k}U(r)\{\hat{j}_\ell(kr) - \hat{n}_\ell(kr)\tan(\eta_\ell(r))\}^2. \tag{B-26}$$

The disadvantage of eqn (B-26) is the discontinuity of the tan function whenever its argument passes through the value of $\pi/2 + \pi n$, where n is any integer. To avoid this problem the equation must be replaced by the one for the phase shift function itself rather than for its tangent.

B.1 Phase Shift for $l = 0$

For $\ell = 0$ we can write

$$\begin{aligned} \hat{j}_{\ell=0}(kr) &= \sin(kr), \\ \hat{n}_{\ell=0}(kr) &= -\cos(kr). \end{aligned} \tag{B-27}$$

Using eqn (B-26) we can now write

$$\frac{\mathrm{d}\tan(\eta_0(r))}{\mathrm{d}r} = -\frac{1}{k}U(r)\{\sin(kr) + \cos(kr)\tan(\eta_0(r))\}^2 \tag{B-28}$$

and therefore

$$\frac{1}{\cos^2(\eta_0(r))}\frac{\mathrm{d}\eta_0(r)}{\mathrm{d}r} = -\frac{1}{k}U(r)\left\{\sin(kr) + \cos(kr)\frac{\sin(\eta_0(r))}{\cos(\eta_0(r))}\right\}^2 \tag{B-29}$$

or

$$\begin{aligned} \frac{\mathrm{d}\eta_0(r)}{\mathrm{d}r} &= -\frac{1}{k}U(r)\{\cos(\eta_0(r))\sin(kr) + \sin(\eta_0(r))\cos(kr)\}^2 \\ &= -\frac{1}{k}U(r)\sin^2(kr + \eta_0(r)). \end{aligned}$$

The variable phase equation for the $\ell = 0$ phase shift may therefore be written as

$$\frac{\mathrm{d}\eta_0(r)}{\mathrm{d}r} = -\frac{1}{k}U(r)\sin^2(kr + \eta_0(r)). \tag{B-30}$$

B.2 Phase Shift for $l > 0$

From eqn (B-26) we finally have the basic equation for the phase shift analysis

$$\frac{\mathrm{d}\,\eta_\ell(r)}{\mathrm{d}r} = -\frac{1}{k}U(r)\{\cos(\eta_\ell(r))\hat{j}_\ell(kr) - \sin(\eta_\ell(r))\hat{n}_\ell(kr)\}^2 \tag{B-31}$$

with boundary condition $\eta_\ell(r = 0) = 0$.

Below we present some useful formulae for the Riccati–Bessel functions. For $\ell = 1$ we can write

$$\hat{j}_{\ell=1}(kr) = \frac{1}{kr}\sin(kr) - \cos(kr)$$

and

$$\hat{n}_{\ell=1}(kr) = -\frac{1}{kr}\cos(kr) - \sin(kr). \tag{B-32}$$

For $\ell > 1$ the Riccati–Bessel functions can be calculated from eqn (B-27) and (B-32) using the recurrence formulae

$$\hat{j}_{\ell+1}(kr) = \frac{2\ell + 1}{kr}\hat{j}_{\ell}(kr) - \hat{j}_{\ell-1}(kr)$$

$$\hat{n}_{\ell+1}(kr) = \frac{2\ell + 1}{kr}\hat{n}_{\ell}(kr) - \hat{n}_{\ell-1}(kr). \tag{B-33}$$

The use of eqn (B-33) leads to some numerical instabilities for large orders. Methods for overcoming these are discussed elsewhere (see, for instance, ref. 6).

Note the numerical calculation of the phase shift function should be started from $r > 0$ to avoid numerical errors. Methods for making the correct choice of the starting point for the calculation of the phase shift function are discussed in ref. 1 and 7.

B.3 Evaluation of Phase Shift

The variable phase approach provides a clear link between the phase shift and the interatomic interaction potential. For instance, as $(\cos\eta_\ell(r)\hat{j}_\ell(kr) - \sin\eta_\ell(r)\hat{n}_\ell(kr))^2$ in eqn (B-31 is always positive, the sign of the derivative of the phase shift function $\mathrm{d}\eta_\ell(r)/\mathrm{d}r$ is determined explicitly by the sign of potential $V(r)$ (note that $V(r) = \frac{\hbar^2 U(r)}{2\mu}$) see eqn (B-2)). In an attractive potential, when $V(r)$ is less than zero, the derivative of the phase shift is positive and correspondingly the phase shift function becomes increasingly more positive as well. In the case of a repulsive potential ($V(r) > 0$) the derivative of the phase shift function is negative and the phase shift becomes progressively more negative. During the integration of eqn (B-30) for an interatomic potential we can observe the effect of different parts of the potential on the phase shift function because the change in the phase shift function in a particular interval of r is caused by the part of the potential in the same interval.

Figure B.1 shows the phase shift function *versus* distance for the collision of two Ne atoms interacting *via* a Lennard-Jones (12, 6) potential, for different relative collision energies and for $\ell = 0$. It may be clearly seen that as the phase shift function, starting from zero, at first becomes negative due to the repulsive part of the potential. It reaches a minimum at $r = 2.75$ Å shown by a dashed line, where the potential equals zero and becomes progressively more positive as it is integrated over the attractive region of the potential.

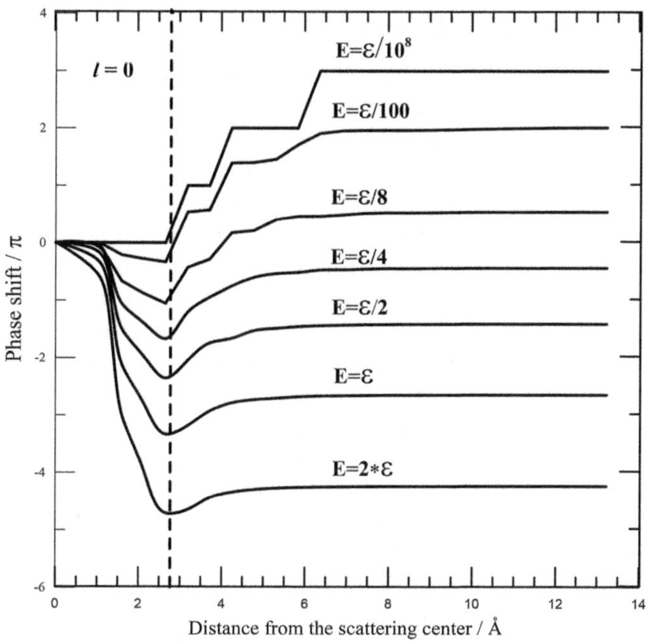

Figure B.1 The phase shifts as a function of interatomic distance between two Ne atoms.

As may be seen from Figure B.1 the phase shift function is not oscillatory as opposed to the wavefunction itself. As $r \to \infty$ the phase shift function yields the absolute value of the phase shift, which greatly simplifies its analysis. Decreasing the relative collision energy E for Ne + Ne system leads to the appearance of a stepwise behavior in the phase shift function (see Figure B.1). The steps become sharper as the collision energy decreases and at extremely low energies the phase shift function approaches the value of 3π at large r. This result confirms Levinson's theorem[8] which can be written as

$$\eta_\ell(E \to 0) = n\pi \tag{B-34}$$

where n is the number of bound states supported by the potential for the chosen orbital angular quantum number ℓ.

The phase shift $\eta_\ell(E \to 0) = 3\pi$ calculated for $\ell = 0$ (see Figure B.1) indicates that the Ne + Ne interatomic potential supports three bound states. For $\ell = 0$, as the energy tends to zero, the negative part of the phase shift function is seen to vanish and the phase shift is determined exclusively by the attractive part of the potential, *i.e.* by the potential well. As ℓ is increased the depth of the effective potential decreases due to the centrifugal repulsion, $\ell(\ell + 1)/2\mu$, and for sufficiently large ℓ values the well will vanish completely (see Figure 2.3). We may therefore expect that the number of bound states will decrease as ℓ increases.

B.4 Levinson's Theorem

Levinson's theorem states that as the collision energy tends to zero (*i.e.* $k \to 0$) the phase shift tends to $n\pi$, where n is the number of bound states supported by the potential.[8,9] The proof of Levinson's theorem is complicated but below we give a simple argument to support its reasonableness.

For a potential, such as a Lennard-Jones potential, that can support bound states, the $v = 1$ wavefunction has one node. The reason for this is that it has to be orthogonal to the $v = 0$ wavefunction which has no nodes. By the same logic the highest bound state wavefunction has $n - 1$ nodes, where the potential supports a total of n bound states. If there were no potential at all, the lowest continuum wavefunction, in the limit of $k \to 0$, would have no nodes. In the presence of a potential, the lowest continuum wavefunction, in the $k \to 0$ limit, has to have one more node than the highest bound state wavefunction, so as to be orthogonal to it. This implies that the lowest continuum wavefunction will have n nodes. In accordance with eqn (B-19) and (B-23), for $\ell = 0$, this wavefunction can be written as $u(r) = A(r)\sin(kr + \delta_0(r))$. This function has nodes when $\sin(kr + \delta_0(r)) = 0$, *i.e.* $kr + \delta_0(r) = \kappa\pi$ where κ is an integer characterizing a node position. If we consider some large value of $r = R$ in the limit of $k \to 0$, we get $\delta_0(R) = \kappa\pi$. As we have seen the maximal value of κ is n and this is reached in the limit of $R \to \infty$ when $\delta(R)$ no longer changes. So finally, phase shift $= \delta_0(R = \infty) = n\pi$. Figure (B.1) confirms these conclusions.

The node at $r = 0$ when $\delta_0(r) = 0$ should not be considered as a node because this is the boundary condition for the phase shift function.

References

1. F. Calogero, *Variable Phase Approach to Potential Scattering*, Academic Press, New York, 1967.
2. A. Ronveaux, Phase equation in quantum mechanics, *Am. J. Phys.*, 1969, **37**, 135.
3. J. L. Dehmer and U. Fano, Phase-amplitude method in atomic physics. I. Basic formulas for an electron in an ionic potential, *Phys. Rev. A: At., Mol., Opt. Phys.*, Aug. 1970, **2**, 304.
4. M. Abromowitz and I. A. Stegun, *Handbook on Mathematical Functions, Applied Mathematics Series 55*, National Bureau of Standards, 10th edn, 1972.
5. A. Jeffrey, *Mathematics for Engineers and Scientists*, Van Nostrand Reinhold, UK, 3rd edn, 1985.
6. A. R. Barnett, *Computational Atomic Physics*, Springer-Verlag, Berlin Heidelberg, 1996, ch. 9, p. 181.
7. V. Babikov, *Variable Phase Approach in Quantum Mechanics*, Nauka, Moscow, 1968 [in Russian].
8. N. Levinson, On the uniqueness of the potential in a Schrodinger equation for a given asymptotic phase, *K. Dan. Vid. Selsk. Mat-Fys. Medd.*, 1949, **25**, 1.
9. N. F. Mott and H. S. W. Massey, *The Theory of Atomic Collisions*, Clarendon Press, Oxford, 2nd edn, 1949.

APPENDIX C

Jacobi Coordinates

Several different coordinate systems may be used to discuss the quantum dynamics of molecular systems.[1,2] The simplest of these is the Jacobi coordinate system. In this appendix we first outline the Jacobi coordinate system for a three atom system and then generalize it to any number of atoms.

C.1 Triatomic System

For a triatomic system, A + BC, we denote the vector positions of the three atoms by $\boldsymbol{r}_1$, $\boldsymbol{r}_2$ and $\boldsymbol{r}_3$.

The first Jacobi coordinate ($\boldsymbol{r} \equiv \boldsymbol{q}_1$) is the coordinate of atom 3 relative to atom 2

$$\boldsymbol{r} = \boldsymbol{r}_3 - \boldsymbol{r}_2. \tag{C-1}$$

The second Jacobi coordinate ($\boldsymbol{R} \equiv \boldsymbol{q}_2$) corresponds to the position of atom 1 relative to the center-of-mass of the diatomic made up from atoms 2 and 3

$$\begin{aligned} \boldsymbol{R} &= \boldsymbol{r}_1 - \frac{m_2\boldsymbol{r}_2 + m_3\boldsymbol{r}_3}{m_2 + m_3} \\ &= \boldsymbol{r}_1 - \frac{m_2\boldsymbol{r}_2 + m_3\boldsymbol{r}_3}{M_{23}}. \end{aligned} \tag{C-2}$$

The final Jacobi coordinate, which we will call $\boldsymbol{R}_c$ ($\equiv \boldsymbol{q}_3$) corresponds to the position of the center-of-mass of the whole system

$$\begin{aligned} \boldsymbol{R}_c &= \frac{m_1\boldsymbol{r}_1 + m_2\boldsymbol{r}_2 + m_3\boldsymbol{r}_3}{m_1 + m_2 + m_3} \\ &= \frac{m_1\boldsymbol{r}_1 + m_2\boldsymbol{r}_2 + m_3\boldsymbol{r}_3}{M}. \end{aligned} \tag{C-3}$$

RSC Theoretical and Computational Chemistry Series No. 7
Theory of Molecular Collisions
By Gabriel G. Balint-Kurti and Alexander P. Palov

Published by the Royal Society of Chemistry, www.rsc.org

The Hamiltonian in the original Cartesian coordinates of the three atoms may be written in the form

$$\begin{aligned}\hat{H} &= \sum_{i=1}^{3} -\frac{\hbar^2}{2m_i}\frac{\partial^2}{\partial \boldsymbol{r}_i^2} + V(\boldsymbol{r}_1, \boldsymbol{r}_2, \boldsymbol{r}_3) \\ &= \hat{T} + V(\boldsymbol{r}_1, \boldsymbol{r}_2, \boldsymbol{r}_3)\end{aligned} \tag{C-4}$$

where $\hat{T}$ is the kinetic energy operator and

$$\frac{\partial^2}{\partial \boldsymbol{r}_i^2} = \nabla_i^2 = \frac{\partial^2}{\partial x_i^2} + \frac{\partial^2}{\partial y_i^2} + \frac{\partial^2}{\partial z_i^2}. \tag{C-5}$$

In order to convert the Hamiltonian from the initial coordinates to Jacobi coordinates we need to convert the differential operators to the new coordinate system. We do this by considering the operation of the differential operators on a test function, $f(\boldsymbol{r}, \boldsymbol{R}, \boldsymbol{R}_\mathrm{c})$, and using the chain rule

$$\frac{\partial f}{\partial x_i} = \frac{\partial x}{\partial x_i}\frac{\partial f}{\partial x} + \frac{\partial X}{\partial x_i}\frac{\partial f}{\partial X} + \frac{\partial X_\mathrm{c}}{\partial x_i}\frac{\partial f}{\partial X_\mathrm{c}} \tag{C-6}$$

where x, X and X_c are the x components of the vectors $\boldsymbol{r}$, $\boldsymbol{R}$ and $\boldsymbol{R}_\mathrm{c}$. From eqn (C-1), (C-2) and (C-3) we have

$$\begin{aligned}&\frac{\partial x}{\partial x_1} = 0, &&\frac{\partial x}{\partial x_2} = -1, &&\frac{\partial x}{\partial x_3} = 1,\\ &\frac{\partial X}{\partial x_1} = 1, &&\frac{\partial X}{\partial x_2} = -\frac{m_2}{M_{23}}, &&\frac{\partial X}{\partial x_3} = -\frac{m_3}{M_{23}},\\ &\frac{\partial X_\mathrm{c}}{\partial x_1} = \frac{m_1}{M}, &&\frac{\partial X_\mathrm{c}}{\partial x_2} = \frac{m_2}{M}, &&\frac{\partial X_\mathrm{c}}{\partial x_3} = \frac{m_3}{M}.\end{aligned} \tag{C-7}$$

We may therefore write

$$\begin{aligned}\frac{\partial f}{\partial x_1} &= \frac{\partial f}{\partial X} + \frac{m_1}{M}\frac{\partial f}{\partial X_\mathrm{c}},\\ \frac{\partial f}{\partial x_2} &= -\frac{\partial f}{\partial x} - \frac{m_2}{M_{23}}\frac{\partial f}{\partial X} + \frac{m_2}{M}\frac{\partial f}{\partial X_\mathrm{c}},\\ \frac{\partial f}{\partial x_3} &= \frac{\partial f}{\partial x} - \frac{m_3}{M_{23}}\frac{\partial f}{\partial X} + \frac{m_3}{M}\frac{\partial f}{\partial X_\mathrm{c}}.\end{aligned} \tag{C-8}$$

We now differentiate eqn (C-6) again

$$\begin{aligned}\frac{\partial^2 f}{\partial x_i^2} = &\left(\frac{\partial x}{\partial x_i}\right)^2 \frac{\partial^2 f}{\partial x^2} + \frac{\partial x}{\partial x_i}\frac{\partial X}{\partial x_i}\frac{\partial^2 f}{\partial x \partial X} + \frac{\partial x}{\partial x_i}\frac{\partial X_\mathrm{c}}{\partial x_i}\frac{\partial^2 f}{\partial x \partial X_\mathrm{c}}\\ &+ \frac{\partial x}{\partial x_i}\frac{\partial X}{\partial x_i}\frac{\partial f}{\partial X \partial x} + \left(\frac{\partial X}{\partial x_i}\right)^2 \frac{\partial^2 f}{\partial X^2} + \frac{\partial X_\mathrm{c}}{\partial x_i}\frac{\partial X}{\partial x_i}\frac{\partial f}{\partial X \partial X_\mathrm{c}}\\ &+ \frac{\partial x}{\partial x_i}\frac{\partial X_\mathrm{c}}{\partial x_i}\frac{\partial f}{\partial x \partial X_\mathrm{c}} + \frac{\partial X}{\partial x_i}\frac{\partial X_\mathrm{c}}{\partial x_i}\frac{\partial f}{\partial X \partial X_\mathrm{c}} + \left(\frac{\partial X_\mathrm{c}}{\partial x_i}\right)^2 \frac{\partial^2 f}{\partial X_\mathrm{c}^2}.\end{aligned} \tag{C-9}$$

Using eqn (C-4), (C-5), (C-7) and (C-9) we obtain, for the x_1, x_2 and x_3 components of the kinetic energy operator acting on a function f

$$-\frac{\hbar^2}{2m_1}\frac{\partial^2 f}{\partial x_1^2} = -\frac{\hbar^2}{2}\left\{\frac{1}{m_1}\frac{\partial^2 f}{\partial X^2} + \frac{1}{M}\frac{\partial^2 f}{\partial X \partial X_c} + \frac{1}{M}\frac{\partial^2 f}{\partial X_c \partial X} + \frac{m_1}{M^2}\frac{\partial^2 f}{\partial X_c^2}\right\}, \tag{C-10}$$

$$\begin{aligned}-\frac{\hbar^2}{2m_2}\frac{\partial^2 f}{\partial x_2^2} = -\frac{\hbar^2}{2}\Bigg\{&\frac{1}{m_2}\frac{\partial^2 f}{\partial x^2} + \frac{1}{M_{23}}\frac{\partial^2 f}{\partial x \partial X} - \frac{1}{M}\frac{\partial^2 f}{\partial x \partial X_c}\\ &+\frac{1}{M_{23}}\frac{\partial^2 f}{\partial x \partial X} + \frac{m_2}{M_{23}^2}\frac{\partial^2 f}{\partial X^2} - \frac{1}{M}\frac{m_2}{M_{23}}\frac{\partial^2 f}{\partial X \partial X_c}\\ &-\frac{1}{M}\frac{\partial^2 f}{\partial x \partial X_c} - \frac{1}{M_{23}}\frac{m_2}{M}\frac{\partial^2 f}{\partial X \partial X_c} + \frac{m_2}{M^2}\frac{\partial^2 f}{\partial X_c^2}\Bigg\}\end{aligned} \tag{C-11}$$

and

$$\begin{aligned}-\frac{\hbar^2}{2m_3}\frac{\partial^2 f}{\partial x_3^2} = -\frac{\hbar^2}{2}\Bigg\{&\frac{1}{m_3}\frac{\partial^2 f}{\partial x^2} - \frac{1}{M_{23}}\frac{\partial^2 f}{\partial x \partial X} + \frac{1}{M}\frac{\partial^2 f}{\partial x \partial X_c}\\ &-\frac{1}{M_{23}}\frac{\partial^2 f}{\partial x \partial X} + \frac{m_3}{M_{23}^2}\frac{\partial^2 f}{\partial X^2} - \frac{1}{M}\frac{m_3}{M_{23}}\frac{\partial^2 f}{\partial X \partial X_c}\\ &+\frac{1}{M}\frac{\partial^2 f}{\partial x \partial X_c} - \frac{1}{M_{23}}\frac{m_3}{M}\frac{\partial^2 f}{\partial X \partial X_c} + \frac{m_3}{M^2}\frac{\partial^2 f}{\partial X_c^2}\Bigg\}.\end{aligned} \tag{C-12}$$

Summing eqn (C-10), (C-11) and (C-12) we obtain the total x component of the kinetic energy operator

$$\begin{aligned}&\left\{-\frac{\hbar^2}{2}\sum_{i=1}^{3}\frac{1}{m_i}\frac{\partial^2}{\partial x_i^2}\right\}f\\ &= -\frac{\hbar^2}{2}\Bigg\{\frac{1}{m_1}\frac{\partial^2 f}{\partial X^2} + \frac{1}{M}\frac{\partial^2 f}{\partial X \partial X_c} + \frac{1}{M}\frac{\partial^2 f}{\partial X_c \partial X} + \frac{m_1}{M^2}\frac{\partial^2 f}{\partial X_c^2}\\ &\quad+\frac{1}{m_2}\frac{\partial^2 f}{\partial x^2} + \frac{1}{M_{23}}\frac{\partial^2 f}{\partial x \partial X} - \frac{1}{M}\frac{\partial^2 f}{\partial x \partial X_c}\\ &\quad+\frac{1}{M_{23}}\frac{\partial^2 f}{\partial x \partial X} + \frac{m_2}{M_{23}^2}\frac{\partial^2 f}{\partial X^2} - \frac{1}{M}\frac{m_2}{M_{23}}\frac{\partial^2 f}{\partial X \partial X_c}\\ &\quad-\frac{1}{M}\frac{\partial^2 f}{\partial x \partial X_c} - \frac{1}{M_{23}}\frac{m_2}{M}\frac{\partial^2 f}{\partial X \partial X_c} + \frac{m_2}{M^2}\frac{\partial^2 f}{\partial X_c^2}\\ &\quad+\frac{1}{m_3}\frac{\partial^2 f}{\partial x^2} - \frac{1}{M_{23}}\frac{\partial^2 f}{\partial x \partial X} + \frac{1}{M}\frac{\partial^2 f}{\partial x \partial X_c}\\ &\quad-\frac{1}{M_{23}}\frac{\partial^2 f}{\partial x \partial X} + \frac{m_3}{M_{23}^2}\frac{\partial^2 f}{\partial X^2} - \frac{1}{M}\frac{m_3}{M_{23}}\frac{\partial^2 f}{\partial X \partial X_c}\\ &\quad+\frac{1}{M}\frac{\partial^2 f}{\partial x \partial X_c} - \frac{1}{M_{23}}\frac{m_3}{M}\frac{\partial^2 f}{\partial X \partial X_c} + \frac{m_3}{M^2}\frac{\partial^2 f}{\partial X_c^2}\Bigg\}.\end{aligned} \tag{C-13}$$

If we look carefully we see that all the off-diagonal cross derivative terms cancel leaving only the "diagonal" terms

$$
\begin{aligned}
\left\{-\frac{\hbar^2}{2}\sum_{i=1}^{3}\frac{1}{m_i}\frac{\partial^2}{\partial x_i^2}\right\}f = -\frac{\hbar^2}{2}\Bigg\{&\left(\frac{1}{m_2}+\frac{1}{m_3}\right)\frac{\partial^2 f}{\partial x^2} \\
&+\left(\frac{m_2}{M_{23}^2}+\frac{m_3}{M_{23}^2}+\frac{1}{m_1}\right)\frac{\partial^2 f}{\partial X^2} \\
&+\left(\frac{m_1}{M^2}+\frac{m_2}{M^2}+\frac{m_3}{M^2}\right)\frac{\partial^2 f}{\partial X_c^2}\Bigg\} \\
= -\frac{\hbar^2}{2}\Bigg\{&\frac{1}{\mu_{23}}\frac{\partial^2 f}{\partial x^2}+\frac{1}{\mu}\frac{\partial^2 f}{\partial X^2}+\frac{1}{M}\frac{\partial^2 f}{\partial X_c^2}\Bigg\}
\end{aligned} \tag{C-14}
$$

where $\mu_{23} = (m_2 m_3)/(m_2 + m_3)$ and $\mu = (m_2 + m_3)m_1/M$.

The potential energy function V depends only on the "internal" coordinates $\boldsymbol{r}$ and $\boldsymbol{R}$ and is independent of $\boldsymbol{R}_c$, the coordinate of the center-of-mass. As there is no coupling between $\boldsymbol{R}_c$ and the internal coordinates in the kinetic energy expression (see eqn (C-14)) the total Hamiltonian becomes a sum of an internal Hamiltonian and one for the motion of the center-of-mass. This results in a separation of the Schrödinger equations for the motion of the center-of-mass from that for the internal motion. The respective wavefunctions then become products of the two separately computed wavefunctions.[3]

Any coordinate system that uses the coordinate of the center-of-mass as one of its coordinates will result in this separation of the wavefunction of the motion of the center-of-mass from the internal wavefunction. The additional advantage of the Jacobi coordinates is the simplicity of the form of the kinetic energy operator in that it contains no cross terms.

C.2 Generalization to an *n* Atom System

The Jacobi coordinates are denoted as $\boldsymbol{q}_k$, $k = 1, 2, 3, \ldots, n$. The following general rule defines the Jacobi coordinates for an arbitrary molecule consisting of n atoms

$$\boldsymbol{q}_1 = \boldsymbol{r}_n - \boldsymbol{r}_{n-1} \qquad \text{for } k = 1, \tag{C-15}$$

$$\boldsymbol{q}_k = \boldsymbol{r}_{n-k} - \frac{(1 - 2\,\delta_{k,n})\,(m_{n-k+1}\boldsymbol{r}_{n-k+1} + \cdots + m_n\boldsymbol{r}_n)}{m_{n-k+1} + \cdots + m_n} \qquad \text{for } k \geq 2 \tag{C-16}$$

where $\boldsymbol{r}_k$ is the position of the k-th atom in a laboratory reference frame, m_k are the masses of the corresponding atoms and $\boldsymbol{r}_0 = 0$. The position of the center-of-mass of the whole system is defined by

$$\boldsymbol{q}_n = \frac{m_1\boldsymbol{r}_1 + \cdots + m_n\boldsymbol{r}_n}{m_1 + \cdots + m_n}. \tag{C-17}$$

In the case of a triatomic molecule $\boldsymbol{q}_1$ corresponds to $\boldsymbol{r}$ as defined in eqn (C-1), $\boldsymbol{q}_2$ to $\boldsymbol{R}$ as in eqn (C-2) and $\boldsymbol{q}_n$ at $n = 3$ to $\boldsymbol{R}_c$ as in eqn (C-3).

Let us now transform the molecular Hamiltonian from laboratory Cartesian coordinates to Jacobi ones. For this we need introduce a **B** matrix defining the transformation of coordinates

$$\boldsymbol{q}_k = \sum_{i=1}^{n} B_{ki}\boldsymbol{r}_i \tag{C-18}$$

For an n atomic molecule we obtain (see eqn (C-15)–(C-18))

$$\begin{aligned}
\boldsymbol{q}_1 &= B_{1,1}\boldsymbol{r}_1 + \cdots + B_{1,n-1}\boldsymbol{r}_{n-1} + B_{1,n}\boldsymbol{r}_n \\
&\rightarrow B_{1,1\ldots n-2} = 0, B_{1,n-1} = -1, B_{1,n} = 1 \\
\boldsymbol{q}_2 &= B_{2,1}\boldsymbol{r}_1 + \cdots + B_{2,n-2}\boldsymbol{r}_{n-2} + B_{2,n-1}\boldsymbol{r}_{n-1} + B_{2,n}\boldsymbol{r}_n \\
&\rightarrow B_{2,1\ldots n-3} = 0, B_{2,n-2} = 1, B_{2,n-1} = -\frac{m_{n-1}}{M_2}, B_{2,n} = -\frac{m_n}{M_2} \\
&\vdots \\
\boldsymbol{q}_k &= B_{k,1}\boldsymbol{r}_1 + \cdots + B_{k,n-k}\boldsymbol{r}_{n-k} + B_{k,n-k+1}\boldsymbol{r}_{n-k+1} + \cdots + B_{k,n}\boldsymbol{r}_n \\
&\rightarrow B_{k,1\ldots n-k-1} = 0, B_{k,n-k} = 1, B_{k,n-k+1} = -\frac{m_{n-k+1}}{M_k}, \ldots, B_{k,n} = -\frac{m_n}{M_k} \\
&\vdots \\
\boldsymbol{q}_n &= B_{n,1}\boldsymbol{r}_1 + \cdots + B_{n,n}\boldsymbol{r}_n \\
&\rightarrow B_{n,1} = \frac{m_1}{M_n}, \ldots, B_{n,n} = \frac{m_n}{M_n}
\end{aligned} \tag{C-19}$$

where

$$M_k = \sum_{i=n-k+1}^{n} m_i.$$

The $n \times n$ **B** matrix can be written in a form

$$B_{ki} = \frac{\partial \boldsymbol{q}_k}{\partial \boldsymbol{r}_i} = \begin{pmatrix} 0 & \ldots & 0 & -1 & 1 \\ 0 & \ldots & 1 & -\frac{m_{n-1}}{M_2} & -\frac{m_n}{M_2} \\ \ldots & \ldots & \ldots & \ldots & \ldots \\ \frac{m_1}{M_n} & \ldots & \ldots & \ldots & \frac{m_n}{M_n} \end{pmatrix}. \tag{C-20}$$

In accordance with the chain rule the gradient of the function $f(\boldsymbol{q}_1, \ldots, \boldsymbol{q}_n)$ is given by

$$\frac{\partial f}{\partial \boldsymbol{r}_i} = \sum_{k=1}^{n} \frac{\partial f}{\partial \boldsymbol{q}_k}\frac{\partial \boldsymbol{q}_k}{\partial \boldsymbol{r}_i} = \sum_{k=1}^{n} B_{ki}\frac{\partial f}{\partial \boldsymbol{q}_k} \tag{C-21}$$

and differentiating again we obtain

$$\frac{\partial^2 f}{\partial \boldsymbol{r}_i^2} = \sum_{l=1}^{n} B_{li}\frac{\partial}{\partial \boldsymbol{q}_l}\sum_{k=1}^{n} B_{ki}\frac{\partial f}{\partial \boldsymbol{q}_k} = \sum_{l=1}^{n}\sum_{k=1}^{n} B_{li}B_{ki}\frac{\partial^2 f}{\partial \boldsymbol{q}_l \boldsymbol{q}_k}. \tag{C-22}$$

We can now write out the kinetic energy operator for the n atom system in generalized Jacobi coordinates

$$\hat{T}f = -\frac{\hbar^2}{2}\sum_{i=1}^{n}\frac{1}{m_i}\frac{\partial^2 f}{\partial \boldsymbol{r}_i^2} = -\frac{h^2}{2}\sum_{i=1}^{n}\sum_{l=1}^{n}\sum_{k=1}^{n}\frac{1}{m_i}B_{li}B_{ki}\frac{\partial^2 f}{\partial \boldsymbol{q}_l \boldsymbol{q}_k}. \tag{C-23}$$

We have seen that for the three atom case all the cross terms in the Jacobi coordinate kinetic energy operator vanish after summation over all the second derivatives (see eqn (C-13) and (C-14)). This also applies to the n atom case and as for the triatomic system the first $(n-1)$ Jacobi coordinates describe the internal variables while q_n is the coordinate of the center-of-mass of the system:

$$\hat{T}f = -\frac{\hbar^2}{2}\sum_{k=1}^{n}\frac{1}{\mu_k}\frac{\partial^2 f}{\partial \boldsymbol{q}_k^2}. \tag{C-24}$$

The reduced masses are defined by the equations

$$\begin{aligned} \frac{1}{\mu_k} &= \frac{1}{M_k} + \frac{1}{m_{n-k}} \qquad \text{for } k < n, \\ \frac{1}{\mu_k} &= \frac{1}{M_k} \qquad \text{for } k = n. \end{aligned} \tag{C-25}$$

As the potential function does not depend on the center-of-mass coordinate, q_n, the Shrödinger equation can be factored into an equation for the internal coordinates (q_1 to q_{n-1}) and a separate equation for the motion of the center-of-mass.

Finally, the Shrödinger equation for the internal motion of an n-atomic molecule may be written as

$$\hat{H}f = -\frac{\hbar^2}{2}\sum_{k=1}^{n-1}\frac{1}{\mu_k}\frac{\partial^2 f}{\partial \boldsymbol{q}_k^2} + V(\boldsymbol{q}_1 \cdots \boldsymbol{q}_{n-1})f. \tag{C-26}$$

References

1. B. T. Sutcliffe and J. Tennyson, A generalized approach to the calculation of ro-vibrational spectra of triatomic molecules, *Mol. Phys.*, 1986, **58**, 1053.
2. B. T. Sutcliffe and J. Tennyson, A general treatment of vibration-rotation coordinates for triatomic molecules, *Int. J. Quantum Chem.*, 1991, **39**, 183.
3. L. Pauling and E. B. Wilson, *Introduction to Quantum Mechanics*, McGraw-Hill Book Co., 1935.

APPENDIX D

Body-Fixed Formulation of Inelastic Scattering Theory

In this appendix we derive the equations for the treatment of inelastic scattering theory in the body-fixed reference frame.

D.1 Total Angular Momentum Eigenfunctions

In order to represent the wavefunction in body-fixed coordinates we need first of all to transform the angular basis functions $\mathcal{Y}_{j\ell}^{JM}(\hat{\boldsymbol{R}}, \hat{\boldsymbol{r}})$ of eqn (3.12) to the body-fixed frame. The spherical harmonic $Y_{\ell,m_\ell}(\theta_R, \phi_R)$ can be written as a Wigner **D** matrix (see eqn 4.17(1) and 4.4(2) of ref. 1)

$$Y_{\ell,m_\ell}(\theta_R, \phi_R) = \left(\frac{2\ell+1}{4\pi}\right)^{\frac{1}{2}} D_{m_\ell,0}^{\ell*}(\phi_R, \theta_R, 0). \tag{D-1}$$

This is in effect the required transformation as the angles θ_R, ϕ_R are both zero in the body-fixed axes as the body-fixed Z axis points in this direction. Incidentally, as the spherical harmonics are normalized, this equation shows that the function on the right hand side of the equation is also normalized.

Upon rotating the coordinate system the diatomic rotational wavefunction $Y_{j,m_j}(\theta_r, \phi_r)$, becomes (see eqn 4.1(1) of ref. 1)

$$Y_{j,\Omega}(\gamma, \phi_r') = \sum_{m_j} Y_{j,m_j}(\theta_r, \phi_r) D_{m_j,\Omega}^{j}(\phi_R, \theta_R, 0) \tag{D-2}$$

where ϕ_r' is the azimuthal angle made by the axis of the diatomic in the body-fixed coordinate system.

RSC Theoretical and Computational Chemistry Series No. 7
Theory of Molecular Collisions
By Gabriel G. Balint-Kurti and Alexander P. Palov

Published by the Royal Society of Chemistry, www.rsc.org

The inverse relationship is (see eqn 4.1(5) of ref. 1)

$$Y_{j,m_j}(\theta_r,\phi_r)=\sum_{\Omega} Y_{j,\Omega}(\gamma,\phi_r')D^{j*}_{m_j,\Omega}(\phi_R,\theta_R,0). \tag{D-3}$$

Using eqn (3.12), (D-1) and (D-3) we can write an expression for the angular functions $\mathcal{Y}^{JM}_{j\ell}(\hat{\boldsymbol{R}},\hat{\boldsymbol{r}})$ in the body-fixed coordinate system

$$\begin{aligned}\mathcal{Y}^{JM}_{j\ell}(\hat{\boldsymbol{R}},\hat{\boldsymbol{r}})=\sum_{m_\ell m_j}\left(\frac{2\ell+1}{4\pi}\right)^{\frac{1}{2}}\sum_{\Omega} Y_{j,\Omega}(\gamma,\phi_r')D^{j*}_{m_j,\Omega}(\phi_R,\theta_R,0)\\ \times D^{\ell*}_{m_\ell,0}(\phi_R,\theta_R,0)(jm_j\ell m_\ell|j\ell JM)\end{aligned} \tag{D-4}$$

The product of the Wigner **D** matrices in eqn (D-4) may be written in the form (see eqn 4.6(1) of ref. 1)

$$\begin{aligned}&D^{j*}_{m_j,\Omega}(\phi_R,\theta_R,0)D^{\ell*}_{m_\ell,0}(\phi_R,\theta_R,0)\\ &=\sum_{L}\sum_{N}(jm_j\ell m_\ell|j\ell LN)D^{L*}_{N,\Omega}(\phi_R,\theta_R,0)(j\Omega\ell 0|j\ell L\Omega)\end{aligned} \tag{D-5}$$

where we have used the fact that $(j\Omega\ell 0|LM)=0$ if $M\neq\Omega$ to eliminate a sum over M. Substituting eqn (D-5) into (D-4) gives

$$\begin{aligned}\mathcal{Y}^{JM}_{j\ell}(\hat{\boldsymbol{R}},\hat{\boldsymbol{r}})=\left(\frac{2\ell+1}{4\pi}\right)^{\frac{1}{2}}\sum_{\Omega} Y_{j,\Omega}(\gamma,\phi_r')\sum_{L}\sum_{N}D^{L*}_{N,\Omega}(\phi_R,\theta_R,0)(j\Omega\ell 0|j\ell L\Omega)\\ \times\left\{\sum_{m_\ell m_j}(jm_j\ell m_\ell|j\ell LN)(jm_j\ell m_\ell|j\ell JM)\right\}.\end{aligned} \tag{D-6}$$

The term in curly brackets is equal to $\delta_{L,J}\delta_{N,M}$ (see eqn 8.1(8) of ref. 1), yielding

$$\begin{aligned}&\mathcal{Y}^{JM}_{j\ell}(\hat{\boldsymbol{R}},\hat{\boldsymbol{r}})\\ &=\left(\frac{2\ell+1}{4\pi}\right)^{\frac{1}{2}}\sum_{\Omega} Y_{j,\Omega}(\gamma,\phi_r')D^{J*}_{M,\Omega}(\phi_R,\theta_R,0)(j\Omega\ell 0|j\ell J\Omega).\end{aligned} \tag{D-7}$$

We now use the symmetry property (see eqn 8.4.3(10) of ref. 1 or eqn (3.5.16) of ref. 2) to change the term $(2\ell+1)$ to $(2J+1)$

$$(2\ell+1)^{\frac{1}{2}}(j\Omega\ell 0|j\ell J\Omega)=(-1)^{j-\Omega}(2J+1)^{\frac{1}{2}}(J\Omega j-\Omega|Jj\ell 0) \tag{D-8}$$

and substituting this into eqn (D-7) gives

$$\begin{aligned}\mathcal{Y}^{JM}_{j\ell}(\hat{\boldsymbol{R}},\hat{\boldsymbol{r}})=\sum_{\Omega}[(-1)^{j-\Omega}(J\Omega j-\Omega|Jj\ell 0)]\\ \times\left\{Y_{j,\Omega}(\gamma,\phi_r')\left(\frac{2J+1}{4\pi}\right)^{\frac{1}{2}}D^{J*}_{M,\Omega}(\phi_R,\theta_R,0)\right\}.\end{aligned} \tag{D-9}$$

D.2 Unitary Transformation Matrix Between Space-Fixed and Body-Fixed Angular Basis Functions

The term in square brackets in eqn (D-9) is an element of a unitary transformation matrix that we denote as **B**

$$B_{\ell,\Omega} = (-1)^{j-\Omega}(J\Omega j - \Omega|Jj\ell 0). \tag{D-10}$$

The term in curly brackets in eqn (D-9) is the body-fixed angular basis which we denote by $\Xi^{JM}_{j\Omega}(\hat{\boldsymbol{R}},\hat{\boldsymbol{r}}')$

$$\Xi^{JM}_{j\Omega}(\hat{\boldsymbol{R}},\hat{\boldsymbol{r}}') = Y_{j,\Omega}(\gamma,\phi_r')\left(\frac{2J+1}{4\pi}\right)^{\frac{1}{2}} D^{J*}_{M,\Omega}(\phi_R,\theta_R,0). \tag{D-11}$$

The **B** matrix transforms the body-fixed orthonormal basis functions $\Xi^{JM}_{j\Omega}(\hat{\boldsymbol{R}},\hat{\boldsymbol{r}}')$ to the space-fixed angular functions $\mathcal{Y}^{JM}_{j\ell}(\hat{\boldsymbol{R}},\hat{\boldsymbol{r}})$. It can be shown to be unitary using the properties of the Clebsch–Gordan coefficients (see eqn 8.1(8) of ref. 1). Employing eqn (D-10) and (D-11) in eqn (D-9) we can write

$$\mathcal{Y}^{JM}_{j\ell}(\hat{\boldsymbol{R}},\hat{\boldsymbol{r}}) = \sum_{\Omega} B_{\ell,\Omega}\,\Xi^{JM}_{j\Omega}(\hat{\boldsymbol{R}},\hat{\boldsymbol{r}}'). \tag{D-12}$$

The unitary properties of the transformation matrix **B** can be expressed in the form

$$\begin{aligned}\sum_{\Omega} B_{\ell,\Omega}B_{\ell',\Omega} &= \delta_{\ell,\ell'},\\ \sum_{\ell} B_{\ell,\Omega}B_{\ell,\Omega'} &= \delta_{\Omega,\Omega'}.\end{aligned} \tag{D-13}$$

D.3 The Coupled Equations

Substituting eqn (D-12) into the expansion of the wavefunction eqn (3.17) we obtain

$$\begin{aligned}\psi(\boldsymbol{R},\boldsymbol{r},E) &= \left(\frac{k_{vj}\mu}{2\pi h^2}\right)^{\frac{1}{2}}\\ &\times \sum_{JM\ell}\{4\pi(2\ell+1)\}^{\frac{1}{2}} i^{\ell} \sum_{v'j'}\sum_{\Omega'}\sum_{\ell'} \frac{\phi^{Jvj\ell}_{v'j'\ell'}(R,E)}{k_{vj}R} B_{\ell',\Omega'}\\ &\times \chi_{v'j'}(r)\,\Xi^{JM}_{j'\Omega'}(\hat{\boldsymbol{R}},\hat{\boldsymbol{r}}')(j\ell JM|jm_j\ell 0)\end{aligned} \tag{D-14}$$

In order to fully transform the radial wavefunction to the body-fixed basis we need to change the ℓ superscript in $\phi^{J\nu j\ell}_{\nu' j' \ell'}(R,E)$ to Ω. This can be accomplished using eqn (D-13) as follows

$$\begin{aligned} i^{\ell}\phi^{J\nu j\ell}_{\nu' j' \ell'}(R,E) &= \sum_{\ell''} \delta_{\ell,\ell''} i^{\ell''} \phi^{J\nu j\ell''}_{\nu' j' \ell'}(R,E) \\ &= \sum_{\Omega} B_{\ell,\Omega} \left\{ \sum_{\ell''} B_{\ell'',\Omega}\, i^{\ell''} \phi^{J\nu j\ell''}_{\nu' j' \ell'}(R,E) \right\}. \end{aligned} \tag{D-15}$$

Inserting this into eqn (D-14) we obtain

$$\begin{aligned} \psi(\boldsymbol{R},\boldsymbol{r},E) = \left(\frac{k_{\nu j}\mu}{2\pi h^2}\right)^{\frac{1}{2}} &\sum_{JM\ell} \{4\pi(2\ell+1)\}^{\frac{1}{2}} \frac{1}{k_{\nu j}R} \\ &\times \sum_{\nu' j' \Omega'} \sum_{\Omega} B_{\ell,\Omega} \left\{ \sum_{\ell''\,\ell'} B_{\ell'',\Omega}\, i^{\ell''} \phi^{J\nu j\ell''}_{\nu' j' \ell'}(R,E) B_{\ell',\Omega'} \right\} \\ &\times \chi_{\nu' j'}(r)\, \Xi^{JM}_{j'\Omega'}(\hat{\boldsymbol{R}},\hat{\boldsymbol{r}}')(j\ell JM|jm_j\ell 0). \end{aligned} \tag{D-16}$$

The term in curly brackets in the last line of eqn (D-16) is the *body-fixed radial wavefunction*. We denote this wavefunction by $\Phi^{J\nu j\Omega}_{\nu' j' \Omega'}(R,E)$

$$\Phi^{J\nu j\Omega}_{\nu' j' \Omega'}(R,E) = \left\{ \sum_{\ell''\,\ell'} B_{\ell'',\Omega}\, i^{\ell''} \frac{\phi^{J\nu j\ell''}_{\nu' j' \ell'}(R,E)}{k_{\nu j}R} B_{\ell',\Omega'} \right\}. \tag{D-17}$$

Substituting eqn (D-17) into eqn (D-16) we obtain for the expansion of the full wavefunction

$$\begin{aligned} \psi(\boldsymbol{R},\boldsymbol{r},E) = \left(\frac{k_{\nu j}\mu}{2\pi h^2}\right)^{\frac{1}{2}} 2\pi^{\frac{1}{2}} &\sum_{JM} \sum_{\nu' j' \Omega'} \sum_{\Omega} \left\{ \sum_{\ell} (2\ell+1)^{\frac{1}{2}} (j\ell JM|jm_j\ell 0) B_{\ell,\Omega} \right\} \\ &\times \Phi^{J\nu j\Omega}_{\nu' j' \Omega'}(R,E)\chi_{\nu' j'}(r)\, \Xi^{JM}_{j'\Omega'}(\hat{\boldsymbol{R}},\hat{\boldsymbol{r}}'). \end{aligned} \tag{D-18}$$

The summation over M can be omitted as the Clebsch–Gordan coefficient $(j\ell JM|jm_j\ell 0)$ is zero unless $M = m_j$. Setting $M = m_j$ and using eqn 8.4.3(10) of ref. 1 or eqn (3.5.16) of ref. 2 we obtain for the summation over ℓ in the curly brackets

$$\begin{aligned} &\sum_{\ell} (2\ell+1)^{\frac{1}{2}} (j\ell JM|jm_j\ell 0) B_{\ell,\Omega} \\ &\quad = (2J+1)^{\frac{1}{2}} (-1)^{-m_j-\Omega} \sum_{\ell} (Jm_j j - m_j|Jj\ell 0)(J\Omega j - \Omega|Jj\ell 0) \\ &\quad = (2J+1)^{\frac{1}{2}} (-1)^{-m_j-\Omega} \delta_{m_j,\Omega} = (2J+1)^{\frac{1}{2}} \delta_{m_j,\Omega} \end{aligned} \tag{D-19}$$

where we have used eqn 8.1.1(8) of ref. 1 or eqn (3.5.5) of ref. 2 and we assume, here and elsewhere, that the angular momentum quantum numbers are all integer. Substituting eqn (D-19) into eqn (D-18) and omitting the summation over M we obtain

$$\psi(\boldsymbol{R},\boldsymbol{r},E) = \left(\frac{k_{vj}\mu}{2\pi h^2}\right)^{\frac{1}{2}} 2\pi^{\frac{1}{2}} \sum_J \sum_{v'j'\Omega'} (2J+1)^{\frac{1}{2}} \\ \Phi^{Jvjm_j}_{v'j'\Omega'}(R,E)\chi_{v'j'}(r)\,\Xi^{Jm_j}_{j'\Omega'}(\hat{\boldsymbol{R}},\hat{\boldsymbol{r}}'). \tag{D-20}$$

We see therefore that, in the case of the body-fixed formulation, the angular basis functions in which the full wavefunction should be expanded are

$$\Xi^{JM}_{j'\Omega'}(\hat{\boldsymbol{R}},\hat{\boldsymbol{r}}') = Y_{j',\Omega'}(\gamma,\phi'_r)\left(\frac{2J+1}{4\pi}\right)^{\frac{1}{2}} D^{J^*}_{M,\Omega'}(\phi_R,\theta_R,0).$$

The quantum number ℓ of the space-fixed formulation has been replaced by Ω in the body-fixed case. Expanding the wavefunction in this basis and substituting into the Schrödinger equation (eqn (3.1)) we obtain, as before, a set of coupled second order differential equations for the body-fixed radial wavefunctions $\Phi^{Jvj\Omega}_{v'j'\Omega'}(R,E)$[3,4]

$$\left[-\frac{\partial^2}{\partial R^2} - k^2_{v'j'} + \frac{(J(J+1)+j'(j'+1)-2\Omega'^2)}{R^2} + U^{J\Omega'}_{v'j';v'j'}(R)\right] \\ \times \Phi^{Jvj\Omega}_{v'j'\Omega'}(R,E) + C^{J+}_{\Omega',\Omega'+1}\,\Phi^{Jvj\Omega}_{v'j'\Omega'+1}(R,E) + C^{J-}_{\Omega',\Omega'-1}\,\Phi^{Jvj\Omega}_{v'j'\Omega'-1}(R,E) \\ + \sum_{v'',j''\neq v',j'} U^{J\Omega'}_{v'j';v''j''}(R)\Phi^{Jvj\Omega}_{v''j''\Omega'}(R,E) = 0. \tag{D-21}$$

where

$$C^{J\pm}_{\Omega,\Omega\pm1} = -\frac{[J(J+1)-\Omega(\Omega\pm1)]^{\frac{1}{2}}[j(j+1)-\Omega(\Omega\pm1)]^{\frac{1}{2}}}{R^2} \tag{D-22}$$

and

$$U^{J\Omega}_{vj;v'j'}(R) = \frac{2\mu}{\hbar^2}\left(\frac{2J+1}{4\pi}\right) \\ \times \langle\chi_{vj}(r)Y_{j,\Omega}(\hat{\boldsymbol{r}}')D^{J*}_{M,\Omega}(\hat{\boldsymbol{R}})|\,\{V(R,r,\gamma) \\ - V_{\text{diatom}}(r)\}\,|\chi_{v'j'}(r)Y_{j',\Omega}(\hat{\boldsymbol{r}}')D^{J*}_{M,\Omega}(\hat{\boldsymbol{R}})\rangle \tag{D-23}$$

where we have used the notation $(\hat{\boldsymbol{r}}' = \gamma,\phi'_r)$ and $(\hat{\boldsymbol{R}} = \phi_R,\theta_R,0)$, and as in the case of eqn (3.19) the angular brackets $\langle.|\,.\,|.\rangle$ imply integration over all four angles and over the diatomic radial coordinate r.

The potential is diagonal in the helicity quantum number Ω and therefore does not couple different values of Ω, however the $C^{J\pm}$ terms in eqn (D-21) do couple Ω' radial wavefunctions to those with $\Omega' \pm 1$.

D.4 Parity

A system remains unchanged when all coordinates are reflected or inverted through a point in space. The wavefunction remains unchanged or changes sign under such an inversion. The behavior of the wavefunction under this symmetry operation is termed the "parity" and is ± 1. In the space-fixed case the parity[5] is given by $(-1)^{J+j+\ell}$ so for a given value of the total angular momentum, J, all calculations can be broken up into two non-interacting sets, one for even and the other for odd parity depending on the value of $(-1)^{j+\ell}$.

In the body-fixed case the situation is more complicated, as the angular basis functions $Y_{j,\Omega}(\gamma,\phi_r')D^{J*}_{M,\Omega}(\phi_R,\theta_R,0)$ do not, on their own, posses a definite parity. If we denote the inversion operator by $\hat{\mathcal{I}}$, then under inversion the four angles transform as follows

$$\hat{\mathcal{I}}\phi_R = \pi + \phi_R, \quad \hat{\mathcal{I}}\theta_R = \pi - \theta_R, \quad \hat{\mathcal{I}}\gamma = \gamma, \quad \hat{\mathcal{I}}\phi_r' = \pi - \phi_r'. \tag{D-24}$$

Note that γ and ϕ_r' are expressed in body-fixed coordinates while θ_R and ϕ_R are in space-fixed coordinates, and hence the difference in behavior under inversion between ϕ_R and ϕ_r'.

The effect of inversion on the spherical harmonic $Y_{j,\Omega}(\gamma,\phi_r')$ is therefore (see eqn 5.1.5(11) and 5.4(11) of ref. 1)

$$\hat{\mathcal{I}}Y_{j,\Omega}(\gamma,\phi_r') = Y_{j,\Omega}(\gamma,\pi-\phi_r') = (-1)^{\Omega}Y_{j,\Omega}(\gamma,\phi_r'-\pi) = Y_{j,-\Omega}(\gamma,\phi_r'), \tag{D-25}$$

while on the Wigner $\mathbf{D}$ matrix it has the effect (see eqn 4.4(1) and 4.3(1) of ref. 1)

$$\hat{\mathcal{I}}D^{J*}_{M,\Omega}(\phi_R,\theta_R,0) = D^{J*}_{M,\Omega}(\pi+\phi_R,\pi-\theta_R,0) = (-1)^J D^{J*}_{M,-\Omega}(\phi_R,\theta_R,0). \tag{D-26}$$

Overall, the angular basis functions therefore behave as

$$\hat{\mathcal{I}}\,Y_{j,\Omega}(\gamma,\phi_r')D^{J*}_{M,\Omega}(\phi_R,\theta_R,0) = (-1)^J Y_{j,-\Omega}(\gamma,\phi_r')D^{J*}_{M,-\Omega}(\phi_R,\theta_R,0). \tag{D-27}$$

Armed with this knowledge, we can now form parity adapted angular basis functions $\Theta^{JMp}_{j\lambda}(\hat{\boldsymbol{r}},\hat{\boldsymbol{R}})$

$$\Theta^{JMp}_{j\lambda}(\hat{\boldsymbol{r}},\hat{\boldsymbol{R}}) = t_\lambda\left(\frac{2J+1}{4\pi}\right)^{\frac{1}{2}}\Big\{Y_{j,\lambda}(\gamma,\phi_r')D^{J*}_{M,\lambda}(\phi_R,\theta_R,0) + (-1)^{J+p}Y_{j,-\lambda}(\gamma,\phi_r')D^{J*}_{M,-\lambda}(\phi_R,\theta_R,0)\Big\} \tag{D-28}$$

where $\lambda = |\Omega|$ and is always positive, $p = 0$ or 1

$$\begin{aligned} p &= 0 \quad \text{even parity,} \\ p &= 1 \quad \text{odd parity} \end{aligned} \tag{D-29}$$

and

$$t_\lambda = \frac{1}{\sqrt{2(1+\delta_{\lambda,0})}}. \tag{D-30}$$

The parity adapted angular basis functions $\Theta_{j\lambda}^{JMp}(\hat{\boldsymbol{r}},\hat{\boldsymbol{R}})$ are orthogonal and normalized. They are said to form an orthonormal set.

The expansion of the wavefunction (eqn (D-20)) in terms of the parity adapted functions becomes

$$\psi(\boldsymbol{R},\boldsymbol{r},E) = \left(\frac{k_{vj}\mu}{2\pi h^2}\right)^{\frac{1}{2}} 2\pi^{\frac{1}{2}} \sum_{J} \sum_{v'j'} \sum_{\lambda'=s} (2J+1)^{\frac{1}{2}} \Phi_{v'j'\lambda'}^{Jvj\lambda}(R,E)\chi_{v'j'}(r)\frac{1}{t_{\lambda'}}\Theta_{j'\lambda'}^{Jm_jp}(\hat{\boldsymbol{r}},\hat{\boldsymbol{R}}) \tag{D-31}$$

where $s = \dfrac{(1-(-1)^{J+p})}{2}$. The use of s to define the lower limit of the summations over λ' takes account of the fact that when $(-1)^{J+p} = -1$ there is no parity adapted angular basis function (see eqn (D-28)) for $\lambda' = 0$.

When the parity adapted basis is used, the form of the coupled differential equations (eqn (D-21)) is still quite similar, but a few details are modified.[4] Thus the equations may now be written as

$$\left[-\frac{\partial^2}{\partial R^2} - k_{v'j'}^2 + \frac{(J(J+1)+j'(j'+1)-\lambda'^2)}{R^2} + U_{v'j';v'j'}^{Jp\lambda'}(R)\right]\Phi_{v'j'\lambda'}^{Jpvj\lambda}(R,E)$$
$$+ C_{\lambda',\lambda'+1}^{Jp+}\Phi_{v'j'\lambda'+1}^{Jpvj\lambda}(R,E) + C_{\lambda',\lambda'-1}^{Jp-}\Phi_{v'j'\lambda'-1}^{Jpvj\lambda}(R,E)$$
$$+ \sum_{v'',j''\neq v',j'} U_{v'j';v''j''}^{Jp\lambda'}(R)\Phi_{v''j''\lambda'}^{Jpvj\lambda}(R,E) = 0 \tag{D-32}$$

The superscript "p"s indicate that the parity adapted basis is being used. The potential matrix elements are now given by

$$U_{vj;v'j'}^{Jp\lambda}(R) = \frac{2\mu}{\hbar^2}\langle\chi_{vj}(r)\Theta_{j\lambda}^{JMp}(\hat{\boldsymbol{r}},\hat{\boldsymbol{R}})|\{V(R,r,\gamma)-V_{\text{diatom}}(r)\}|\chi_{v'j'}(r)\Theta_{j'\lambda}^{JMp}(\hat{\boldsymbol{r}},\hat{\boldsymbol{R}})\rangle. \tag{D-33}$$

Recalling that λ is always positve, the $C_{\lambda,\lambda\pm1}^{Jp+}$ are equal to the $C_{\Omega,\Omega\pm1}^{J\pm}$ of eqn (D-22) unless one of the subscripts is equal to zero. In this latter case we have

$$C_{0,1}^{Jp+} = \sqrt{2}\,C_{0,1}^{J+},\quad C_{1,0}^{Jp-} = \sqrt{2}\,C_{1,0}^{J-} \qquad \text{if } (-1)^{J+p} = +1, \tag{D-34}$$

$$C_{0,1}^{J+} = 0, \qquad C_{1,0}^{J-} = 0 \qquad \text{if } (-1)^{J+p} = -1. \tag{D-35}$$

Eqn (D-12) gives the relationship between the space-fixed and the body-fixed angular basis functions. In order to find the relationship between the parity adapted body-fixed angular basis functions and the space-fixed functions we need to establish the behavior of the transformation matrix elements $B_{\ell,\Omega}$ when the sign of Ω is reversed. From eqn (D-10) we have

$$\begin{aligned}B_{\ell,-\Omega} &= (-1)^{j+\Omega}(J-\Omega j\Omega|Jj\ell 0)\\ &= (-1)^{j+\Omega}(-1)^{J+j-\ell}(J\Omega j-\Omega|Jj\ell 0)\\ &= (-1)^{J+j+\ell}B_{\ell,\Omega}\\ &= (-1)^{p}B_{\ell,\Omega}\end{aligned} \tag{D-36}$$

where we have used eqn 8.4.3(11) of ref. 1 or eqn (3.5.17) of ref. 2.

Using eqn (D-36) in eqn (D-12) we obtain

$$\begin{aligned}\mathcal{Y}^{JM}_{j\ell}(\hat{\boldsymbol{R}},\hat{\boldsymbol{r}}) &= \sum_{\Omega} B_{\ell,\Omega}\,\Xi^{JM}_{j\Omega}(\hat{\boldsymbol{R}},\hat{\boldsymbol{r}}')\\ &= \sum_{\lambda=s}^{\min(J,j)}\left\{B_{\ell,\Omega}\,\Xi^{JM}_{j\Omega}(\hat{\boldsymbol{R}},\hat{\boldsymbol{r}}') + B_{\ell,-\Omega}\,\Xi^{JM}_{j-\Omega}(\hat{\boldsymbol{R}},\hat{\boldsymbol{r}}')\right\}\\ &= \sum_{\lambda=s}^{\min(J,j)} B_{\ell,\Omega}\left\{\Xi^{JM}_{j\Omega}(\hat{\boldsymbol{R}},\hat{\boldsymbol{r}}') + (-1)^{p}\,\Xi^{JM}_{j-\Omega}(\hat{\boldsymbol{R}},\hat{\boldsymbol{r}}')\right\}.\end{aligned} \tag{D-37}$$

Comparing eqn (D-37) with eqn (D-28) we see that the transformation will be a little different for even and odd values of J

$$\mathcal{Y}^{JM}_{j\ell}(\hat{\boldsymbol{R}},\hat{\boldsymbol{r}}) = \begin{cases}\displaystyle\sum_{\lambda=s}^{\min(J,j)} \frac{1}{t_\lambda}B_{\ell,\Omega}\Theta^{JMp}_{j\lambda}(\hat{\boldsymbol{r}},\hat{\boldsymbol{R}}) & \text{for even values of } J\\ \displaystyle\sum_{\lambda=s}^{\min(J,j)} \frac{1}{t_\lambda}B_{\ell,\Omega}\Theta^{JM(p+1)}_{j\lambda}(\hat{\boldsymbol{r}},\hat{\boldsymbol{R}}) & \text{for odd values of } J.\end{cases}$$

We can unite these two equations in the form

$$\mathcal{Y}^{JM}_{j\ell}(\hat{\boldsymbol{R}},\hat{\boldsymbol{r}}) = \sum_{\lambda=s}^{\min(J,j)} \frac{1}{t_\lambda}B_{\ell,\Omega}\Theta^{JM(p+J)}_{j\lambda}(\hat{\boldsymbol{r}},\hat{\boldsymbol{R}}) \tag{D-38}$$

where the angular function has even parity if the superscript $(p+J)$ is even and odd parity when it is odd.

D.5 Boundary Conditions, T Matrix and Cross Sections

Second order differential equations of the form of eqn (3.18) and (D-21) have two linearly independent solutions. We are always interested in just one of these, *i.e.* the regular solution which remains finite at the origin. The centrally important **S** and **T** matrices are obtained by examining the long range form of the computed radial wavefunctions and comparing it with the required asymptotic form or boundary condition at long range (eqn (3.20)). It is best to do the analysis of the long range form of the wavefunction in the space-fixed coordinate system. This is due to the presence of the helicity coupling terms of the type $C^{Jp+}_{\lambda',\lambda'+1}$ in eqn (D-32). These are generally much longer ranged than the potential coupling in the equivalent space-fixed coupled radial differential equations (eqn (3.18)). So the analysis of the boundary conditions and the extraction of the **T** and **S** is best done in the space-fixed coordinate representation. The wavefunction is easily converted from the body-fixed to the space-fixed representation using the inverse of eqn (D-17). On the other hand, the relationship between the **T** matrix and the cross sections, especially the differential cross section, is much simpler in the body-fixed representation.

In order to establish the form of the cross section in body-fixed coordinates we insert the asymptotic form of the space-fixed radial wavefunction (eqn (3.20)) into eqn (D-20), while also making use of eqn (D-17)

$$\begin{aligned}\psi(\boldsymbol{R},\boldsymbol{r},E)\widetilde{_{R\to\infty}}&\left(\frac{k_{vj}\mu}{2\pi h^2}\right)^{\frac{1}{2}}2\pi^{\frac{1}{2}}\sum_J\sum_{v'j'\Omega'}(2J+1)^{\frac{1}{2}}\frac{1}{k_{vj}R}\sum_{\ell''\,\ell'}B_{\ell'',m_j}\,i^{\ell''}\\&\times\Bigg[\hat{j}_{\ell'}(k_{v'j'}R)\delta_{v,v'}\delta_{j,j'}\delta_{\ell'',\ell'}-T^J_{v'j'\ell',vj\ell''}\\&\times\frac{1}{2i}\left(\frac{k_{vj}}{k_{v'j'}}\right)^{\frac{1}{2}}e^{i\left(k_{v'j'}R-\frac{\ell''\pi}{2}\right)}\Bigg]B_{\ell',\Omega'}\,\chi_{v'j'}(r)\,\Xi^{Jm_j}_{j'\Omega'}(\hat{\boldsymbol{R}},\hat{\boldsymbol{r}}').\end{aligned}\tag{D-39}$$

The first term in the square brackets, involving the spherical Riccati–Bessel function $\hat{j}_{\ell'}(k_{v'j'}R)$, yields a plane wave after all the summations (see eqn (3.23) and (3.24)). Also using eqn (D-11) to expand the body-fixed angular basis functions we obtain

$$\begin{aligned}\psi(\boldsymbol{R},\boldsymbol{r},E)\widetilde{_{R\to\infty}}&\left(\frac{k_{vj}\mu}{2\pi h^2}\right)^{\frac{1}{2}}\Bigg\{e^{ik_{vj}Z}\chi_{vj}(r)\,Y_{j,m_j}(\theta_r,\phi_r)-\frac{1}{2i}\sum_J\sum_{v'j'\Omega'}\\&\times(2J+1)\left[\sum_{\ell''\,\ell'}B_{\ell',\Omega'}\,T^J_{v'j'\ell',vj\ell''}B_{\ell'',m_j}\right]\left(\frac{1}{k_{vj}\,k_{v'j'}}\right)^{\frac{1}{2}}\\&\frac{e^{i(k_{v'j'}R)}}{R}\chi_{v'j'}(r)\,Y_{j',\Omega'}(\gamma,\phi'_r)D^{J*}_{m_j,\Omega'}(\phi_R,\theta_R,0).\Bigg\}\end{aligned}\tag{D-40}$$

The term in square brackets is the body-fixed **T** matrix with $\Omega=m_j$

$$T^J_{v'j'\Omega',vj(\Omega=m_j)}=\sum_{\ell''\,\ell'}B_{\ell',\Omega'}\,T^J_{v'j'\ell',vj\ell''}B_{\ell'',m_j}.\tag{D-41}$$

Comparing eqn (D-40) with eqn (3.8), using eqn (D-41) and eqn 4.3(1) of ref. 1, we can obtain an expression for the scattering amplitude

$$
\begin{aligned}
f_{v'j'\Omega' \leftarrow vjm_j}(\theta_R, \phi_R) \\
&= \frac{i}{2}\left(\frac{1}{k_{vj}\,k_{v'j'}}\right)^{\frac{1}{2}} \sum_J (2J+1) T^J_{v'j'\Omega',vj(\Omega=m_j)} D^{J*}_{m_j,\Omega'}(\phi_R, \theta_R, 0) \\
&= \frac{i}{2}\left(\frac{1}{k_{vj}\,k_{v'j'}}\right)^{\frac{1}{2}} \sum_J (2J+1) T^J_{v'j'\Omega',vj(\Omega=m_j)} \mathrm{e}^{im_j\phi_R} d^J_{m_j,\Omega'}(\theta_R)
\end{aligned}
\tag{D-42}
$$

where $d^J_{m_j,\Omega'}(\theta_R)$ is a Wigner (small) **d** matrix element.

Inserting this into eqn (3.9) we obtain an expression for the differential cross section

$$
\begin{aligned}
\sigma_{v'j'\Omega' \leftarrow vjm_j}(\theta_R, \phi_R, E) &= \frac{k_{v'j'}}{k_{vj}} \left| f_{v'j'\Omega' \leftarrow vjm_j}(\theta_R, \phi_R) \right|^2 \\
&= \frac{1}{4\,k^2_{vj}} \left| \sum_J (2J+1) T^J_{v'j'\Omega',vj(\Omega=m_j)} d^J_{m_j,\Omega'}(\theta_R) \right|^2
\end{aligned}
$$

or, summarizing

$$
\sigma_{v'j'\Omega' \leftarrow vjm_j}(\theta_R, \phi_R, E) = \frac{1}{4\,k^2_{vj}} \left| \sum_J (2J+1) T^J_{v'j'\Omega',vj(\Omega=m_j)} d^J_{m_j,\Omega'}(\theta_R) \right|^2 . \tag{D-43}
$$

An expression for the integral cross section may now be obtained by integrating over all angles θ_R and ϕ_R. The orthogonality condition for the small Wigner **d** matrices is (see eqn 4.10(6) of ref. 1)

$$
\int_0^{\pi} \mathrm{d}\theta_R\, d^J_{m_j,\Omega'}(\theta_R) d^{J'}_{m_j,\Omega'}(\theta_R) = \frac{2}{2J+1}\delta_{JJ'} . \tag{D-44}
$$

The integration over the azimuthal angle ϕ_R yields a factor of 2π and the integral cross section may therefore be written as

$$
\sigma_{v'j'\Omega' \leftarrow vjm_j}(E) = \frac{\pi}{k^2_{vj}} \sum_J (2J+1) \left| T^J_{v'j'\Omega',vjm_j} \right|^2 . \tag{D-45}
$$

References

1. D. A. Varshalovich, A. N. Moskalev and V. K. Khersonskii, *Quantum Theory of Angular Momentum*, World Scientific, Singapore, 1988.
2. A. R. Edmonds, *Angular Momentum in Quantum Mechanics*, Princeton University Press, Princeton, New Jersey, 1960.

3. G. C. Schatz and A. Kuppermann, Quantum mechanical reactive scattering for three-dimensional atom plus diatom systems. I. Theory, *J. Chem. Phys.*, 1976, **65**, 4642.
4. G. G. Balint-Kurti and M. Shapiro, Photofragmentation of triatomic molecules. Theory of angular and state distribution of product fragments, *Chem. Phys.*, 1981, **61**, 137.
5. W. H. Miller, Coupled equations and the minimum principle for collisions of an atom and a diatomic molecule, including rearrangements, *J. Chem. Phys.*, 1969, **50**, 407.

APPENDIX E

Integral Equations and Green's Function

In this appendix we outline the derivation of the integral equations of scattering theory and introduce the concept of a Green's function. The equations we are interested in solving take the general form of eqn (3.18). For the current discussion we will simplify this equation by omitting the vibrational quantum numbers v and also the total angular momentum quantum number J

$$\left[-\frac{\partial^2}{\partial R^2} + \frac{\ell'(\ell'+1)}{R^2} - k_{j'}^2 + U_{j'\ell';,j'\ell'}(R)\right]\phi_{j'\ell'}^{j\ell}(R,E)$$
$$= -\sum_{j'',\ell''\neq j',\ell'} U_{j'\ell';j''\ell''}(R)\phi_{j''\ell''}^{j\ell}(R,E). \tag{E-1}$$

We will first start with the simplest case where we place all the potential energy matrix elements on the right hand side of the equation, giving

$$\left[-\frac{\partial^2}{\partial R^2} + \frac{\ell'(\ell'+1)}{R^2} - k_{j'}^2\right]\phi_{j'\ell'}^{j\ell}(R,E) = -\sum_{j'',\ell''} U_{j'\ell';j''\ell''}(R)\phi_{j''\ell''}^{j\ell}(R,E). \tag{E-2}$$

Note that all the potential energy matrix elements go to zero at large separations.

To simplify the notation we will consider the equation

$$\left[-\frac{\partial^2}{\partial R^2} + \frac{\ell(\ell+1)}{R^2} - k^2\right]\phi(R) = F(R). \tag{E-3}$$

where $F(R)$ is equal to the right hand side of eqn (E-2).

RSC Theoretical and Computational Chemistry Series No. 7
Theory of Molecular Collisions
By Gabriel G. Balint-Kurti and Alexander P. Palov

Published by the Royal Society of Chemistry, www.rsc.org

The integral equations make use of the solutions of the "homogeneous" equation obtained by setting the right hand side to zero. The homogeneous equation is

$$\left[-\frac{\partial^2}{\partial R^2}+\frac{\ell(\ell+1)}{R^2}-k^2\right]u(R)=0 \tag{E-4}$$

Such an equation always has two linearly independent solutions, in this case namely the spherical Riccati–Bessel functions $\hat{j}_\ell(kR)$ and $\hat{n}_\ell(kR)$ (see, for instance, eqn (2.12)). The two solutions we will be interested in are:
1) the regular solution $u_1(R)=\hat{j}_\ell(kR)$ which is finite (*i.e.* zero) at the origin and has the asymptotic form

$$u_1(R)\ \widetilde{_{R\to\infty}}\ \sin\left(kR-\frac{\ell\pi}{2}\right) \tag{E-5}$$

and 2) the irregular solution $u_2(R)=-\frac{1}{k}(\hat{n}_\ell(kR)-i\hat{j}_\ell(kR))$ which becomes infinite at the origin and asymptotically has the form of a radially out-going spherical wave

$$u_2(R)\ \widetilde{_{R\to\infty}}\ \frac{1}{k}e^{i(kR-\frac{\ell\pi}{2})}. \tag{E-6}$$

From eqn (B-5) we see that the Wronskian of these two solutions is

$$W=u_1(R)\frac{du_2(R)}{dR}-\frac{du_1(R)}{dR}u_2(R)=-1. \tag{E-7}$$

Let us now return to the inhomogeneous equations, eqn (E-3). Consider the wavefunction

$$\phi(R)=u_1(R)\int_R^\infty u_2(R')F(R')\,dR'+u_2(R)\int_0^R u_1(R')F(R')\,dR'. \tag{E-8}$$

Using the notation $\phi'(R)=\dfrac{d\phi}{dR}$ we now differentiate $\phi(R)$ to obtain

$$\begin{aligned}\phi'(R)&=u_1'(R)\int_R^\infty u_2(R')F(R')\,dR'+u_2'(R)\int_0^R u_1(R')F(R')\,dR'\\&\quad-u_1(R)u_2(R)F(R)+u_2(R)u_1(R)F(R)\\&=u_1'(R)\int_R^\infty u_2(R')F(R')\,dR'+u_2'(R)\int_0^R u_1(R')F(R')\,dR'\end{aligned} \tag{E-9}$$

and differentiating this again we get

$$\begin{aligned}\phi''(R)&=u_1''(R)\int_R^\infty u_2(R')F(R')\,dR'+u_2''(R)\int_0^R u_1(R')F(R')\,dR'\\&\quad+\{u_1(R)u_2'(R)-u_1'(R)u_2(R)\}F(R)\\&=u_1''(R)\int_R^\infty u_2(R')F(R')\,dR'+u_2''(R)\int_0^R u_1(R')F(R')\,dR'-F(R).\end{aligned} \tag{E-10}$$

We now wish to prove that the function $\phi(R)$ defined in eqn (E-8) is a solution of the second order differential equation eqn (E-3). Substituting eqn (E-10) into eqn (E-3) we obtain

$$
\begin{aligned}
&- u_1''(R)\int_R^\infty u_2(R')F(R')\,\mathrm{d}R' - u_2''(R)\int_0^R u_1(R')F(R')\,\mathrm{d}R' \\
&+ F(R) + \left[\frac{\ell(\ell+1)}{R^2} - k^2\right]\phi(R) = F(R).
\end{aligned}
\tag{E-11}
$$

We can now cancel $F(R)$ on the two sides of the equation and replace both $u_1''(R)$ and $u_2''(R)$ using eqn (E-4)

$$
\begin{aligned}
&- \left[\frac{\ell(\ell+1)}{R^2} - k^2\right] u_1(R)\int_R^\infty u_2(R')F(R')\,\mathrm{d}R' \\
&- \left[\frac{\ell(\ell+1)}{R^2} - k^2\right] u_2(R)\int_0^R u_1(R')F(R')\,\mathrm{d}R' + \left[\frac{\ell(\ell+1)}{R^2} - k^2\right]\phi(R) = 0.
\end{aligned}
\tag{E-12}
$$

Now combining the first two terms in this equation we can write

$$
\begin{aligned}
&- \left[\frac{\ell(\ell+1)}{R^2} - k^2\right] \\
&\times \left\{u_1(R)\int_R^\infty u_2(R')F(R')\,\mathrm{d}R' + u_2(R)\int_0^R u_1(R')F(R')\,\mathrm{d}R'\right\} \\
&+ \left[\frac{\ell(\ell+1)}{R^2} - k^2\right]\phi(R) = 0.
\end{aligned}
\tag{E-13}
$$

From eqn (E-8) we see that the term in curly brackets is equal to $\phi(R)$ and that consequently all the terms on the left hand side of the equation cancel. This proves that the function $\phi(R)$ defined in eqn (E-8) is a solution of the differential equation in eqn (E-3). Because of the choice of the function $u_2(R)$, the wavefunction has the required asymptotic form, as discussed in Chapter 4 (see eqn (4.30)).

We may rewrite the solution, eqn (E-8), in terms of a Green's function

$$
\phi(R) = \int_0^\infty G(R,R')F(R')\,\mathrm{d}R' \tag{E-14}
$$

where

$$
G(R,R') = \begin{cases} u_1(R)u_2(R') & \text{for } R < R' \\ u_2(R)u_1(R') & \text{for } R > R'. \end{cases} \tag{E-15}
$$

The solutions in eqn (E-8) and (E-14) are "particular" solutions to eqn (E-3) ref. 1. In order to satisfy any particular boundary condition we may add, on the right hand side, any solution of the homogeneous equation (E-4). Consider the situation in which there is no potential, *i.e.* $F(R)$ in eqn (E-3) is zero. In this case the equation becomes the homogeneous equation and the particular solution of eqn (E-14) (or eqn (E-8)) is identically zero. This would clearly be incorrect. We need, therefore, to add the appropriate solution of the homogeneous equation to the right hand side of eqn (E-14) in order to obtain the general solution to the problem with the correct boundary conditions. The general solution therefore becomes

$$\phi(R) = u_1(R) + \int_0^\infty G(R,R')F(R')\,\mathrm{d}R' \tag{E-16}$$

Reverting now to the more cumbersome notation of eqn (E-2), it is clear that the Green's function as well as the the solutions to the homogeneous equations need to carry the labels j, ℓ and E. In terms of this Green's function the solution to eqn (E-2) can be written as

$$\begin{aligned}\phi^{j\ell}_{j'\ell'}(R,E) = u^1_{j\ell}(R,E)\delta_{j,j'}\delta_{\ell,\ell'} - \int_0^\infty G_{j'\ell'}(R,R';E)\\ \times \sum_{j'',\ell''} U_{j'\ell';j''\ell''}(R')\phi^{j\ell}_{j''\ell''}(R',E)\,\mathrm{d}R'.\end{aligned} \tag{E-17}$$

It is important to note that $G_{j'\ell'}(R,R';E)$ rather than $G_{j\ell}(R,R';E)$ appears in eqn (E-17). This is needed so as to obtain the correct boundary conditions. $G_{j\ell}(R,R';E)$ is defined as

$$G_{j\ell}(R,R';E) = \begin{cases} u^1_{j\ell}(R,E)u^2_{j\ell}(R',E) & \text{for } R < R' \\ u^2_{j\ell}(R,E)u^1_{j\ell}(R',E) & \text{for } R > R' \end{cases} \tag{E-18}$$

and $u^1_{j\ell}(R,E)$ and $u^2_{j\ell}(R,E)$ are solutions of the homogeneous equation derived by setting the right hand side of eqn (E-2) to zero having the asymptotic behavior

$$\begin{aligned} u^1_{j\ell}(R,E) &\widetilde{\scriptstyle R\to\infty} \sin\left(k_jR - \frac{\ell\pi}{2}\right)\\ u^2_{j\ell}(R,E) &\widetilde{\scriptstyle R\to\infty} \frac{1}{k_j}\mathrm{e}^{i(k_jR-\frac{\ell\pi}{2})}.\end{aligned} \tag{E-19}$$

Eqn (E-17) is an integral equation for the scattering wavefunction $\phi^{j\ell}_{j'\ell'}(R,E)$.

E.1 The Distorted Wave Green's Function

The analysis above has been based on eqn (E-2). We could equally well have based the discussion on eqn (E-1) and the corresponding homogeneous equation obtained by setting the right hand side of the equation to zero. The main difference is that the solutions of the homogeneous equation are no longer analytic but must rather be obtained numerically. In this case eqn (E-17) and (E-18) still apply but $u^1_{j\ell}(R,E)$ and $u^2_{j\ell}(R,E)$ are now solutions of the homogeneous equation derived by setting the right hand side of eqn (E-1) to zero and having the asymptotic forms

$$u^1_{j\ell}(R,E)\widetilde{_{R\to\infty}}\sin\left(k_jR-\frac{\ell\pi}{2}+\eta_{j\ell}\right)$$
$$u^2_{j\ell}(R,E)\widetilde{_{R\to\infty}}\frac{1}{k_j}\mathrm{e}^{i\left(k_jR-\frac{\ell\pi}{2}+\eta_{j\ell}\right)} \tag{E-20}$$

where the phase shift $\eta_{j\ell}$ appears because of the presence of a potential in the left hand side of eqn (E-1).

Reference

1. A. Jeffrey, *Mathematics for Engineers and Scientists*, Van Nostrand Reinhold (UK), 3rd edn, 1985.

APPENDIX F

Semiclassical or JWKB Approximation

F.1 Semiclassical Treatment of One-Dimensional Schrödinger Equation

For heavy collision partners when the de Broglie wavelength of the colliding particles is small compared to the dimensions or range of the interaction potential a semiclassical approach, commonly known as the JWKB (Jeffreys–Wentzel–Kramers–Brillouin) approximations, will generally be applicable.[1–4] In this appendix we will outline the basic concepts of the approach and derive the formulae needed to compute the phase shift in a simple case.

We will first start with a one-dimensional problem and will then extend the approach to encompass the one-dimensional radial Schrödinger equation. The one-dimensional Schrödinger equation is

$$-\frac{\hbar^2}{2\mu}\frac{\mathrm{d}^2}{\mathrm{d}x^2}\psi(x) + V(x)\psi(x) = E\psi(x) \tag{F-1}$$

or

$$\left[\frac{\hbar^2}{2\mu}\frac{\mathrm{d}^2}{\mathrm{d}x^2} + (E - V(x))\right]\psi(x) = 0. \tag{F-2}$$

We first consider a region of the x such that $E > V(x)$, *i.e.* the classically allowed region of space.

RSC Theoretical and Computational Chemistry Series No. 7
Theory of Molecular Collisions
By Gabriel G. Balint-Kurti and Alexander P. Palov

Published by the Royal Society of Chemistry, www.rsc.org

Let us write the solution of eqn (F-2) as $\psi(x) = e^{iS(x)/\hbar}$, where $S(x)$ will be expanded in a series in powers of $\hbar$, *i.e.*, $S(x) = S_0(x) + \hbar S_1(x) + \cdots$. Substituting this into eqn (F-2) we obtain

$$-\frac{1}{2\mu}\left[S_0'^2(x) + 2\hbar S_0'(x)S_1'(x) + \hbar^2 S_1'^2(x)\right] + \frac{i\hbar}{2\mu}\left[S_0''(x) + \hbar S_1''(x)\right] + (E - V(x)) = 0. \tag{F-3}$$

Separating powers of $\hbar$ we obtain for the zeroth and first powers in $\hbar$

$$-S_0'^2(x) + 2\mu(E - V(x)) = 0 \tag{F-4}$$

$$-2S_0'(x)S_1'(x) + iS_0''(x) = 0. \tag{F-5}$$

Let us first "solve" eqn (F-4). The local value of the momentum is $p(x) = \sqrt{2\mu(E - V(x))}$. We define the local wavevector $q(x) = p(x)/\hbar$, so the general solution to eqn (F-4) is

$$S_0(x) = \pm\hbar \int q(x)\,dx. \tag{F-6}$$

Let us now look at eqn (F-5) for the solution for the next term, $S_1(x)$.

$$\begin{aligned} S_1'(x) &= \frac{i}{2}\frac{1}{S_0'(x)}S_0''(x) \\ &= \frac{i}{2}\frac{1}{q(x)}q'(x) \\ &= \frac{i}{2}\frac{d\ln q(x)}{dx} \\ &= i\frac{d\ln\sqrt{q(x)}}{dx} \end{aligned} \tag{F-7}$$

and therefore

$$S_1(x) = i\ln\sqrt{q(x)} + \text{ (an arbitrary constant)}. \tag{F-8}$$

Using eqn (F-6) and (F-7) in the exponential form of the wavefunction, we can write

$$\psi(x) = A\frac{1}{\sqrt{q(x)}}e^{i\int q(x)\,dx} + B\frac{1}{\sqrt{q(x)}}e^{-i\int q(x)\,dx}. \tag{F-9}$$

In the classically forbidden region the general solution is

$$\psi(x) = A\frac{1}{\sqrt{|q(x)|}}e^{-\int|q(x)|\,dx} + B\frac{1}{\sqrt{|q(x)|}}e^{\int|q(x)|\,dx}. \tag{F-10}$$

In order to obtain eqn (F-4) from eqn (F-3) we needed to neglect the term $\frac{i\hbar}{2\mu}S_0''(x)$ in eqn (F-3). This neglect is valid if $\hbar\,|S_0''(x)/S_0'^2(x)| << 1$, and this provides us with various validity conditions for the semiclassical approximation. Using eqn (F-6) we obtain a criterion for the validity of the semiclassical approximation. Below we express this criterion in several different, but equivalent, forms

$$\left|\frac{\mathrm{d}}{\mathrm{d}x}\left(\frac{\hbar}{S_0'(x)}\right)\right| = \left|\frac{\mathrm{d}}{\mathrm{d}x}\left(\frac{1}{q(x)}\right)\right| = \left|\frac{\mathrm{d}}{\mathrm{d}x}\left(\frac{\lambda_{\mathrm{DB}}}{2\pi}\right)\right|$$
$$= \left|\frac{\hbar}{p^2}\frac{\mathrm{d}p(x)}{\mathrm{d}x}\right| = \left|\frac{\hbar\mu}{p^3}\frac{\mathrm{d}V(x)}{\mathrm{d}x}\right| = \left|\frac{\mu\lambda_{\mathrm{DB}}^3}{8\hbar^2\pi^3}\frac{\mathrm{d}V(x)}{\mathrm{d}x}\right| << 1 \tag{F-11}$$

where the de Broglie wavelength is $\lambda_{\mathrm{DB}} = \frac{h}{p}$.

Thus the validity condition for the semiclassical approximation can be stated in several ways. One way of stating it is that the change in the de Broglie wavelength λ_{DB} must be small over a distance equal to itself.[1] Another way of stating it is that the change in the potential over a wavelength must be very small. The neglect of the third term (S_2) in the expansion of $S(x)$ in powers of $\hbar$ leads to the additional condition that $\frac{\mathrm{d}V}{\mathrm{d}x}$ must tend to zero as x increases.[1]

It can be seen from eqn (F-11) that the approximation breaks down in the vicinity of a classical turning point x_0 where $p(x_0) = \sqrt{2\mu(E - V(x_0))} = 0$. So a problem we encounter is how to connect the solution in the classically forbidden region with that in the classically allowed region. The best way is to actually solve the Schrödinger equation in the vicinity of the classical turning point.

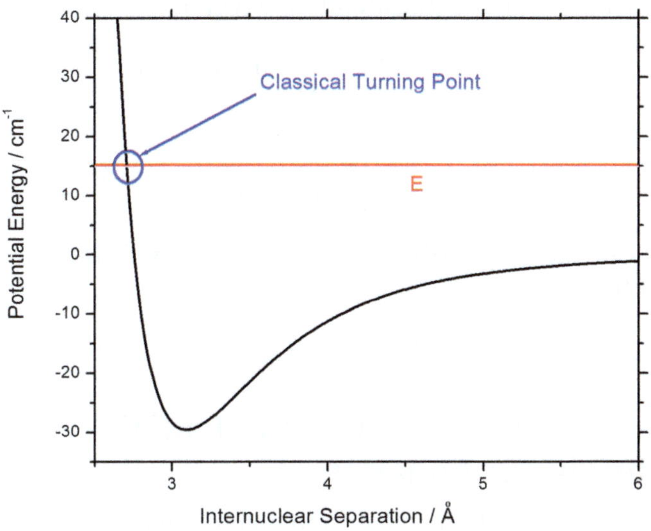

Figure F.1 Prototypical atom–atom interaction potential showing the classical turning point.

From Figure F.1 we see that a natural way of solving the Schrödinger equation in the region of the classical turning point would be to approximate the potential as a linear function in this region. The linearly independent solutions in this case are the regular ($Ai(\zeta)$) and irregular ($Bi(\zeta)$) Airy functions[5] where the argument of the functions is $\zeta = \frac{2}{3}a(x - x_0)^{\frac{3}{2}}$ and $q(x) = a\sqrt{(x - x_0)}$ in the close vicinity of the classical turning point.[6] We will not be interested in the irregular Airy function because this goes to infinity as we proceed further to the left into the classically forbidden region.

A uniform approximation in which the model linear potential problem, whose solution is the regular Airy function, is mapped onto to the full problem is discussed by Berry and Mount.[3] For a single turning point problem as illustrated in Figure F.1 the solution may be written in the form

$$\psi(x) = \left[\frac{\zeta(x)}{q^2(x)}\right]^{\frac{1}{4}} Ai(\zeta(x)) \tag{F-12}$$

where

$$\zeta(x) = \left[\frac{3}{2}\int_{x_0}^{x} |q(x')|\mathrm{d}x'\right]^{\frac{2}{3}} \tag{F-13}$$

and

$$q(x) = \frac{\sqrt{2\mu(E - V(x))}}{\hbar}. \tag{F-14}$$

This wavefunction passes smoothly through the classical turning point region. It has the correct regular behavior in the classical forbidden region and at large x it goes over to the required sinusoidal behavior.

As x increases into the classically allowed region,

$$\psi(x) \underset{x \to \infty}{\approx} A\frac{1}{\sqrt{q(x)}} \sin\left(\omega + \frac{\pi}{4}\right) \tag{F-15}$$

where

$$\omega = \int_{x_0}^{x} |q(x')|\mathrm{d}x'. \tag{F-16}$$

As x goes to the left, further into the classically forbidden region, the wavefunction becomes

$$\psi(x) \underset{x \to -\infty}{\approx} A\frac{1}{2\sqrt{q(x)}}\mathrm{e}^{-|\omega|}. \tag{F-17}$$

F.2 Semiclassical Treatment of Radial Schrödinger Equation

The use of the Airy function equation to obtain the connection formula for the semiclassical wavefunction assumes that the range of the argument, x, runs from $-\infty$ to $+\infty$. In contrast, the argument r in the radial Schrödinger equation has a range restricted to between 0 and $+\infty$. In this context an important obstaclc also arises from the appearance of the centrifugal potential, $\frac{\ell(\ell+1)\hbar^2}{2\mu r^2}$, which goes to infinity at $r = 0$. For instance, according to Berry and Mount[3] the JWKB approximation for this case does not predict the vanishing of the phase shift when the potential is zero.

In order to obtain correct values of the phase shifts in these circumstances Langer[3,7] suggested the substitution $\ell(\ell+1) \to \left(\ell + \frac{1}{2}\right)^2$. As a consequence $q(x)$ in eqn (F-13) is replaced by

$$q(r) = \frac{\sqrt{2\mu(E - V(r)) - \frac{\hbar^2\left(\ell + \frac{1}{2}\right)^2}{r^2}}}{\hbar} \tag{F-18}$$

and the asymptotic form of the semiclassical radial wavefunction becomes

$$\psi(r) \underset{r\to\infty}{\approx} A \frac{1}{\sqrt{q(r)}} \sin\left(\omega + \frac{\pi}{4}\right) \tag{F-19}$$

where

$$\omega = \frac{1}{\hbar} \int_{r_0}^{r} \sqrt{2\mu(E - V(r')) - \frac{\hbar^2\left(\ell + \frac{1}{2}\right)^2}{r'^2}}\, \mathrm{d}r'. \tag{F-20}$$

The semiclassical approximation to the phase shift is given by comparing the phase of the wavefunction in eqn (F-19) with that of a similar wavefunction in the absence of a potential

$$\eta_\ell = \frac{1}{\hbar} \lim_{r\to\infty} \left\{ \int_{r_0}^{r} \sqrt{2\mu(E - V(r')) - \frac{\hbar^2\left(\ell + \frac{1}{2}\right)^2}{r'^2}}\, \mathrm{d}r' - \int_{r_{f0}}^{r} \sqrt{2\mu E - \frac{\hbar^2\left(\ell + \frac{1}{2}\right)^2}{r'^2}}\, \mathrm{d}r' \right\} \tag{F-21}$$

where r_{f0} is the classical turning point for the modified centrifugal potential $\frac{\hbar^2\left(\ell + \frac{1}{2}\right)^2}{r'^2}$. By construction we see that in the absence of a potential the phase shift is zero, as it should be.

References

1. L. D. Landau and E. M. Lifshitz, *Quantum Mechanics: Non-Relativistic Theory*, Butterworth-Heinemann, Oxford, 3rd edn, 1977 (Reprinted with corrections 1991.)
2. M. S. Child, *Molecular Collision Theory*, Academic Press, 1974. Corrected Dove Publication 1996.

3. M. V. Berry and K. E. Mount, Semiclassical approximations in wave mechanics, *Rep. Prog. Phys.*, 1972, **35**, 315.
4. M. S. Child, *Semiclassical Mechanics with Molecular Applications*, Oxford University Press, 1991.
5. M. Abromowitz and I. A. Stegun, *Handbook on Mathematical Functions, Applied Mathematics Series 55*, National Bureau of Standards, 10th edn, 1972.
6. P. M. Morse and H. Feschbach, *Methods of Theoretical Physics*, McGraw-Hill Book Co. Inc., New York, 1st edn, 1953. See Section 9.3.
7. R. E. Langer, On the connection formulas and the solutions of the wave equation, *Phys. Rev.*, 1937, **51**, 669.

APPENDIX G

Hyperspherical Coordinates and the Schrödinger Equation

Quantum reactive scattering calculations are often performed using hyperspherical coordinates.[1–6] These coordinates are preferred to the Jacobi coordinates introduced in Appendix A and used in Chapter 3 as they attempt to treat reactants and products on a more equal basis. There are several different forms of hyperspherical coordinates in the literature. In this appendix we will discuss two specific forms of these coordinates and for consistency we will base our definitions on those of Pack and Parker.[1] All the different hyperspherical coordinate systems start by defining mass-scaled Jacobi coordinates. The use of these coordinates already introduces some simplifications into the quantum scattering equations.

G.1 Mass-Scaled Jacobi Coordinates

We consider here reactions of the type A + BC → AB + C. The Jacobi coordinates for a triatomic system of this type depend on which arrangement "channel" we are considering (see Appendix A). We will denote the three possible arrangement channels by the symbols α (A + BC), β (B + CA) and γ (C + AB).

The Hamiltonian for an atom–diatom system in the center-of-mass reference frame can be written in the form (see eqn (3.1) to (3.3))

$$\begin{aligned}\hat{H} &= -\frac{\hbar^2}{2\mu_{\boldsymbol{R}_\alpha}}\nabla^2_{\boldsymbol{R}_\alpha} - \frac{\hbar^2}{2\mu_{\boldsymbol{r}_\alpha}}\nabla^2_{\boldsymbol{r}_\alpha} + V(R_\alpha, r_\alpha, \gamma_\alpha)\\ &= -\frac{\hbar^2}{2\mu_{R_\alpha}}\left\{\frac{1}{R_\alpha^2}\frac{\partial}{\partial R_\alpha}\left(R_\alpha^2\frac{\partial}{\partial R_\alpha}\right)\right\} - \frac{\hat{\boldsymbol{\ell}}_\alpha^2}{2\mu_{R_\alpha}R_\alpha^2}\\ &\quad - \frac{\hbar^2}{2\mu_{r_\alpha}}\left\{\frac{1}{r_\alpha^2}\frac{\partial}{\partial r_\alpha}\left(r_\alpha^2\frac{\partial}{\partial r_\alpha}\right)\right\} - \frac{\hat{\boldsymbol{j}}_\alpha^2}{2\mu_{r_\alpha}r_\alpha^2} + V(R_\alpha, r_\alpha, \gamma_\alpha)\end{aligned} \tag{G-1}$$

RSC Theoretical and Computational Chemistry Series No. 7
Theory of Molecular Collisions
By Gabriel G. Balint-Kurti and Alexander P. Palov

Published by the Royal Society of Chemistry, www.rsc.org

where the subscript α denotes the arrangement channel, μ_{R_α} is the reduced mass of atom A relative to the BC diatom, μ_{r_α} is the reduced mass of the BC diatom, $\boldsymbol{r}_\alpha$ is the position vector of atom C relative to atom B and $\boldsymbol{R}_\alpha$ is the position vector of atom A relative to the center-of-mass of the diatom BC.[1,6,7]

The Jacobi angle γ_α is defined as the angle between these two vectors

$$\cos\gamma_\alpha = \frac{\boldsymbol{R}_\alpha \cdot \boldsymbol{r}_\alpha}{R_\alpha r_\alpha}. \tag{G-2}$$

The orbital angular momentum of atom A relative to the BC diatomic is denoted by $\hat{\boldsymbol{\ell}}_\alpha$ and the rotational angular momentum of the diatomic by $\hat{\boldsymbol{j}}_\alpha$. Clearly similar equations can be written for the other two arrangement channels.

We denote the mass-scaled Jacobi coordinates for the arrangement channel α by $\tilde{\boldsymbol{R}}_\alpha$ and $\tilde{\boldsymbol{r}}_\alpha$. They are related to the unscaled coordinates by a simple scaling factor d_α[8–17]

$$\begin{aligned} \tilde{\boldsymbol{R}}_\alpha &= d_\alpha \boldsymbol{R}_\alpha, \\ \tilde{\boldsymbol{r}}_\alpha &= d_\alpha^{-1} \boldsymbol{r}_\alpha \end{aligned} \tag{G-3}$$

where the dimensionless scaling factor d_α is

$$d_\alpha = \left[\frac{m_\mathrm{A}}{\mu}\left(1 - \frac{m_\mathrm{A}}{M}\right)\right]^{1/2}, \tag{G-4}$$

$M = m_\mathrm{A} + m_\mathrm{B} + m_\mathrm{C}$ is the total mass of the triatomic system and the three-body reduced mass, μ, is

$$\mu = \left(\frac{m_\mathrm{A} m_\mathrm{B} m_\mathrm{C}}{M}\right)^{1/2}. \tag{G-5}$$

Note that μ is the same for all of the arrangement channels.

Substituting eqn (G-3) in eqn (G-1) and using eqn (G-4) and (G-5) we obtain the Hamiltonian in terms of mass-scaled Jacobi coordinates

$$\begin{aligned} \hat{H} &= -\frac{\hbar^2}{2\mu}\left\{\nabla^2_{\tilde{\boldsymbol{R}}_\alpha} + \nabla^2_{\tilde{\boldsymbol{r}}_\alpha}\right\} + V(\tilde{R}_\alpha, \tilde{r}_\alpha, \gamma_\alpha) \\ &= -\frac{\hbar^2}{2\mu}\left\{\frac{1}{\tilde{R}_\alpha^2}\frac{\partial}{\partial \tilde{R}_\alpha}\left(\tilde{R}_\alpha^2 \frac{\partial}{\partial \tilde{R}_\alpha}\right)\right\} - \frac{\hat{\boldsymbol{\ell}}_\alpha^2}{2\mu \tilde{R}_\alpha^2} - \frac{\hbar^2}{2\mu}\left\{\frac{1}{r_\alpha^2}\frac{\partial}{\partial \tilde{r}_\alpha}\left(\tilde{r}_\alpha^2 \frac{\partial}{\partial \tilde{r}_\alpha}\right)\right\} \\ &\quad - \frac{\hat{\boldsymbol{j}}_\alpha^2}{2\mu \tilde{r}_\alpha^2} + V(\tilde{R}_\alpha, \tilde{r}_\alpha, \gamma_\alpha). \end{aligned} \tag{G-6}$$

The Hamiltonian now contains only one reduced mass. It can be viewed as describing the motion of a single particle in a six-dimensional space and is similar in this sense to the Hamiltonian for atom–atom scattering.

Another advantage of introducing mass-scaled Jacobi coordinates is that the transformation between different arrangement channels may now be described by a "kinematic rotation".[6,8–14,16,17] If we define a six component vector using the space-fixed scaled Jacobi vectors $\tilde{\boldsymbol{R}}_\alpha$ and $\tilde{\boldsymbol{r}}_\alpha$

$$\begin{pmatrix} \tilde{\boldsymbol{R}}_\alpha \\ \tilde{\boldsymbol{r}}_\alpha \end{pmatrix} = \begin{pmatrix} \tilde{X}_\alpha \\ \tilde{Y}_\alpha \\ \tilde{Z}_\alpha \\ \tilde{x}_\alpha \\ \tilde{y}_\alpha \\ \tilde{z}_\alpha \end{pmatrix} \tag{G-7}$$

then transformation from, for instance, arrangement channel α to γ can be described by operating with a unitary transformation matrix $\mathbf{T}(\chi_{\gamma\alpha})$

$$\begin{pmatrix} \tilde{\boldsymbol{R}}_\gamma \\ \tilde{\boldsymbol{r}}_\gamma \end{pmatrix} = \mathbf{T}(\chi_{\gamma\alpha}) \begin{pmatrix} \tilde{\boldsymbol{R}}_\alpha \\ \tilde{\boldsymbol{r}}_\alpha \end{pmatrix} \tag{G-8}$$

where $\mathbf{T}(\chi_{\gamma\alpha})$ is a matrix

$$\mathbf{T}(\chi_{\gamma\alpha}) = \begin{pmatrix} \cos\chi_{\gamma\alpha} & 0 & 0 & \sin\chi_{\gamma\alpha} & 0 & 0 \\ 0 & \cos\chi_{\gamma\alpha} & 0 & 0 & \sin\chi_{\gamma\alpha} & 0 \\ 0 & 0 & \cos\chi_{\gamma\alpha} & 0 & 0 & \sin\chi_{\gamma\alpha} \\ -\sin\chi_{\gamma\alpha} & 0 & 0 & \cos\chi_{\gamma\alpha} & 0 & 0 \\ 0 & -\sin\chi_{\gamma\alpha} & 0 & 0 & \cos\chi_{\gamma\alpha} & 0 \\ 0 & 0 & -\sin\chi_{\gamma\alpha} & 0 & 0 & \cos\chi_{\gamma\alpha} \end{pmatrix} \tag{G-9}$$

and the obtuse kinematic angle $\chi_{\gamma\alpha}$ is defined by the equations

$$\begin{aligned} \cos\chi_{\gamma\alpha} &= -\frac{\mu}{d_\gamma d_\alpha m_\mathrm{B}} \\ \sin\chi_{\gamma\alpha} &= -\frac{1}{d_\gamma d_\alpha}. \end{aligned} \tag{G-10}$$

Note also the relationships $\chi_{\alpha\alpha} = 0$, $\chi_{\gamma\alpha} = -\chi_{\alpha\gamma}$ and $\chi_{\alpha\beta} + \chi_{\beta\gamma} + \chi_{\gamma\alpha} = 2\pi$.[1]

G.2 Body-Fixed Frame and Arrangement Channel Hyperspherical Coordinates

We can now rotate the frame of reference to a body-fixed frame in which the Z axis points along the direction $\tilde{\boldsymbol{R}}_\alpha$ and the triatomic molecules lies in the ZX plane. The formalism for this change of frame is discussed fully in Appendix D (see also Refs 1,7). The wavefunction for the overall tumbling motion of the system then becomes the Wigner $\mathbf{D}$ matrix $D^{J^*}_{M,\Omega}(\phi_{R_\alpha}, \theta_{R_\alpha}, \phi'_{r_\alpha})$ (see eqn D-7), where the Euler angles $\phi_{R_\alpha}, \theta_{R_\alpha}, \phi'_{r_\alpha}$ are those required to rotate the axes so that the new body-fixed ZX axes are in the molecular plane with the Z axis pointed along the $\tilde{\boldsymbol{R}}_\alpha$ direction. Note that we have now performed an extra rotation of the axis system so as to bring the X axis of the body-fixed system into the molecular plane.[1,7] This effectively moves the body-fixed

ϕ'_{r_α} variable associated with the rotation of the diatomic from the spherical harmonic to the Wigner **D** function, but has no effect on the complete wavefunction of the system. We see therefore from eqn (D-11) in Appendix D that the body-fixed angular basis is

$$\Xi^{JM}_{j\Omega}(\hat{\tilde{\boldsymbol{R}}}_\alpha, \hat{\tilde{\boldsymbol{r}}}'_\alpha) = Y_{j,\Omega}(\gamma, 0)\left(\frac{2J+1}{4\pi}\right)^{\frac{1}{2}} D^{J^*}_{M,\Omega}(\phi_{R_\alpha}, \theta_{R_\alpha}, \phi'_{r_\alpha}). \qquad \text{(G-11)}$$

Besides the four angles in this angular basis, there are still two radial variables, R_α and r_α. We now replace these† with a "hyperradius" ρ and a further angle α_α. The new and old coordinates are related to each other by the relationships

$$\begin{aligned}\tilde{r}_\alpha &= \rho \sin \alpha_\alpha,\\ \tilde{R}_\alpha &= \rho \cos \alpha_\alpha\end{aligned} \qquad \text{(G-12)}$$

where $0 \leq \rho \leq \infty$ and $0 \leq \alpha_\alpha \leq \pi/2$.

The variables ρ and α_α are called Delves coordinates. The hyperradius, ρ, does not depend on the arrangement channel and therefore we have not placed a subscript α on it. In contrast, the hyperangle α_α does depend on the arrangement channel. The inverse relationship to eqn (G-12) is

$$\begin{aligned}\rho &= (\tilde{R}^2_\alpha + \tilde{r}^2_\alpha)^{1/2}\\ \alpha_\alpha &= \tan^{-1}\left(\frac{\tilde{r}_\alpha}{\tilde{R}_\alpha}\right).\end{aligned} \qquad \text{(G-13)}$$

The volume element of the six-dimensional space expressed in hyperspherical Delves body-fixed coordinates can be written, with the help of three Euler angles ϕ_{R_α} and θ_{R_α} and ϕ'_{r_α}, as

$$\mathrm{d}\tilde{\boldsymbol{r}}_\alpha \mathrm{d}\tilde{\boldsymbol{R}}_\alpha = \rho^5 \mathrm{d}\rho \frac{1}{4} \sin^2 2\alpha_\alpha \mathrm{d}\alpha_\alpha \mathrm{d}\phi_{R_\alpha} \sin\theta_{R_\alpha} \mathrm{d}\theta_{R_\alpha} \mathrm{d}\phi'_\alpha \sin\gamma_\alpha \mathrm{d}\gamma_\alpha. \qquad \text{(G-14)}$$

In terms of these new coordinates the Hamiltonian becomes

$$\hat{H} = -\frac{\hbar^2}{2\mu\rho^5}\frac{\partial}{\partial\rho}\rho^5\frac{\partial}{\partial\rho} + \frac{\Delta^2_\alpha}{2\mu\rho^2} + V(\rho, \alpha_\alpha, \gamma_\alpha) \qquad \text{(G-15)}$$

where Δ^2_α is the square of the "grand angular momentum operator"[1,11,12]

$$\Delta^2_\alpha = -\frac{\hbar^2}{\sin^2 2\alpha_\alpha}\frac{\partial}{\partial\alpha_\alpha}\sin^2 2\alpha_\alpha\frac{\partial}{\partial\alpha_\alpha} + \frac{\hat{\boldsymbol{\ell}}_\alpha^{\,2}}{\cos^2\alpha_\alpha} + \frac{\hat{\boldsymbol{j}}_\alpha^{\,2}}{\sin^2\alpha_\alpha}. \qquad \text{(G-16)}$$

The orbital angular momentum $\boldsymbol{\ell}_\alpha$ and the angular momentum of the rotation of the diatomic $\boldsymbol{j}_\alpha$ can be expressed either through four space-fixed or body-fixed angles. If we choose to work in a body-fixed coordinate system then we can use the identities $\hat{\boldsymbol{\ell}}_\alpha^{\,2} = \{\hat{\boldsymbol{J}} - \hat{\boldsymbol{j}}_\alpha\}^2 = \hat{\boldsymbol{J}}^2 + \hat{\boldsymbol{j}}_\alpha^{\,2} - 2j_{\alpha z}J_{z_\alpha} - (j^-_\alpha J^+ + j^+_\alpha J^-)$ (see ref. 7 eqn (A10)) and apply the body-fixed basis set of eqn (G-11).

†We follow here the definitions of Delves[18,19] as used also by Pack and Parker.[1] These differ by a factor of two from those of Kuppermann[7,20] and Johnson.[6]

We can now expand the wavefunction[‡] in terms of "vibrational" wavefunctions, $\chi^J_{\nu j\Omega}(\alpha;\rho)$ in the angular variable α. These basis functions depend parametrically on the hyperradius ρ

$$\psi(\tilde{\boldsymbol{R}}_\alpha,\tilde{\boldsymbol{r}}_\alpha,E)=2\sum_{JM}\sum_{\alpha'\nu'j'\Omega'}\sum_{\Omega}\left\{\sum_{\ell}(2\ell+1)^{\frac{1}{2}}(j\ell JM|jm_j\ell 0)B_{\ell,\Omega}\right\}\times\frac{\Gamma^{J\alpha\nu j\Omega}_{\alpha'\nu'j'\Omega'}(\rho,E)}{\rho^{\frac{5}{2}}}\frac{\chi^J_{\alpha'\nu'j'\Omega'}(\alpha_\alpha;\rho)}{\sin 2\alpha_\alpha}\Xi^{JM}_{j'\Omega'}(\hat{\tilde{\boldsymbol{R}}}_\alpha,\hat{\tilde{\boldsymbol{r}}}'_\alpha) \tag{G-17}$$

The vibrational functions $\chi^J_{\alpha\nu j\Omega}(\alpha;\rho)$ are solutions of the equation

$$\left\{\frac{\hbar^2}{2\mu\rho_s^2}\left[-\frac{\partial^2}{\partial\alpha_\alpha}+\frac{j_\alpha(j_\alpha+1)}{\sin^2\alpha_\alpha}+\frac{J(J+1)+j_\alpha(j_\alpha+1)-2\Omega^2}{\cos^2\alpha_\alpha}\right]+V^{\text{diatom}}_\alpha(\rho_s\sin\alpha_\alpha)-\mathcal{E}_{\nu j}(\rho_s)\right\}\chi^J_{\alpha\nu j\Omega}(\alpha_\alpha;\rho_s)=0 \tag{G-18}$$

where we imagine that we solve the problem by dividing ρ up into segments and use one of the standard methods of solving coupled differential equations as described in Chapter 4. ρ_s indicates a particular value of ρ in the center of segment s. The potential $V^{\text{diatom}}_\alpha(\rho_s\sin\alpha_\alpha)=V^{\text{diatom}}_\alpha(r_{\alpha_\alpha})$ is the asymptotic diatomic potential at a large R_α value. Note that ν, j and Ω should always be associated with the channel subscript α, which has been omitted in several places for brevity.

Substituting the expansion of eqn (G-17) into the Schrödinger equation we obtain the following set of coupled differential equations for $\Gamma^{J\nu j\Omega}_{\nu'j'\Omega'}(\rho,E)$ within ρ in a segment s

$$\begin{aligned}&\left\{-\frac{\partial^2}{\partial\rho^2}+\frac{2\mu}{\hbar^2}\left[\frac{\rho_s^2}{\rho^2}\mathcal{E}_{\nu j}(\rho_s)-E\right]+\frac{15}{4\rho^2}\right\}\Gamma^{J\alpha\nu j\Omega}_{\alpha'\nu'j'\Omega'}(\rho,E)\\&=-\sum_{\nu''j''}\left\langle\chi^J_{\alpha'\nu'j'\Omega'}(\alpha_\alpha;\rho)\Xi^{JM}_{j'\Omega'}(\hat{\tilde{\boldsymbol{R}}}_\alpha,\hat{\tilde{\boldsymbol{r}}}'_\alpha)\right|U\left|\chi^J_{\alpha'\nu''j''\Omega'}(\alpha_\alpha;\rho)\Xi^{JM}_{j''\Omega'}(\hat{\tilde{\boldsymbol{R}}}_\alpha,\hat{\tilde{\boldsymbol{r}}}'_\alpha)\right\rangle\\&\times\Gamma^{J\alpha\nu j\Omega}_{\alpha'\nu''j''\Omega'}(\rho,E)-\left[\frac{(J(J+1)+j'(j'+1)-\Omega'^2)}{\rho^2\cos^2\alpha_\alpha}\right]\Gamma^{J\alpha\nu j\Omega}_{\alpha'\nu'j'\Omega'}(\rho,E)\\&-\frac{C^{J+}_{\Omega',\Omega'+1}}{\rho^2\cos^2\alpha_\alpha}\Gamma^{J\alpha\nu j\Omega}_{\alpha'\nu'j'\Omega'+1}(\rho,E)-\frac{C^{J-}_{\Omega',\Omega'-1}}{\rho^2\cos^2\alpha_\alpha}\Gamma^{J\alpha\nu j\Omega}_{\alpha'\nu'j'\Omega'-1}(\rho,E)=0\end{aligned} \tag{G-19}$$

where $U=\dfrac{2\mu}{\hbar^2}\left[V-\dfrac{\rho_s^2}{\rho^2}V^{\text{diatom}}_\alpha(\rho_s\sin\alpha_\alpha)\right]$ and its matrix elements are diagonal in the helicity quantum number Ω. $C^{J+}_{\Omega',\Omega'+1}$ and $C^{J-}_{\Omega',\Omega'-1}$ are defined

---

[‡]Note that the expansion in eqn (G-17) should include an extra, constant, factor of $\left[\dfrac{k\mu_\alpha}{2\pi\hbar^2}\right]^{\frac{1}{2}}$ so as to agree with the energy scale normalisation use in eqn (2.1) and elsewhere. This factor has been omitted here to simplify the notation.

in Appendix D eqn (D-22). The factors of $\frac{1}{\rho^{\frac{5}{2}}}$ and $\frac{2}{\sin 2\alpha_{\alpha}}$ in the expansion of eqn (G-17) have led to a considerable simplification of the Hamiltonian operator (see eqn (G-15) and (G-16)).

The set of equations in eqn (G-19) are the body-fixed arrangement channel coupled differential equations in hyperspherical coordinates. Parity remains a good quantum number and this can be accounted for in the same way as in appendix D. The equivalent space-fixed equations are given in eqn (161) of ref. 1.

G.3 Adiabatically Adjusting Principal Axes Hyperspherical (APH) Coordinates

As we see, the arrangement channel or Delves hyperspherical coordinates depend on the particular arrangement channel chosen. They do not therefore treat all channels in an equal manner. If these coordinates are used, complicated arrangements must be made to transform the wavefunction from one arrangement channel to the other to account for a chemical reaction process[3,7] and to maintain the continuity of the wavefunction and its first derivative. We seek a "universal" coordinate system which is independent of the arrangement channel. In the arrangement channel hyperspherical coordinate system discussed in Section G.2 above the Euler angles were chosen so as to bring the body-fixed Z axis into coincidence with the Jacobi $\boldsymbol{R}$ vector, which connected a particular atom to the center-of-mass of the diatom. In a universal coordinate system the direction of the axes must not be governed by any particular choice of arrangement channel or Jacobi coordinates. One way of satisfying this requirement is to use the principal axes of inertia of the system[21] to determine the orientation of the body-fixed axes.[1,6,10] The effect of this is that for small values of the hyperradius the orientation of the body-fixed Z axis does not coincide with the Jacobi coordinates of any of the channels. For large values of the hyperradius, however, one of the coordinates will point along the Jacobi $\boldsymbol{R}$ vector, depending on the value of the hyperangle. In the discussion which follows we choose the adiabatically adjusting, principal axes hyperspherical (APH) coordinates.[1] In these coordinates, as the hyperradius increases the body-fixed Z axis points along the Jacobi $\boldsymbol{R}$ vector of one of the arrangement channels.

The main idea behind the APH coordinate system is to specify an arbitrary angle $\chi_{Q\alpha}$ for the kinematic rotation which transforms one set of arrangement channel Jacobi coordinates into another (see eqn (G-8)). Only particular angles will in fact be appropriate to perform this transformation. So by choosing an arbitrary continuous angle we are creating a new and unphysical arrangement "channel", Q

$$\begin{pmatrix} \tilde{\boldsymbol{R}}_Q \\ \tilde{\boldsymbol{r}}_Q \end{pmatrix} = \mathbf{T}(\chi_{Q\alpha}) \begin{pmatrix} \tilde{\boldsymbol{R}}_\alpha \\ \tilde{\boldsymbol{r}}_\alpha \end{pmatrix}. \tag{G-20}$$

Using eqn (G-9) and (G-20) we can show that the squares of the new vectors $\tilde{\boldsymbol{R}}_Q$ and $\tilde{\boldsymbol{r}}_Q$ take the form

$$\tilde{R}_Q^2 = \tilde{R}_\alpha^2 \cos^2 \chi_{Q\alpha} + \tilde{r}_\alpha^2 \sin^2 \chi_{Q\alpha} + \left(\tilde{\boldsymbol{R}}_\alpha \cdot \tilde{\boldsymbol{r}}_\alpha\right) \sin 2\chi_{Q\alpha} \tag{G-21}$$

and

$$\tilde{r}_Q^2 = \tilde{R}_\alpha^2 \sin^2 \chi_{Q\alpha} + \tilde{r}_\alpha^2 \cos^2 \chi_{Q\alpha} - \left(\tilde{\boldsymbol{R}}_\alpha \cdot \tilde{\boldsymbol{r}}_\alpha\right) \sin 2\chi_{Q\alpha}. \tag{G-22}$$

The value of the kinematic angle $\chi_{Q\alpha}$ is chosen, for each molecular geometry, so as to maximize the magnitude of $\boldsymbol{R}_Q$. This condition is satisfied if (see ref. 1 eqn (41))

$$\tan(2\chi) = \frac{2\,\tilde{\boldsymbol{R}}_\alpha \cdot \tilde{\boldsymbol{r}}_\alpha}{\tilde{R}_\alpha^2 - \tilde{r}_\alpha^2} \tag{G-23}$$

It turns out that with this choice, coordinates $\tilde{\boldsymbol{R}}_Q$ and $\tilde{\boldsymbol{r}}_Q$ are mutually orthogonal and they also constitute two of the principal axes of inertia of the system (see ref. 1 eqn (45)). The body-fixed Z axis is taken to lie along $\tilde{\boldsymbol{R}}_Q$ and is associated with the smallest moment of inertia $\mu\tilde{\boldsymbol{r}}_Q^2$; the body-fixed X axis lies along the $\tilde{\boldsymbol{r}}_Q$ direction and is the inertial axis with the second lowest moment of inertia $\mu\tilde{\boldsymbol{R}}_Q^2$. The body-fixed Y axis lies perpendicular to the molecular plane and is associated with the largest moment of inertia $\mu(\tilde{\boldsymbol{r}}_Q^2 + \tilde{\boldsymbol{R}}_Q^2)$.

The fact that $\tilde{\boldsymbol{R}}_Q$ and $\tilde{\boldsymbol{r}}_Q$ are mutually orthogonal effectively rules out the use of the Jacobi angle as one of the internal angular variables in the arrangement channel hyperspherical coordinate treatment. This angle is replaced in the formalism by the now continuous kinematic angle $\chi = \chi_{Q\alpha}$.

The APH hyperspherical coordinates that describe the body-fixed geometry of the molecule are therefore the hyperradius ρ describing the size of the triatomic system and the two angles θ and $\chi = \chi_{Q\alpha}$ that describe its shape

$$\begin{aligned} \rho &= \left[\tilde{R}_Q^2 + \tilde{r}_Q^2\right]^{1/2} \\ \theta &= \pi/2 - 2\tan^{-1}(r_Q/R_Q) \end{aligned} \tag{G-24}$$

where $0 \le \rho \le \infty$, $0 \le \theta \le \pi/2$ and $-\pi \le \chi \le \pi$. The coordinates may be viewed as spherical polar coordinates and cover the upper half of a sphere in the internal coordinate space.

The APH coordinate system treats all three arrangement channels equivalently. Normally it is used in only the strong interaction region of coordinate space. As ρ increases we need first to switch over to the body-fixed arrangement channel hyperspherical coordinates, then to body-fixed arrangement channel Jacobi coordinates and finally to space-fixed Jacobi coordinates which are used to apply the boundary conditions and compute the **S** matrix.

Below we give some formulae connecting the internal APH coordinates with the mass-scaled Jacobi ones

$$\rho = [\tilde{R}_\alpha^2 + \tilde{r}_\alpha^2]^{1/2}, \tag{G-25}$$

$$\tan\theta = \frac{\left[(2\tilde{R}_\alpha\tilde{r}_\alpha\cos\gamma_\alpha)^2 + (\tilde{R}_\alpha^2 - \tilde{r}_\alpha^2)^2\right]^{1/2}}{2\tilde{R}_\alpha\tilde{r}_\alpha\sin\gamma_\alpha} \tag{G-26}$$

and

$$\tan(2\chi) = \frac{2\tilde{R}_\alpha\tilde{r}_\alpha\cos\gamma_\alpha}{\tilde{R}_\alpha^2 - \tilde{r}_\alpha^2}. \tag{G-27}$$

Note that the angle θ is responsible for the triatomic shape modifications from an equilateral triangle for $\theta = 0$ to a collinear geometry at $\theta = \pi/2$ while the χ angle mainly describes the ratio of $\tilde{r}_\alpha$ to $\tilde{R}_\alpha$ for a fixed ρ value.

The body-fixed mass-scaled arrangement channel Jacobi coordinates can be expressed in terms of the APH inner coordinates by the relationships

$$\tilde{R}_\alpha = \frac{\rho}{\sqrt{2}}[1 + \sin\theta\cos(2\chi)]^{1/2}, \tag{G-28}$$

$$\tilde{r}_\alpha = \frac{\rho}{\sqrt{2}}[1 - \sin\theta\cos(2\chi)]^{1/2} \tag{G-29}$$

and

$$\cos\gamma_\alpha = \frac{\sin\theta\sin(2\chi)}{\sqrt{1 - \sin^2\theta\cos^2(2\chi)}}. \tag{G-30}$$

The volume element of the six-dimensional APH body-fixed coordinate system can be written in terms of the three Euler angles α_Q, β_Q and γ_Q and the three internal hyperspherical coordinates[1,22]

$$\mathrm{d}\tilde{\boldsymbol{r}}_\alpha\,\mathrm{d}\tilde{\boldsymbol{R}}_\alpha = \rho^5\mathrm{d}\rho \times \frac{1}{8}\sin 2\theta\,\mathrm{d}\theta\,\mathrm{d}\chi\,\mathrm{d}\alpha_Q\,\sin\beta_Q\,\mathrm{d}\beta_Q\,\mathrm{d}\gamma_Q. \tag{G-31}$$

An integral of an arbitrary function F over all space can be written as[1]

$$\int_{-\infty}^{\infty} F\mathrm{d}\tilde{\boldsymbol{R}}_\alpha\mathrm{d}\tilde{\boldsymbol{r}}_\alpha = \frac{1}{16}\int_0^\infty \rho^5\mathrm{d}\rho\int_{-\pi}^{\pi}\mathrm{d}\chi\int_0^{\pi/2}\sin 2\theta\,\mathrm{d}\theta \times \int_0^{2\pi}\mathrm{d}\alpha_Q\int_0^{\pi}\sin\beta_Q\,d\beta_Q\int_0^{2\pi}\mathrm{d}\gamma_Q F. \tag{G-32}$$

The extra factor of $\frac{1}{2}$ in front of the right hand side of the equation is due to the fact that allowing χ to extend over the range $-\pi \le \chi \le \pi$ in fact covers all space twice over. This range is used in order to allow us to properly represent the effects of inversion on the system.[1]

The APH Hamiltonian is a sum of the kinetic energy operator and potential energy

$$\hat{H} = \hat{T} + V(\rho, \theta, \chi) \tag{G-33}$$

where in order to evaluate V the APH inner coordinates, ρ, θ and χ, can be converted to Jacobi coordinates using eqn (G-28) to (G-30).

The kinetic energy operator can be obtained using the general formulae of Podolsky[23] for transforming derivatives from Cartesian to curvilinear coordinates. Performing the required operations leads to the following kinetic energy operator[1,22,24]

$$\hat{T} = -\frac{\hbar^2}{2\mu\rho^5}\frac{\partial}{\partial\rho}\rho^5\frac{\partial}{\partial\rho} - \frac{\Delta_Q^2}{2\mu\rho^2} + \frac{J_x^2}{\mu\rho^2(1+\sin\theta)} + \frac{J_y^2}{2\mu\rho^2\sin^2\theta} + \frac{J_z^2}{\mu\rho^2(1-\sin\theta)} - \frac{i\hbar\cos\theta}{\mu\rho^2\sin^2\theta}J_y\frac{\partial}{\partial\chi} \tag{G-34}$$

where Δ_Q^2 presents the square of the APH angular momentum operator

$$\Delta_Q^2 = \hbar^2\left(\frac{4}{\sin 2\theta}\frac{\partial}{\partial\theta}\sin 2\theta\frac{\partial}{\partial\theta} + \frac{1}{\sin^2\theta}\frac{\partial^2}{\partial\chi^2}\right) \tag{G-35}$$

and the operators J_x, J_y and J_z, are the components of the total angular momentum operator in the space-fixed reference frame.

We can now define the "rotational constants"

$$\begin{aligned} A(\rho,\theta) &= \frac{1}{\mu\rho^2(1+\sin\theta)} \\ B(\rho,\theta) &= \frac{1}{2\mu\rho^2\sin^2\theta} \\ C(\rho,\theta) &= \frac{1}{\mu\rho^2(1-\sin\theta)} \end{aligned} \tag{G-36}$$

and rewrite the Hamiltonian in the form

$$\begin{aligned} \hat{T} &= -\frac{\hbar^2}{2\mu\rho^5}\frac{\partial}{\partial\rho}\rho^5\frac{\partial}{\partial\rho} - \frac{\Delta_Q^2}{2\mu\rho^2} + A\hat{J}_x^2 + B\hat{J}_y^2 + C\hat{J}_z^2 - \frac{i\hbar\cos\theta}{\mu\rho^2\sin^2\theta}\hat{J}_y\frac{\partial}{\partial\chi} \\ &= -\frac{\hbar^2}{2\mu\rho^5}\frac{\partial}{\partial\rho}\rho^5\frac{\partial}{\partial\rho} - \frac{\Delta_Q^2}{2\mu\rho^2} + \frac{(A+B)}{2}\hat{J}^2 + \left[C - \frac{(A+B)}{2}\right]\hat{J}_z^2 \\ &\quad + \frac{(A-B)}{4}\left(\hat{J}_+^2 + \hat{J}_-^2\right) - \frac{i\hbar\cos\theta}{\mu\rho^2\sin^2\theta}\left\{\frac{1}{2i}(\hat{J}_- - \hat{J}_+)\right\}\frac{\partial}{\partial\chi}. \end{aligned} \tag{G-37}$$

The expansion of the wavefunction in the body-fixed reference frame now takes the form

$$\Psi^{JM\Omega n} = 4\sum_{n'\Omega'}\frac{\psi_{\Omega'n'}^{J\Omega n}(\rho,E)}{\rho^{\frac{5}{2}}}\Phi_{\Omega'n'}^{J}(\theta,\chi;\rho)D_{M,\Omega'}^{J^*}(\alpha_Q,\beta_Q,\gamma_Q). \tag{G-38}$$

Note that this wavefunction is expressed entirely in APH hyperspherical coordinates. It is useful only within the strong interaction region. Before we can apply the necessary boundary conditions these functions must be re-expressed in arrangement channel mass-scaled Jacobi coordinates. The reader is referred to the article by Pack and Parker[1] for the details of this transformation.

The hyperradius ρ is, as before, divided into segments. At the center of each such segment we solve for the two-dimensional hyperspherical surface functions $\Phi^J_{\Omega'n'}(\theta,\chi;\rho)$ which are parametrically dependent on ρ and obey the differential equations

$$\left\{-\frac{\hbar^2}{2\mu\rho_s^2}\left[\frac{4}{\sin 2\theta}\frac{\partial}{\partial\theta}\sin 2\theta\frac{\partial}{\partial\theta}+\frac{1}{\sin^2\theta}\frac{\partial^2}{\partial\chi^2}\right]+\frac{\mathrm{A}+\mathrm{B}}{2}\hbar^2 J(J+1)+\frac{15\hbar^2}{8\mu\rho_s^2}\right.$$
$$\left.+\left[\mathrm{C}-\frac{(\mathrm{A}+\mathrm{B})}{2}\right]\hbar^2\Omega^2+V(\rho_s,\theta,\chi)-\mathcal{E}^J_{\Omega n}(\rho_s)\right\}\Phi^J_{\Omega n}(\theta,\chi;\rho_s)=0. \tag{G-39}$$

In order to find the differential equations obeyed by the radial functions $\psi^{J\Omega n}_{\Omega'n'}(\rho,E)$ we need to find the matrix elements of the different parts of the Hamiltonian (eqn (G-37)) between the angular basis functions $\Phi^J_{\Omega'n'}(\theta,\chi;\rho)D^{J*}_{M,\Omega'}(\alpha_Q,\beta_Q,\gamma_Q)$. The matrix elements involving the $\hat{J}_+$ and $\hat{J}_-$ operators are

$$\left\langle\Phi^J_{\Omega'n'}(\theta,\chi;\rho)D^{J*}_{M,\Omega'}(\alpha_Q,\beta_Q,\gamma_Q)\right|-\frac{i\hbar\cos\theta}{\mu\rho^2\sin^2\theta}J_y\frac{\partial}{\partial\chi}\left|\right.$$
$$\left.\times\,\Phi^J_{\Omega''n''}(\theta,\chi;\rho)D^{J*}_{M,\Omega''}(\alpha_Q,\beta_Q,\gamma_Q)\right\rangle$$
$$=-\frac{i\hbar}{\mu\rho^2}\left\langle\Phi^J_{\Omega'n'}(\theta,\chi;\rho)\right|\frac{\cos\theta}{\sin^2\theta}\frac{\partial}{\partial\chi}\left|\Phi^J_{\Omega''n''}(\theta,\chi;\rho)\right\rangle$$
$$\times\left\langle D^{J*}_{M,\Omega'}(\alpha_Q,\beta_Q,\gamma_Q)\right|\left\{\frac{1}{2i}(\hat{J}_--\hat{J}_+)\right\}\left|D^{J*}_{M,\Omega''}(\alpha_Q,\beta_Q,\gamma_Q)\right\rangle$$
$$=-\frac{\hbar^2}{2\mu\rho^2}\left\langle\Phi^J_{\Omega'n'}(\theta,\chi;\rho)\right|\frac{\cos\theta}{\sin^2\theta}\frac{\partial}{\partial\chi}\left|\Phi^J_{\Omega''n''}(\theta,\chi;\rho)\right\rangle$$
$$\times\left\{\bar{\lambda}_-(J,\Omega'')\delta_{\Omega',\Omega''+1}-\bar{\lambda}_+(J,\Omega'')\delta_{\Omega',\Omega''-1}\right\} \tag{G-40}$$

where

$$\bar{\lambda}_\pm(J,\Omega)=\left[(J\pm\Omega+1)(J\mp\Omega)\right]^{\frac{1}{2}}. \tag{G-41}$$

The bar on $\bar{\lambda}_\pm$ indicates that it has the opposite sign convention to the usual one (for a clear explanation of this see ref. 1 eqn (171) and the paragraph above eqn (169) and also Zare[25] Section 3.4)

$$\left\langle\Phi^J_{\Omega'n'}(\theta,\chi;\rho)D^{J*}_{M,\Omega'}(\alpha_Q,\beta_Q,\gamma_Q)\right|\frac{(\mathrm{A}-\mathrm{B})}{4}\left(\hat{J}_+^2+\hat{J}_-^2\right)\left|\right.$$
$$\left.\times\,\Phi^J_{\Omega''n''}(\theta,\chi;\rho)D^{J*}_{M,\Omega''}(\alpha_Q,\beta_Q,\gamma_Q)\right\rangle$$
$$=\frac{\hbar^2}{4}\left\langle\Phi^J_{\Omega'n'}(\theta,\chi;\rho)\right|(\mathrm{A}-\mathrm{B})\left|\Phi^J_{\Omega''n''}(\theta,\chi;\rho)\right\rangle$$
$$\times\left\{\bar{\lambda}_+(J,\Omega''+1)\bar{\lambda}_+(J,\Omega'')\delta_{\Omega',\Omega''-2}\right.$$
$$\left.+\,\bar{\lambda}_-(J,\Omega''-1)\bar{\lambda}_-(J,\Omega'')\delta_{\Omega',\Omega''+2}\right\}. \tag{G-42}$$

Now substituting the expansion of eqn (G-38) into the Schrödinger equation, and making use of eqn (G-39)–(G-42) we obtain a set of coupled differential equations for the function $\psi^{J\Omega n}_{\Omega' n'}(\rho, E)$

$$\begin{aligned}
&\left\{-\frac{\partial^2}{\partial\rho^2} + \frac{2\mu}{\hbar^2}\left(\frac{\rho_s^2}{\rho^2}\mathcal{E}^J_{\Omega' n'}(\rho_s) - E\right)\right\}\psi^{J\Omega n}_{\Omega' n'}(\rho, E) \\
&= \left\{-\frac{2\mu}{\hbar^2}\sum_{n''}\left\langle \Phi^J_{\Omega n'}(\theta,\chi;\rho_s)\right| U \left|\Phi^J_{\Omega n''}(\theta,\chi;\rho_s)\right\rangle \delta_{\Omega',\Omega''}\right. \\
&+ \frac{1}{\rho^2}\sum_{\Omega'',n''}\left\langle \Phi^J_{\Omega' n'}(\theta,\chi;\rho)\right| \frac{\cos\theta}{\sin^2\theta}\frac{\partial}{\partial\chi}\left|\Phi^J_{\Omega'' n''}(\theta,\chi;\rho)\right\rangle \\
&\times \left\{\bar{\lambda}_-(J,\Omega'')\delta_{\Omega',\Omega''+1} - \bar{\lambda}_+(J,\Omega'')\delta_{\Omega',\Omega''-1}\right\} \\
&- \frac{\mu}{2}\left\langle \Phi^J_{\Omega' n'}(\theta,\chi;\rho)\right| (\mathrm{A}-\mathrm{B}) \left|\Phi^J_{\Omega'' n''}(\theta,\chi;\rho)\right\rangle \\
&\times \left\{\bar{\lambda}_+(J,\Omega''+1)\bar{\lambda}_+(J,\Omega'')\delta_{\Omega',\Omega''-2}\right. \\
&\left.\left. + \bar{\lambda}_-(J,\Omega''-1)\bar{\lambda}_-(J,\Omega'')\delta_{\Omega',\Omega''+2}\right\}\right\}\psi^{J\Omega n}_{\Omega'' n''}(\rho, E) \qquad \text{(G-43)}
\end{aligned}$$

where

$$U = \left[V - \frac{\rho_s^2}{\rho^2}V(\rho_s,\theta,\chi)\right]. \qquad \text{(G-44)}$$

References

1. R. T. Pack and G. A. Parker, Quantum reactive scattering in three dimensions using hyperspherical (APH) coordinates. Theory, *J. Chem. Phys.*, 1987, **87**, 3888.
2. V. Aquilanti, S. Cavalli and D. De Fazio, Hyperquantization algorithm. I. Theory for triatomic systems, *J. Chem. Phys.*, 1998, **109**, 3792.
3. D. Skouteris, J. F. Castillo and D. E. Manolopoulos, ABC: a quantum reactive scattering program, *Comput. Phys. Commun.*, 2000, **133**, 128.
4. P. Honvault and J.-M. Launay, A quantum-mechanical study of the dynamics of the $O(^1D) + H_2 \rightarrow OH + H$ insertion reaction, *J. Chem. Phys.*, 2000, **114**, 1057.
5. J. Crawford and G. A. Parker, State-to-state three-atom time-dependent reactive scattering in hyperspherical coordinates, *J. Chem. Phys.*, 2013, **138**, 054313.
6. B. R. Johnson, On hyperspherical coordinates and mapping the internal configurations of a three body system, *J. Chem. Phys.*, 1980, **73**, 5051.
7. G. C. Schatz and A. Kuppermann, Quantum mechanical reactive scattering for three-dimensional atom plus diatom systems. I. Theory, *J. Chem. Phys.*, 1976, **65**, 4642.
8. M. A. Eliason and J. O. Hirschfelder, General collision theory treatment for the rate of bimolecular, gas phase reactions, *J. Chem. Phys.*, 1959, **30**, 1426.

9. J. O. Hirschfelder and J. S. Dahler, The kinetic energy of relative motion, *Proc. Natl. Acad. Sci. U. S. A.*, 1956, **42**, 363.
10. R. C. Whitten and F. T. Smith, Symmetric representation for three-body problems. II. Motion in space, *J. Math. Phys.*, 1968, **9**, 1103.
11. F. T. Smith, Participation of vibration in exchange reactions, *J. Chem. Phys.*, 1959, **31**, 1352.
12. F. T. Smith, Generalized angular momentum in many-body collisions, *Phys. Rev.*, 1960, **120**, 1058.
13. D. W. Jepsen and J. O. Hirschfelder, Set of co-ordinate systems which diagonalize the kinetic energy of relative motion, *Proc. Natl. Acad. Sci. U. S. A.*, 1959, **45**, 249.
14. K.-H. (Tom) Yang, J. O. Hirschfelder and B. R. Johnson, Interaction of molecules with electromagnetic fields. II. The multipole operators and dynamics of molecules with moving nuclei in electromagnetic fields, *J. Chem. Phys.*, 1981, **75**, 2321.
15. B. R. Johnson, J. O. Hirschfelder and K.-H. Yang, Interaction of atoms, molecules, and ions with constant electric and magnetic fields, *Rev. Mod. Phys.*, 1983, **55**, 109.
16. F. T. Smith, A symmetric representation for three-body problems. I. Motion in a plane, *J. Math. Phys.*, 1962, **3**, 735.
17. R. C. Whitten, Potentials for three-body systems, *J. Math. Phys.*, 1969, **10**, 1631.
18. L. M. Delves, Tertiary and general-order collisions, *Nucl. Phys.*, 1958–1959, **9**, 391.
19. L. M. Delves, Tertiary and general-order collisions (II), *Nucl. Phys.*, 1960, **20**, 275.
20. A. Kuppermann, A useful mapping of triatomic potential energy surfaces, *Chem. Phys. Lett.*, 1975, **32**, 374.
21. H. Goldstein, *Classical Mechanics*, Addison-Wesley, Reading, Massachusetts, USA, 1950.
22. B. R. Johnson, The quantum dynamics of three particles in hyperspherical coordinates, *J. Chem. Phys.*, 1983, **79**, 1916.
23. B. Podolsky, Quantum-mechanically correct form of Hamiltonian function for conservative systems, *Phys. Rev.*, 1928, **32**, 812.
24. B. R. Johnson, The classical dynamics of three particles in hyperspherical coordinates, *J. Chem. Phys.*, 1983, **79**, 1906.
25. R. N. Zare, *Angular Momentum*. John Wiley and Sons, New York, 1988. See also list of corrections at web address: http://web.stanford.edu/group/Zarelab/Errata.pdf.

APPENDIX H

Formalism for Time-Dependent Quantum Dynamics

In this appendix we discuss in detail two aspects of time-dependent quantum dynamics which are somewhat mathematically involved and which we therefore felt were better consigned to an appendix. The topics covered in the appendix are: (1) helicity decomposition of the wavepacket and (2) analysis of wavepacket to extract the **S** matrix.

H.1 Helicity Decomposition of the Wavepacket

As the total angular momentum is a good, *i.e.* conserved, quantum number, the wavepacket is taken to have a defined value of the total angular momentum. In the body-fixed reference frame our wavepacket takes the form $\Phi_{\Omega'}^{Jvj\Omega}(R,r,\gamma)$ (see eqn (7.15)). Ω is the quantum number for the initial value of the body-fixed Z component of the total angular momentum and Ω' labels the components of the wavepacket which end up having different values of this quantum number as a result of the scattering dynamics. Despite the fact that we seek to use a "grid representation" where the value of the wavefunction is specified by its value at a set of predetermined grid points in the three variables, it will prove easiest to derive the form of the Hamiltonian operator by initially expanding the angular dependence of the wavepacket over a set of angular basis functions. The basis functions we use are eigenfunctions of the rotational angular momentum operator of the diatomic molecule. These are normalized associated Legendre polynomials[1] in the Jacobi angle γ. We will later use a Gauss–Legendre angular grid,[2] which is the one naturally associated with this set of basis functions. The normalized associated Legendre polynomials are just the "θ" part of the spherical harmonics.

RSC Theoretical and Computational Chemistry Series No. 7
Theory of Molecular Collisions
By Gabriel G. Balint-Kurti and Alexander P. Palov

Published by the Royal Society of Chemistry, www.rsc.org

For an atom–diatom system the Schrödinger equation in a space-fixed center-of-mass reference frame is (see eqn (3.1), note that we have omitted the argument "E" from the wavefunction to simplify the notation)

$$\hat{H}\psi(\boldsymbol{R},\boldsymbol{r}) = \left\{ -\frac{\hbar^2}{2\mu_R}\nabla_R^2 - \frac{\hbar^2}{2\mu_r}\nabla_r^2 + V(R,r,\gamma) \right\} \psi(\boldsymbol{R},\boldsymbol{r}) = E\,\psi(\boldsymbol{R},\boldsymbol{r}). \tag{H-1}$$

We wish to transform this to a body-fixed reference frame. This is done by rotating the wavefunction using a Wigner **D** matrix[1,3,4] (see eqn (D-3))

$$\begin{aligned} \psi^{Jvjm_j}(\boldsymbol{R},\boldsymbol{r}) &= \sum_{\Omega''} \frac{1}{Rr} \Phi_{\Omega''}^{Jvj\Omega}(R,r,\gamma)\tilde{D}^{J*}_{m_j,\Omega''}(\phi_R,\theta_R,\phi_r') \\ &= \sum_{\Omega''}\sum_{j''} \frac{1}{Rr}\phi^{Jvj\Omega}_{j'',\Omega''}(R,r)\Theta_{j'',\Omega''}(\gamma)\tilde{D}^{J*}_{m_j,\Omega''}(\phi_R,\theta_R,\phi_r') \end{aligned} \tag{H-2}$$

where $\psi^{Jvjm_j}(\boldsymbol{R},\boldsymbol{r})$ is a space-fixed wavefunction, m_j is the initial quantum number for the Z component of the diatomic rotational angular momentum in the space-fixed frame and is equal to Ω as the body-fixed Z axis coincides initially with the space-fixed axis (the assumption being that the initial relative velocity is along the space-fixed Z axis), the function $\tilde{D}^J_{m_j,\Omega}(\phi_R,\theta_R,\phi_r')$ is a normalized Wigner **D** matrix (*i.e.* $\tilde{D}^J_{m_j,\Omega}(\phi_R,\theta_R,\phi_r') = \frac{\sqrt{2J+1}}{4\pi}D^J_{m_j,\Omega}(\phi_R,\theta_R,\phi_r')$), the set of such functions forms an orthonormal set (see ref. 4 eqn (4.10.3)) and $\Theta_{j,\Omega}(\gamma)$ is a normalized associated Legendre Polynomial.[1]

Substituting eqn (H-2) into eqn (H-1), left multiplying by $\tilde{D}^J_{m_j,\Omega'}(\phi_R,\theta_R,\phi_r')$ $\Theta_{j',\Omega'}(\gamma)$ and intergrating over all angles yields the equation

$$\begin{aligned} &\left\{ -\frac{\hbar^2}{2\mu_R}\frac{\partial^2}{\partial R^2} - \frac{\hbar^2}{2\mu_r}\frac{\partial^2}{\partial r^2} + \frac{\hbar^2 j(j+1)}{2\mu_r r^2} + \right\} \phi^{Jvj\Omega}_{j',\Omega'}(R,r) \\ &\quad + \sum_{j''} V^J_{j',j''}(R,r)\phi^{Jvj\Omega}_{j'',\Omega'}(R,r) \\ &\quad + \frac{1}{2\mu_R R^2}\sum_{\Omega''}\sum_{j''} \langle \tilde{D}^{J*}_{m_j,\Omega'}\Theta_{j',\Omega'} |\, \hat{\ell}^2 \,| \tilde{D}^{J*}_{m_j,\Omega''}\Theta_{j'',\Omega''}\rangle\, \phi^{Jj\Omega}_{j'',\Omega''}(R,r) \\ &= E\,\phi^{Jvj\Omega}_{j',\Omega'}(R,r) \end{aligned} \tag{H-3}$$

where $\hat{\ell}$ is the orbital angular momentum operator, $\hat{j}$ is the rotational angular momentum operator for the diatomic molecule, $V^J_{j',j''} = \langle \Theta_{j'\Omega}(\gamma) |\, V(R,r,\gamma)\, | \Theta_{j''\Omega}(\gamma)\rangle$ and we have used the fact that the normalized Legendre polynomials are eigenfunctions of the $\hat{j}^2$ operator.

We now express the orbital angular momentum as the difference between the total angular momentum and the rotational angular momentum, $\hat{\ell} = \hat{J} - \hat{j}$, leading to the relationship

$$\hat{\ell}^2 = \left(\hat{J} - \hat{j}\right)^2 = \hat{J}^2 + \hat{j}^2 - 2\hat{j}\cdot\hat{J}. \tag{H-4}$$

We can write

$$\hat{j}\cdot\hat{J} = \hat{j}_z\hat{J}_z + \hat{j}_x\hat{J}_x + \hat{j}_y\hat{J}_y. \tag{H-5}$$

Now defining the ladder operators (see ref. 1 eqn (1.17) and (1.31))

$$\begin{aligned}\hat{J}_\pm &= \hat{J}_x \pm i\,\hat{J}_y,\\ \hat{j}_\pm &= \hat{j}_x \pm i\,\hat{j}_y.\end{aligned} \tag{H-6}$$

we can write the inverse relationships as

$$\begin{aligned}\hat{J}_x &= \frac{1}{2}\{\hat{J}_+ + \hat{J}_-\}\\ \hat{J}_y &= \frac{1}{2i}\{\hat{J}_+ - \hat{J}_-\}\end{aligned} \tag{H-7}$$

with similar equations for $\hat{j}_x$ and $\hat{j}_y$.

With these definitions we find that

$$\begin{aligned}\hat{j}_x\hat{J}_x + \hat{j}_y\hat{J}_y &= \frac{1}{4}\left\{(\hat{j}_+ + \hat{j}_-)(\hat{J}_+ + \hat{J}_-) - (\hat{j}_+ - \hat{j}_-)(\hat{J}_+ - \hat{J}_-)\right\}\\ &= \frac{1}{2}\left\{\hat{j}_+\hat{J}_- + \hat{j}_-\hat{J}_+\right\}\end{aligned} \tag{H-8}$$

and substituting eqn (H-5) and (H-8) into eqn (H-4) we obtain

$$\hat{\ell}^2 = (\hat{J} - \hat{j})^2 = \hat{J}^2 + \hat{j}^2 - 2\hat{j}_z\hat{J}_z - \{\hat{j}_+\hat{J}_- + \hat{j}_-\hat{J}_+\}. \tag{H-9}$$

The Wigner **D** matrix is an eigenfunction of the total angular momentum ($\hat{J}$) and of its body-fixed Z component. At this stage, however, we must take some care. We are working in the body-fixed coordinate frame and the actions of the ladder operators in this frame are reversed! (See Zare[1] eqn (3.4.9). This also agrees with the discussion in Appendix G of eqn (G-41)).

We may therefore write

$$\hat{J}_\pm \tilde{D}^{J*}_{m_j,\Omega}(\phi_R,\theta_R,\phi'_r) = \hbar\lambda_\mp(J,\Omega)\tilde{D}^{J*}_{m_j,(\Omega\mp 1)}(\phi_R,\theta_R,\phi'_r) \tag{H-10}$$

where

$$\lambda_\pm(J,\Omega) = [J(J+1) - \Omega(\Omega\pm 1)]^{\frac{1}{2}}. \tag{H-11}$$

The action of the ladder operator $\hat{j}_\pm$ on the eigenfunctions $\Theta_{j'',\Omega''}$ takes the normal form

$$\hat{j}_\pm\Theta_{j,\Omega}(\gamma) = \hbar\lambda_\pm(j,\Omega)\Theta_{j,\Omega\pm 1}(\gamma). \tag{H-12}$$

Now substituting eqn (H-9), (H-10) and (H-12) into eqn (H-3) we obtain

$$\begin{aligned}&\left\{-\frac{\hbar^2}{2\mu_R}\frac{\partial^2}{\partial R^2} - \frac{\hbar^2}{2\mu_r}\frac{\partial^2}{\partial r^2} + \frac{\hbar^2 j(j+1)}{2}\left\{\frac{1}{\mu_r r^2} + \frac{1}{\mu_R R^2}\right\}\right.\\ &\quad\left.+\frac{\hbar^2\left(J(J+1) - 2\Omega'^2\right)}{2\mu_R R^2}\right\}\phi^{Jvj\Omega}_{j',\Omega'}(R,r) + \sum_{j''} V^J_{j',j''}(R,r)\phi^{Jvj\Omega}_{j'',\Omega'}(R,r)\\ &\quad-\frac{\hbar^2}{2\mu_R}\left\{C^{J+}_{\Omega',\Omega'+1}\,\phi^{Jvj\Omega}_{j',\Omega'+1}(R,r) + C^{J-}_{\Omega',\Omega'-1}\,\phi^{Jvj\Omega}_{j',\Omega'-1}(R,r)\right\}\\ &= E\,\phi^{Jvj\Omega}_{j',\Omega'}(R,r)\end{aligned} \tag{H-13}$$

where

$$C^{J\pm}_{\Omega,\Omega\pm1} = -\frac{[J(J+1)-\Omega(\Omega\pm1)]^{\frac{1}{2}}[j(j+1)-\Omega(\Omega\pm1)]^{\frac{1}{2}}}{R^2}. \quad \text{(H-14)}$$

See also Appendix D eqn (D-21) and (D-22) and note that $\lambda_+(j,\Omega-1) = \lambda_-(j,\Omega)$ and $\lambda_-(j,\Omega+1) = \lambda_+(j,\Omega)$.

If we had just expanded the wavefunction in the form

$$\psi^{J\nu jm_j}(\boldsymbol{R},\boldsymbol{r}) = \sum_{\Omega''} \frac{1}{Rr}\Phi^{J\nu j(\Omega=m_j)}_{\Omega''}(R,r,\gamma)\tilde{D}^{J*}_{m_j,\Omega''}(\phi_R,\theta_R,\phi'_r) \quad \text{(H-15)}$$

without using the expansion in terms of normalized Legendre polynomials in eqn (H-2), using the same arguments as above we would have obtained the set of coupled equations

$$\begin{aligned}&\left\{-\frac{\hbar^2}{2\mu_R}\frac{\partial^2}{\partial R^2} - \frac{\hbar^2}{2\mu_r}\frac{\partial^2}{\partial r^2} + \frac{\hat{\boldsymbol{j}}^2}{2}\left\{\frac{1}{\mu_r r^2}+\frac{1}{\mu_R R^2}\right\}\right.\\ &\quad\left. + \frac{\hbar^2\left(J(J+1)-2\Omega'^2\right)}{2\mu_R R^2} + V(R,r,\gamma)\right\}\Phi^{J\nu j\Omega}_{\Omega'}(R,r,\gamma)\\ &\quad - \frac{\hbar}{2\mu_R R^2}\left\{\lambda_+(J,\Omega')\,\hat{\boldsymbol{j}}_-\,\Phi^{J\nu j\Omega}_{\Omega'+1}(R,r,\gamma) + \lambda_-(J,\Omega')\,\hat{\boldsymbol{j}}_+\,\Phi^{J\nu j\Omega}_{\Omega'-1}(R,r,\gamma)\right\}\\ &= E\,\Phi^{J\nu j\Omega}_{\Omega'}(R,r,\gamma). \end{aligned} \quad \text{(H-16)}$$

Note that the term $\left[\lambda_+(J,\Omega')\,\hat{\boldsymbol{j}}_-\,\Phi^{J\nu j\Omega}_{\Omega'+1}(R,r,\gamma)\right]$ is a contribution from the wavepacket associated with $(\Omega'+1)$ to the (Ω') wavepacket due to the action of the $\hat{\boldsymbol{j}}_-$ operator.

The left hand side of eqn (H-16) gives the result of the action of the Hamiltonian on the wavepacket resulting in the "new" Ω' component of the resulting function.

H.2 Analysis of Wavepacket to Extract the S Matrix

Our analysis of the wavepacket is predicated on the idea that by watching the wavepacket as it passes by a point in the asymptotic region of the exit channel we can extract from it the flux going into a particular final channel and hence also the **S** matrix. Figure 7.1 shows a schematic of a potential energy surface for a chemical reaction with an "analysis" line drawn across the product exit channel. We take a cut through the wavepacket along the analysis line at each time step. By projecting this cut onto the product fragment (diatomic) eigenfunctions and then taking the half Fourier transform over time we are able to evaluate the **S** matrix elements. We now discuss the details of how this is done.[5–7]

We denote the time-dependent wavepacket by $\Phi^{J\nu j\Omega}_{\Omega'}(R_\beta,r_\beta,\gamma_\beta,t)$, (see eqn (H-16)) where the superscripts $\Omega = m_j$, ν and j correspond to the quantum state of the reactants. The analysis line is set at $R_\beta = R_{\beta\infty}$. The cut, taken along this line at each time step is then $\Phi^{J\nu j\Omega}_{\Omega'}(R_{\beta\infty},r_\beta,\gamma_\beta,t)$ (the subscript β denotes the use of product Jacobi coordinates). We now expand the cut through the wavepacket in terms of the product vibration–rotation fragment eigenfunctions

$$\Phi_{\Omega'}^{J\nu j\Omega}(R_{\beta\infty}, r_\beta, \gamma_\beta, t) = \sum_{\nu',j'} C^J_{\nu j\Omega\to\nu' j'\Omega'}(t)\, \chi_{\nu' j'}(r_\beta)\, \Theta_{j'\Omega'}(\gamma_\beta) \tag{H-17}$$

where $\Theta_{j'\Omega'}(\gamma_\beta)$ are normalized associated Legendre polynomials and $\chi_{\nu' j'}(r_\beta)$ are the product vibrational eigenfunctions (see eqn (I-25)).

The coefficients $C^J_{\nu j\Omega\to\nu' j'\Omega'}(t)$ are evaluated using the integral

$$C^J_{\nu j\Omega\to\nu' j'\Omega'}(t) = \int \chi^*_{\nu' j'}(r_\beta)\, \Theta^*_{j'\Omega'}(\gamma_\beta)\, \Phi_{\Omega'}^{J\nu j\Omega}(R_{\beta\infty}, r_\beta, \gamma_\beta, t)\, \mathrm{d}r_\beta \sin\gamma_\beta\, \mathrm{d}\gamma_\beta. \tag{H-18}$$

We now take the half Fourier transform of the coefficients $C^J_{\nu j\Omega\to\nu' j'\Omega'}(t)$

$$A^J_{\nu j\Omega\to\nu' j'\Omega'}(E) = \frac{1}{2\pi}\int_0^\infty \mathrm{e}^{iEt/\hbar} C^J_{\nu j\Omega\to\nu' j'\Omega'}(t)\, \mathrm{d}t. \tag{H-19}$$

Substituting eqn (H-18) into eqn (H-19) and inserting a Dirac delta function together with an integration to localize R_β at the analysis line $R_{\beta\infty}$ we can write

$$\begin{aligned} A^J_{\nu j\Omega\to\nu' j'\Omega'}(E) &= \frac{1}{2\pi}\int_0^\infty \mathrm{e}^{iEt/\hbar} \int \chi^*_{\nu' j'}(r_\beta)\, \Theta^*_{j'\Omega'}(\gamma_\beta) \\ &\times \int_0^\infty \mathrm{d}R_\beta\, \delta(R_\beta - R_{\beta\infty}) \Phi_{\Omega'}^{J\nu j\Omega}(R_\beta, r_\beta, \gamma_\beta, t)\, \mathrm{d}r_\beta \sin\gamma_\beta\, \mathrm{d}\gamma_\beta\, \mathrm{d}t. \end{aligned} \tag{H-20}$$

The wavepacket at time t may now be written as the initial wavepacket propagated from $t = 0$ to t

$$\begin{aligned} A^J_{\nu j\Omega\to\nu' j'\Omega'}(E) &= \frac{1}{2\pi}\int_0^\infty \mathrm{e}^{iEt/\hbar} \int \chi^*_{\nu' j'}(r_\beta)\, \Theta^*_{j'\Omega'}(\gamma_\beta) \\ &\int_0^\infty \mathrm{d}R_\beta\, \delta(R_\beta - R_{\beta\infty}) \mathrm{e}^{-i\hat{H}t/\hbar} \Phi^{J\nu j\Omega}(R_\alpha, r_\alpha, \gamma_\alpha, t=0)\, \mathrm{d}r_\beta \sin\gamma_\beta\, \mathrm{d}\gamma_\beta\, \mathrm{d}t \end{aligned} \tag{H-21}$$

where $\Phi^{J\nu j\Omega}(R_\alpha, r_\alpha, \gamma_\alpha, t=0)$ is the initial wavepacket.

We now use a "bra–ket" notation to indicate integration over spatial coordinates and therefore rewrite eqn (H-21) as

$$\begin{aligned} A^J_{\nu j\Omega\to\nu' j'\Omega'}(E) &= \frac{1}{2\pi}\int_0^\infty \mathrm{e}^{iEt/\hbar}\, \mathrm{d}t \\ &\langle \chi_{\nu' j'}(r_\beta)\, \Theta_{j'\Omega'}(\gamma_\beta)\, \delta(R_\beta - R_{\beta\infty}) |\, \mathrm{e}^{-i\hat{H}t/\hbar}\, | \Phi^{J\nu j\Omega}(R_\alpha, r_\alpha, \gamma_\alpha, t=0) \rangle . \end{aligned} \tag{H-22}$$

We now define the set of continuum radial functions $\Phi^{J,\beta-}_{v'j'\Omega'}(R_\beta, r_\beta, \gamma_\beta, E)$. These functions are analogous to those in eqn (H-15), but additionally possess well defined outgoing boundary conditions. They are "energy normalized" (see eqn (2.1) and Appendix A) and correspond asymptotically to a pure outgoing radial wave in product channel "$v'j'\Omega'$" and incoming waves in all possible initial channels (note that this includes both reactant and product channels, see also eqn (3.21))

$$\begin{aligned}
&\Phi^{J,\beta-}_{v'j'\Omega'}(E)\\
&\underset{R_\beta\to\infty}{\sim} \left(\frac{\mu^\beta}{2\pi\hbar^2 k^\beta_{v'j'}}\right)^{\frac{1}{2}} \left\{ e^{ik^\beta_{v'j'}R_\beta}\chi^\beta_{v'j'\Omega'} - \sum_{v''j''} (S^{J,\beta,\beta}_{v''j''\Omega'',v'j'\Omega'})^* \right.\\
&\quad \left. \times \left(\frac{k^\beta_{v'j'}}{k^\beta_{v''j''}}\right)^{\frac{1}{2}} e^{-ik^\beta_{v''j''}R_\beta}\chi^\beta_{v''j''\Omega''} \right\}\\
&\underset{R_\alpha\to\infty}{\sim} -\left(\frac{\mu^\alpha}{2\pi\hbar^2}\right)^{\frac{1}{2}} \sum_{v''j''\Omega''} \frac{(S^{J,\beta,\alpha}_{v'j'\Omega',v''j''\Omega''})^*}{(k^\alpha_{v''j''})^{1/2}} e^{-ik^\alpha_{v''j''}R_\alpha}\chi^\alpha_{v''j''\Omega''}
\end{aligned} \tag{H-23}$$

where we have used the notation

$$\chi^\beta_{v'j'\Omega'} = \chi_{v'j'}(r_\beta)\Theta_{j'\Omega'}(\gamma_\beta) \tag{H-24}$$

as an abbreviation for the diatomic fragment vibrational–rotational eigenfunctions (see eqn (H-2) and (I-25)). $\mathbf{S}^{J,\beta,\beta}$ is an non-reactive scattering **S** matrix between product vibrational–rotational states and $\mathbf{S}^{J,\alpha,\beta}$ is the reactive **S** matrix in which we are principally interested.

As was mentioned above, the set of continuum functions $\Phi^{J,\beta-}_{v'j'\Omega'}(E)$ form a complete orthonormal basis set for energy E with the stated boundary conditions, *i.e.* we may write

$$\begin{aligned}
&\langle \Phi^{J,\beta-}_{v'j'\Omega'}(E) | \Phi^{J,\beta-}_{v''j''\Omega''}(E')\rangle = \delta(E-E')\delta_{v',v''}\delta_{j',j''}\delta_{\Omega',\Omega''}\\
&\int_{-\infty}^{\infty} dE' \sum_{v''j''\Omega''} \langle \Phi^{J,\beta-}_{v'j'\Omega'}(E) | \Phi^{J,\beta-}_{v'''j'''\Omega'''}(E')\rangle = 1,
\end{aligned} \tag{H-25}$$

where the angular brackets $\langle \cdots | \cdots \rangle$ imply integration over all spatial variables R_i, r_i and γ_i with $i = \alpha$ or β. Thus we can now expand the function $\Phi^{Jvj\Omega}(R_\alpha, r_\alpha, \gamma_\alpha, t=0)$ as a linear combination of the functions $\Phi^{J,\beta-}_{v'''j'''\Omega'''}(E')$. Because $\Phi^{Jvj\Omega}(R_\alpha, r_\alpha, \gamma_\alpha, t=0)$ is not an energy eigenfunction and contains contributions from a range of energies its expansion over the energy eigenfunctions $\Phi^{J,\beta-}_{v'''j'''\Omega'''}(E')$ involves an integration over all possible energies as well as a summation over all possible quantum numbers

$$\Phi^{J\nu j\Omega}(R_\alpha, r_\alpha, \gamma_\alpha, t=0) = \int_{-\infty}^{\infty} \mathrm{d}E' \sum_{v''j''\Omega''} \langle \Phi^{J,\beta-}_{v''j''\Omega''}(E') \,|\Phi^{J\nu j\Omega}(R'_\alpha, r'_\alpha, \gamma'_\alpha, t=0)\rangle\, \Phi^{J,\beta-}_{v''j''\Omega''}(E') \tag{H-26}$$

where the angular brackets $\langle\cdots|\cdots\rangle$ mean integration over spatial variables $R'_\alpha, r'_\alpha, \gamma'_\alpha$.

We now substitute this expansion for the last term in eqn (H-22) to obtain

$$\begin{aligned} A^J_{\nu j\Omega\to\nu'j'\Omega'}(E) = \frac{1}{2\pi}\int_0^\infty \mathrm{e}^{iEt/\hbar}\,\mathrm{d}t \int_{-\infty}^{\infty} \mathrm{d}E' \sum_{v''j''\Omega''} \\ \times \langle \Phi^{J,\beta-}_{v''j''\Omega''}(E')|\, \Phi^{J\nu j\Omega}(R'_\alpha, r'_\alpha, \gamma'_\alpha, t=0)\rangle \\ \times \langle \chi_{v'j'}(r_\beta)\,\Theta_{j'\Omega'}(\gamma_\beta)\,\delta(R_\beta - R_{\beta\infty})|\Phi \mathrm{e}^{-i\hat{H}t/\hbar}|\Phi^{J,\beta-}_{v''j''\Omega''}(E')\rangle \end{aligned} \tag{H-27}$$

where the last angular brackets $\langle\cdots|\cdots\rangle$ mean integration over spatial variables $R_\beta, r_\beta, \gamma_\beta$.

$\Phi^{J,\beta-}_{v''j''\Omega''}(E')$ is an eigenfunction of the Hamiltonian so we can replace $\mathrm{e}^{-i\hat{H}t/\hbar}\,\Phi^{J,\beta-}_{v''j''\Omega''}(E')\rangle$ by $\mathrm{e}^{-iE't/\hbar}\,\Phi^{J,\beta-}_{v''j''\Omega''}(E')\rangle$ to get

$$\begin{aligned} A^J_{\nu j\Omega\to\nu'j'\Omega'}(E) = \frac{1}{2\pi}\int_0^\infty \mathrm{e}^{i(E-E')t/\hbar}\,\mathrm{d}t \int_{-\infty}^{\infty} \mathrm{d}E' \sum_{v''j''\Omega''} \\ \times \langle \Phi^{J,\beta-}_{v''j''\Omega''}(E')|\, \Phi^{J\nu j\Omega}(R'_\alpha, r'_\alpha, \gamma'_\alpha, t=0)\rangle \\ \times \langle \chi_{v'j'}(r_\beta)\,\Theta_{j'\Omega'}(\gamma_\beta)\,\delta(R_\beta - R_{\beta\infty})|\Phi^{J,\beta-}_{v''j''\Omega''}(E')\rangle. \end{aligned} \tag{H-28}$$

In the first set of angular brackets in eqn (H-28), the brackets $\langle\cdots|\cdots\rangle$ imply integration over the spatial variables $R'_\alpha, r'_\alpha, \gamma'_\alpha$. The wavepacket at time $t=0$ (*i.e.* the initial wavepacket) is located in the asymptotic large R_α region, so the continuum function $\Phi^{J,\beta-}_{v''j''\Omega''}(E')$ will take on the lower of the two asymptotic forms shown in eqn (H-23). If we use this and also eqn (7.13), (7.14) and (H-24) we can write the first angular brackets in eqn (H-28) as

$$
\begin{aligned}
&\langle \Phi^{J,\beta-}_{v''j''\Omega''}(E')|\Phi^{Jvj\Omega}(R'_\alpha, r'_\alpha, \gamma'_\alpha, t=0)\rangle \\
&= -\left(\frac{\mu^\alpha}{2\pi\hbar^2}\right)^{\frac{1}{2}} \int \mathrm{d}R'_\alpha\, \mathrm{d}r'_\alpha \sin\gamma'_\alpha\, \mathrm{d}\sin\gamma'_\alpha \sum_{v'''j'''\Omega'''} \\
&\quad \times \frac{(S^{J,\beta,\alpha}_{v'''j'''\Omega''',v''j''\Omega''})}{(k^\alpha_{v'''j'''})^{1/2}} \mathrm{e}^{ik^\alpha_{v'''j'''}R_\alpha} \chi^\alpha_{v'''j'''\Omega'''} \Phi^{Jvj\Omega}(R'_\alpha, r'_\alpha, \gamma'_\alpha, t=0) \\
&= -\left(\frac{\mu^\alpha}{2\pi\hbar^2}\right)^{\frac{1}{2}} 2\pi \sum_{v'''j'''\Omega'''} \frac{(S^{J,\beta,\alpha}_{v'''j'''\Omega''',v''j''\Omega''})}{(k^\alpha_{v'''j'''})^{1/2}} \delta_{v,v'''}\delta_{j,j'''}\delta_{\Omega,\Omega'''} \\
&\quad \times \left\{\frac{1}{2\pi}\int_{-\infty}^{\infty} \psi_0(R'_\alpha)\mathrm{e}^{ik^\alpha_{v'''j'''}R'_\alpha}\, \mathrm{d}R'_\alpha\right\} \\
&= -\left(\frac{\mu^\alpha 2\pi}{\hbar^2}\right)^{\frac{1}{2}} \frac{S^{J,\beta,\alpha}_{vj\Omega,v''j''\Omega''}}{(k^\alpha_{vj})^{1/2}} g(-k^\alpha_{vj})
\end{aligned}
\tag{H-29}
$$

where $\psi_0(R'_\alpha)$ is the is the Jacobi R dependence of the initial wavepacket (see eqn (7.13)) and $g(k)$ is its Fourier transform (see eqn (7.14)).

We turn now to the second of the angular brackets $\langle\cdots|\cdots\rangle$ in eqn (H-28). Recalling that the angular brackets $\langle\cdots|\cdots\rangle$ denote integration over spatial variables $R_\beta, r_\beta, \gamma_\beta$. The brackets contain the Dirac delta function $\delta(R_\beta - R_{\beta\infty})$. The presence of delta function means that the function $\Phi^{J,\beta-}_{v''j''\Omega''}(E')$ takes on the first of the two asymptotic forms given in eqn (H-23). Bearing all this in mind, and using eqn (H-24) we can write

$$
\begin{aligned}
&\langle \chi_{v'j'}(r_\beta)\,\Theta_{j'\Omega'}(\gamma_\beta)\,\delta(R_\beta - R_{\beta\infty})|\Phi^{J,\beta-}_{v''j''\Omega''}(E')\rangle \\
&= \int_0^\infty \mathrm{d}R_\beta\,\delta(R_\beta - R_{\beta\infty}) \int \chi^*_{v'j'}(r_\beta)\,\Theta^*_{j'\Omega'}(\gamma_\beta)\,\Phi^{J,\beta-}_{v''j''\Omega''}(E')\,\mathrm{d}r_\beta \sin\gamma_\beta\,\mathrm{d}\gamma_\beta \\
&= \int_0^\infty \mathrm{d}R_\beta\,\delta(R_\beta - R_{\beta\infty}) \int \chi^*_{v'j'}(r_\beta)\,\Theta^*_{j'\Omega'}(\gamma_\beta) \left(\frac{\mu^\beta}{2\pi\hbar^2 k^\beta_{v''j''}}\right)^{\frac{1}{2}} \\
&\quad \times \left\{ \mathrm{e}^{ik^\beta_{v''j''}R_\beta}\chi^\beta_{v''j''\Omega''} - \sum_{v'''j'''\Omega'''} (S^{J,\beta,\beta}_{v'''j'''\Omega''',v''j''\Omega''})^* \right. \\
&\quad \left. \times \left(\frac{k^\beta_{v''j''}}{k^\beta_{v'''j'''}}\right)^{\frac{1}{2}} \mathrm{e}^{-ik^\beta_{v'''j'''}R_\beta}\chi^\beta_{v'''j'''\Omega'''} \right\} \mathrm{d}r_\beta \sin\gamma_\beta\,\mathrm{d}\gamma_\beta \\
&= \int_0^\infty \mathrm{d}R_\beta\,\delta(R_\beta - R_{\beta\infty}) \left(\frac{\mu^\beta}{2\pi\hbar^2 k^\beta_{v''j''}}\right)^{\frac{1}{2}} \\
&\quad \times \left\{ \mathrm{e}^{ik^\beta_{v'j'}R_\beta}\delta_{v',v''}\delta_{j',j''}\delta_{\Omega',\Omega''} - (S^{J,\beta,\beta}_{v'j'\Omega',v''j''\Omega''})^* \left(\frac{k^\beta_{v''j''}}{k^\beta_{v'j'}}\right)^{\frac{1}{2}} \mathrm{e}^{-ik^\beta_{v'j'}R_\beta} \right\}.
\end{aligned}
\tag{H-30}
$$

We can now substitute eqn (H-29) and (H-30) into eqn (H-28) to obtain

$$\begin{aligned}
A^{J}_{vj\Omega\to v'j'\Omega'}(E) = &-\frac{1}{2\pi}\int_0^\infty \mathrm{d}t \int_{-\infty}^{\infty} \mathrm{d}E' \sum_{v''j''\Omega''} \mathrm{e}^{i(E-E')t/\hbar} \\
&\times \int_0^\infty \mathrm{d}R_\beta\, \delta(R_\beta - R_{\beta\infty}) \left(\frac{\mu^\beta}{2\pi\hbar^2 k^\beta_{v''j''}}\right)^{\frac{1}{2}} \\
&\times \left\{ \mathrm{e}^{ik^\beta_{v'j'}R_\beta}\delta_{v',v''}\delta_{j',j''}\delta_{\Omega',\Omega''} - (S^{J,\beta,\beta}_{v'j'\Omega',v''j''\Omega''})^* \right. \\
&\times \left. \left(\frac{k^\beta_{v''j''}}{k^\beta_{v'j'}}\right)^{\frac{1}{2}} \mathrm{e}^{-ik^\beta_{v'j'}R_\beta} \right\} \\
&\times \left(\frac{\mu^\alpha 2\pi}{\hbar^2}\right)^{\frac{1}{2}} \frac{S^{J,\beta,\alpha}_{vj\Omega,v''j''\Omega''}}{(k^\alpha_{vj})^{1/2}} g(-k^\alpha_{vj}).
\end{aligned} \tag{H-31}$$

We now define the integrals $F_\pm(E,R)$

$$F_\pm(E,R) = \frac{1}{2\pi}\int_0^\infty \mathrm{d}t \int_{-\infty}^{\infty} \mathrm{d}E' \mathrm{e}^{i(E-E')t/\hbar} \frac{\mathrm{e}^{\pm ik^\beta_{v'j'}R_\beta}}{\sqrt{k^\beta_{v'j'}(E')}} \tag{H-32}$$

where we have explicitly noted that $k^\beta_{v'j'}$ is dependent on the total energy E' (see just below eqn (3.18) or eqn (3.8)) and we note that we wish to evaluate the integral at asymptotically large values of R.

The integral over time is

$$\begin{aligned}
&\frac{1}{2\pi}\int_0^\infty \mathrm{d}t\, \mathrm{e}^{i(E-E')t/\hbar} \\
&= \frac{1}{2\pi}\lim_{t\to\infty}\left\{\int_{t'=0}^{t} \mathrm{d}t' \mathrm{e}^{i(E-E')t'/\hbar}\right\} \\
&= \frac{1}{2\pi}\lim_{t\to\infty}\left\{\frac{\hbar}{i(E-E')}\left[\mathrm{e}^{i(E-E')t/\hbar} - 1\right]\right\} \\
&= \frac{1}{2}\lim_{t\to\infty}\left\{\frac{1}{\pi}\frac{\sin[(E-E')t/\hbar]}{(E-E')/\hbar} + \frac{i}{\pi}\frac{[1-\cos[(E-E')t/\hbar]]}{(E-E')/\hbar}\right\}.
\end{aligned} \tag{H-33}$$

But if we use the identities $\delta(x) = \lim_{t\to\infty}\left\{\frac{1}{\pi}\frac{\sin xt}{x}\right\}$[8] and $\delta(x/\hbar) = \hbar\delta(x)$ we can rewrite eqn (H-33) as

$$\frac{1}{2\pi}\int_0^\infty \mathrm{d}t\, \mathrm{e}^{i(E-E')t/\hbar} = \frac{\hbar}{2}\delta(E-E') + \frac{1}{2}\lim_{t\to\infty}\left\{\frac{i}{\pi}\frac{[1-\cos[(E-E')t/\hbar]]}{(E-E')/\hbar}\right\}. \tag{H-34}$$

Substituting this into eqn (H-32) we obtain

$$
\begin{aligned}
F_{\pm}(E,R) \\
&= \int_{-\infty}^{\infty} \mathrm{d}E' \left\{ \frac{\hbar}{2}\delta(E - E') + \frac{1}{2}\lim_{t\to\infty}\left[\frac{i}{\pi}\frac{[1 - \cos[(E - E')t/\hbar]]}{(E - E')/\hbar}\right]\right\} \frac{\mathrm{e}^{\pm ik^{\beta}_{\nu' j'}R_{\beta}}}{\sqrt{k^{\beta}_{\nu' j'}(E')}} \\
&= \frac{\hbar}{2}\frac{\mathrm{e}^{\pm ik^{\beta}_{\nu' j'}R_{\beta}}}{\sqrt{k^{\beta}_{\nu' j'}(E)}} + \frac{i}{2\pi}\lim_{t\to\infty}\left\{\int_{-\infty}^{\infty} \mathrm{d}E' \left[\frac{[1 - \cos[(E - E')t/\hbar]]}{(E - E')/\hbar}\right]\frac{\mathrm{e}^{\pm ik^{\beta}_{\nu' j'}R_{\beta}}}{\sqrt{k^{\beta}_{\nu' j'}(E')}}\right\}.
\end{aligned}
\tag{H-35}
$$

Setting $\omega = (E - E')/\hbar$ we can write

$$
\begin{aligned}
f(\omega) &= \frac{[1 - \cos[(E - E')t/\hbar]]}{(E - E')/\hbar} \\
&= \frac{[1 - \cos\omega t]}{\omega}.
\end{aligned}
\tag{H-36}
$$

In Figure H.1 we plot the function $f(\omega)$ for two values of t. We see that the function has sharp peaks close to $\omega = 0$. As $t \to \infty$ these two peaks become progressively more pronounced and get progressively closer to $\omega = 0$.

Let us now consider the second term in eqn (H-35). The presence of the function $f(\omega)$ (*i.e.* the term in square brackets in eqn (H-37) below) and of the limit $t \to \infty$ means that we can replace $\frac{1}{\sqrt{k^{\beta}_{\nu' j'}(E')}}$ by $\frac{1}{\sqrt{k^{\beta}_{\nu' j'}(E)}}$

$$
\begin{aligned}
&\lim_{t\to\infty}\left\{\int_{-\infty}^{\infty} \mathrm{d}E' \left[\frac{[1 - \cos[(E - E')t/\hbar]]}{(E - E')/\hbar}\right]\frac{\mathrm{e}^{\pm ik^{\beta}_{\nu' j'}R_{\beta}}}{\sqrt{k^{\beta}_{\nu' j'}(E')}}\right\} \\
&\quad = \frac{1}{\sqrt{k^{\beta}_{\nu' j'}(E)}}\lim_{t\to\infty}\left\{\int_{-\infty}^{\infty} \mathrm{d}E' \left[\frac{[1 - \cos[(E - E')t/\hbar]]}{(E - E')/\hbar}\right]\mathrm{e}^{\pm ik^{\beta}_{\nu' j'}R_{\beta}}\right\}.
\end{aligned}
\tag{H-37}
$$

The term $k^{\beta}_{\nu' j'}$ appearing in the exponential in eqn (H-37) can now be rewritten in terms of a function $k(\omega)$ defined as

$$
\begin{aligned}
k(\omega)^2 &= \left[k^{\beta}_{\nu' j'}(E')\right]^2 \\
&= \frac{2\mu^{\beta}}{\hbar^2}\left(E' - \varepsilon_{\nu' j'}\right) \\
&= \frac{2\mu^{\beta}}{\hbar^2}\left(E - \varepsilon_{\nu' j'}\right) - \frac{2\mu^{\beta}}{\hbar^2}(E - E') \\
&= \left[k^{\beta}_{\nu' j'}(E)\right]^2 - \frac{2\mu^{\beta}\omega}{\hbar}.
\end{aligned}
\tag{H-38}
$$

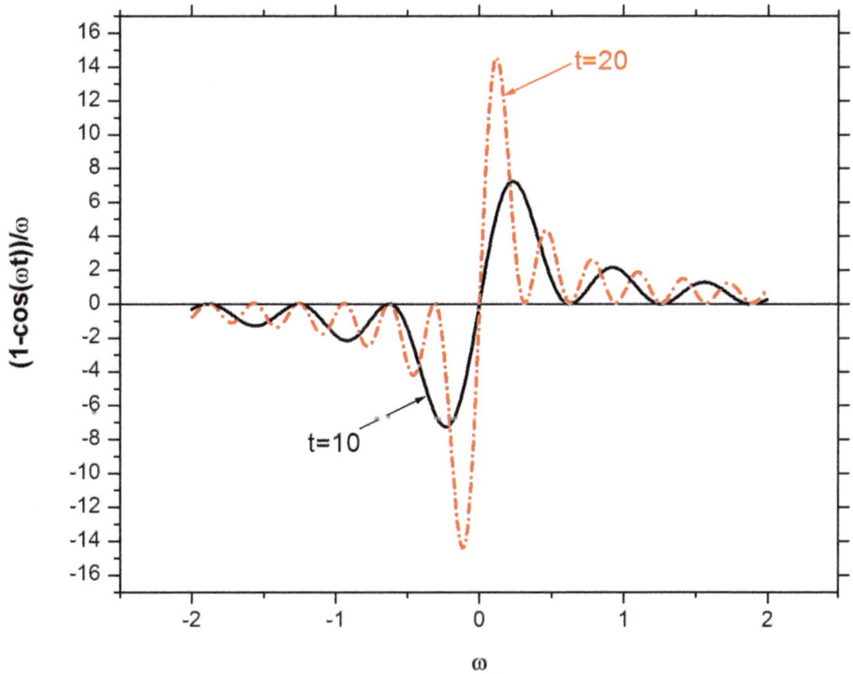

Figure H.1 Plot of the function $f(\omega) = (1 - \cos \omega t)/\omega$ for two values of t. The solid line shows $t = 10$, the dot-dashed line $t = 20$.

Taking account of the fact that $\omega \to 0$ in the limit $t \to \infty$, we can further write

$$\begin{aligned} k(\omega) &= \left[\left[k^{\beta}_{v'j'}(E)\right]^2 - \frac{2\mu^{\beta}\omega}{\hbar}\right]^{\frac{1}{2}} \\ &= k^{\beta}_{v'j'}(E)\left[1 - \frac{2\mu^{\beta}\omega}{\hbar\left[k^{\beta}_{v'j'}(E)\right]^2}\right]^{\frac{1}{2}} \\ &\approx k^{\beta}_{v'j'}(E)\left[1 - \frac{\mu^{\beta}\omega}{\hbar\left[k^{\beta}_{v'j'}(E)\right]^2}\right] \\ &\approx k^{\beta}_{v'j'}(E) - \frac{\mu^{\beta}\omega}{\hbar k^{\beta}_{v'j'}(E)}. \end{aligned} \tag{H-39}$$

Substituting eqn (H-39) into eqn (H-37), changing the variable of integration to ω (note that $dE' = -\hbar d\omega$) and splitting the integral into two parts we can write

$$
\begin{aligned}
&\lim_{t\to\infty}\left\{\int_{-\infty}^{\infty} \mathrm{d}E' \left[\frac{[1-\cos[(E-E')t/\hbar]]}{(E-E')/\hbar}\right] \frac{\mathrm{e}^{\pm ik^{\beta}_{v'j'}R_\beta}}{\sqrt{k^{\beta}_{v'j'}(E')}}\right\}\\
&= \frac{\hbar \mathrm{e}^{\pm ik^{\beta}_{v'j'}(E)R_\beta}}{\sqrt{k^{\beta}_{v'j'}(E)}} \lim_{t\to\infty}\left\{\int_{\omega=-\infty}^{0} \mathrm{d}\omega \left[\frac{[1-\cos\omega t]}{\omega}\right] e^{\mp i\mu^\beta \omega R_\beta/\left(\hbar k^{\beta}_{v'j'}(E)\right)}\right.\\
&\left.+\int_{\omega=0}^{\infty} \mathrm{d}\omega \left[\frac{[1-\cos\omega t]}{\omega}\right] \mathrm{e}^{\mp i\mu^\beta \omega R_\beta/\left(\hbar k^{\beta}_{v'j'}(E)\right)}\right\}\\
&= \frac{\hbar \mathrm{e}^{\pm ik^{\beta}_{v'j'}(E)R_\beta}}{\sqrt{k^{\beta}_{v'j'}(E)}} \lim_{t\to\infty}\left\{-\int_{\omega=0}^{\infty} \mathrm{d}\omega \left[\frac{[1-\cos\omega t]}{\omega}\right] \mathrm{e}^{\pm i\mu^\beta \omega R_\beta/\left(\hbar k^{\beta}_{v'j'}(E)\right)}\right.\\
&\left.+\int_{\omega=0}^{\infty} \mathrm{d}\omega \left[\frac{[1-\cos\omega t]}{\omega}\right] \mathrm{e}^{\mp i\mu^\beta \omega R_\beta/\left(\hbar k^{\beta}_{v'j'}(E)\right)}\right\}\\
&= \frac{\hbar \mathrm{e}^{\pm ik^{\beta}_{v'j'}(E)R_\beta}}{\sqrt{k^{\beta}_{v'j'}(E)}} \lim_{t\to\infty}\left\{\int_{\omega=0}^{\infty} \mathrm{d}\omega \left[\frac{[1-\cos\omega t]}{\omega}\right] \left[\mathrm{e}^{\mp i\mu^\beta \omega R_\beta/\left(\hbar k^{\beta}_{v'j'}(E)\right)}\right.\right.\\
&\left.\left.-\mathrm{e}^{\pm i\mu^\beta \omega R_\beta/\left(\hbar k^{\beta}_{v'j'}(E)\right)}\right]\right\}\\
&= \mp\frac{2i\hbar \mathrm{e}^{\pm ik^{\beta}_{v'j'}(E)R_\beta}}{\sqrt{k^{\beta}_{v'j'}(E)}}\\
&\times \lim_{t\to\infty}\left\{\int_{\omega=0}^{\infty} \mathrm{d}\omega[1-\cos\omega t]\left[\frac{\sin\left(\mu^\beta \omega R_\beta/\left(\hbar k^{\beta}_{v'j'}(E)\right)\right)}{\omega}\right]\right\}\\
&= \mp\frac{2i\hbar \mathrm{e}^{\pm ik^{\beta}_{v'j'}(E)R_\beta}}{\sqrt{k^{\beta}_{v'j'}(E)}} \lim_{t\to\infty}\left\{\int_{\omega=0}^{\infty} \mathrm{d}\omega \frac{\sin\left(\mu^\beta \omega R_\beta/\left(\hbar k^{\beta}_{v'j'}(E)\right)\right)}{\omega}\right.\\
&\left.-\int_{\omega=0}^{\infty} \mathrm{d}\omega \cos\omega t\left[\frac{\sin\left(\mu^\beta \omega R_\beta/\left(\hbar k^{\beta}_{v'j'}(E)\right)\right)}{\omega}\right]\right\}.
\end{aligned}
\tag{H-40}
$$

To evaluate the first integral in the last line of eqn (H-40) we use the fact that the value of R is chosen to be asymptotically large and that for such large values of R the integrand is a representation of the Dirac delta function $\delta(\omega) = \lim_{R\to\infty}\left\{\frac{1}{\pi}\frac{\sin\omega R}{\omega}\right\}$.[8] Using this the integral yields

$$\lim_{R\to\infty}\left\{\int_{\omega=0}^{\infty} d\omega \frac{\sin\left(\mu^{\beta}\omega R_{\beta}/\left(\hbar k_{v'j'}^{\beta}(E)\right)\right)}{\omega}\right\} = \frac{\pi}{2}. \tag{H-41}$$

The second integral in eqn (H-40) is

$$\lim_{t\to\infty}\left\{\int_{\omega=0}^{\infty} d\omega \cos\omega t \left[\frac{\sin\left(\mu^{\beta}\omega R_{\beta}/\left(\hbar k_{v'j'}^{\beta}(E)\right)\right)}{\omega}\right]\right\} = 0. \tag{H-42}$$

For any fixed value of R, the term $\cos\omega t$ oscillates rapidly between $+1$ and -1 thus causing the integral to be zero.

Substituting eqn (H-42) and (H-41) into eqn (H-40) we obtain

$$\lim_{t\to\infty}\left\{\int_{-\infty}^{\infty} dE' \left[\frac{[1-\cos[(E-E')t/\hbar]]}{(E-E')/\hbar}\right] \frac{e^{\pm ik_{v'j'}^{\beta}R_{\beta}}}{\sqrt{k_{v'j'}^{\beta}(E')}}\right\} = \mp\frac{i\pi\hbar e^{\pm ik_{v'j'}^{\beta}(E)R_{\beta}}}{\sqrt{k_{v'j'}^{\beta}(E)}} \tag{H-43}$$

and substituting this result back into eqn (H-35) we obtain

$$F_{\pm}(E,R) = \frac{\hbar}{2}\frac{e^{\pm ik_{v'j'}^{\beta}R_{\beta}}}{\sqrt{k_{v'j'}^{\beta}(E)}} \pm \frac{\hbar}{2}\frac{e^{\pm ik_{v'j'}^{\beta}(E)R_{\beta}}}{\sqrt{k_{v'j'}^{\beta}(E)}} \tag{H-44}$$

so we finally get the result

$$\begin{aligned} F_{+}(E,R) &= \frac{\hbar}{\sqrt{k_{v'j'}^{\beta}}} e^{ik_{v'j'}^{\beta}R_{\beta}}, \\ F_{-}(E,R) &= 0. \end{aligned} \tag{H-45}$$

Now substituting this back into eqn (H-31) we obtain

$$\begin{aligned} A_{vj\Omega\to v'j'\Omega'}^{J}(E) &= -\frac{1}{\hbar}\left(\frac{\mu^{\alpha}\mu^{\beta}}{k_{vj}^{\alpha}k_{v'j'}^{\beta}}\right)^{\frac{1}{2}} S_{v'j'\Omega',vj\Omega}^{J,\beta,\alpha}\, g(-k_{vj}^{\alpha}) \\ &\quad\times \int_{0}^{\infty} dR_{\beta}\,\delta(R_{\beta}-R_{\beta\infty}) e^{ik_{v'j'}^{\beta}R_{\beta}} \\ &= -\frac{1}{\hbar}\left(\frac{\mu^{\alpha}\mu^{\beta}}{k_{vj}^{\alpha}k_{v'j'}^{\beta}}\right)^{\frac{1}{2}} S_{v'j'\Omega',vj\Omega}^{J,\beta,\alpha}\, g(-k_{vj}^{\alpha}) e^{ik_{v'j'}^{\beta}R_{\beta\infty}}. \end{aligned} \tag{H-46}$$

H.2.1 Trick Analysis

The initial wavepacket should, in principal, be located in a region where both the potential and the centrifugal potential are zero. In practice, especially for $J > 0$, it will be difficult or impossible to do this. In order to allow for this we have developed a technique[9] named "Trick Analysis Method" which corrects for this situation. We assume that the initial wavepacket is situated sufficiently far out that the action of the Hamiltonian will not lead to any inelastic scattering. We then create an effective one-dimensional potential by averaging the potential over the initial two-dimensional vibrational–rotational wavefunction and propagate the R_α dependent part of the initial wavepacket $\psi_0(R_\alpha)$ (see eqn (7.13)) outwards or backwards to larger values of R_α. We then take the Fourier transform of the back-propagated wavepacket to find the "correct" value of $g(k)$ (see eqn (7.14) and (H-29)).

H.3 Computing the S Matrix in the Space-Fixed Reference Frame

Because of the long range centrifugal coupling present in the body-fixed reference frame it is best, or even necessary, to perform the asymptotic analysis of the wavepacket in the space-fixed representation (see Appendix D Section D.5, [Appendix A of ref. 6] and ref. 10). In order to transform the $\mathbf{A}^J$ matrix of eqn (H-19) and (H-46) to the space-fixed frame we use the **B** matrix of Appendix D eqn (D-10)

$$A^J_{vj\ell\to v'j'\ell'}(E) = \sum_{\Omega\,\Omega'} B^\alpha_{\ell,\Omega}\, A^J_{vj\Omega\to v'j'\Omega'}(E) B^\beta_{\ell',\Omega'} \tag{H-47}$$

where the labels α and β on the **B** matrices indicate that the matrices refer to reactants and products respectively.

The space-fixed **S** matrix is then obtained using an equation similar to eqn (H-46)

$$S^{J,\beta,\alpha}_{v'j'\ell',vj\ell} = -\hbar\left(\frac{k^\alpha_{vj}k^\beta_{v'j'}}{\mu^\alpha\mu^\beta}\right)^{\frac{1}{2}} \frac{A^J_{vj\ell\to v'j'\ell'}(E)}{g(-k^\alpha_{vj})} e^{-ik^\beta_{v'j'}R_{\beta\infty}}. \tag{H-48}$$

This equation would be adequate if the analysis line ($R_{\beta\infty}$) were chosen to be in a region where both the potential and the centrifugal potential were zero. In practice, however, this is never possible. This is especially so when $J > 0$. In the product exit channel the local value of the wave vector at the analysis line is given by

$$k^\beta_{v'j'} = \sqrt{2\,\mu^\beta\left(E - \left(\varepsilon_{v'j'} + \frac{\ell'(\ell'+1)\hbar^2}{2\,\mu^\beta R^2_{\beta\infty}}\right)\right)} \tag{H-49}$$

where the term $\dfrac{\ell'(\ell'+1)\hbar^2}{2\,\mu^\beta R^2_{\beta\infty}}$ arises to allow for the fact that the centrifugal potential may not be zero at the position of the analysis line. Because of this, a correction, $\delta\eta_{v'j'\ell'}$, should be applied to the phase of the **S** matrix

$$\delta\eta_{v'j'\ell'} = \int_{R_{\beta\infty}}^{\infty} \mathrm{d}R\left\{\sqrt{2\,\mu^\beta\left(E - \varepsilon_{v'j'} - \frac{\ell'(\ell'+1)}{2\,\mu^\beta R^2}\right)} - \sqrt{2\,\mu^\beta\left(E - \varepsilon_{v'j'}\right)}\right\} \tag{H-50}$$

while similar correction is also needed in the entrance channel

$$\delta\eta_{\nu j\ell} = \int_{R_0}^{\infty} \mathrm{d}R \left\{ \sqrt{2\,\mu^\alpha \left(E - \varepsilon_{\nu j} - \frac{\ell(\ell+1)}{2\,\mu^\alpha R^2}\right)} - \sqrt{2\,\mu^\alpha\,(E - \varepsilon_{\nu j})}. \right\} \tag{H-51}$$

The integrals in eqn (H-50) and (H-51) can be performed analytically to yield (see ref. 11 integrals number 2.265 and 2.266)

$$\begin{aligned} \delta\eta &= \int_{R^*}^{\infty} \mathrm{d}R \left\{ \sqrt{a - \frac{b}{R^2}} - \sqrt{a} \right\} \\ &= -\sqrt{a(R^*)^2 - b} + \sqrt{b}\arcsin\left(-\sqrt{\frac{b}{a(R^*)^2}}\right) + \sqrt{a}R^* \end{aligned} \tag{H-52}$$

where $R^* = R_{\beta\infty}$ or $R^* = R_0$, $a = 2\mu^\beta(E - \varepsilon_{\nu' j'})$ or $a = 2\mu^\alpha(E - \varepsilon_{\nu j})$ and $b = \ell'(\ell' + 1)$ or $b = \ell(\ell + 1)$ for eqn (H-50) and (H-51) respectively. The final expression for the **S** matrix elements, together with the phase corrections (eqn (H-50) and (H-51)), is therefore

$$S^{J,\beta,\alpha}_{\nu' j' \ell', \nu j \ell} = -\hbar \left(\frac{k^\alpha_{\nu j} k^\beta_{\nu' j'}}{\mu^\alpha \mu^\beta}\right)^{\frac{1}{2}} \frac{A^J_{\nu j \ell \to \nu' j' \ell'}(E)}{g(-k^\alpha_{\nu j})} \mathrm{e}^{-i\left(k^\beta_{\nu' j'} R_{\beta\infty} + \delta\eta_{\nu' j' \ell'} + \delta\eta_{\nu j \ell}\right)}. \tag{H-53}$$

Having calculated the **S** matrix in the space-fixed basis we now transform back to the body-fixed basis using the same transformation matrices, $\mathbf{B}^\alpha$ and $\mathbf{B}^\beta$, as before (see eqn (H-47) and the comments below it) to yield the result

$$S^{J,\beta,\alpha}_{\nu' j' \Omega', \nu j \Omega} = \sum_{\ell\,\ell'} B^\beta_{\ell',\Omega'} S^{J,\beta,\alpha}_{\nu' j' \ell', \nu j \ell} B^\alpha_{\ell,\Omega}. \tag{H-54}$$

References

1. R. N. Zare, *Angular Momentum*, John Wiley and Sons, New York, 1988. See also list of corrections at web address: http://web.stanford.edu/group/Zarelab/Errata.pdf.
2. W. A. Vetterling, W. H. Press, S. A. Teukolsky and B. P. Flannery, *Numerical Recipes in Fortran*, Cambridge University Press, Cambridge, 2nd edn, 1992.
3. A. R. Edmonds, *Angular Momentum in Quantum Mechanics*, Princeton University Press, Princeton, New Jersey, 1960.
4. D. A. Varshalovich, A. N. Moskalev and V. K. Khersonskii, *Quantum Theory of Angular Momentum*, World Scientific, Singapore, 1988.
5. S. K. Gray and G. G. Balint-Kurti, Quantum dynamics with real wave packets, including application to three-dimensional ($J = 0$) D + H_2 → HD + H reactive scattering, *J. Chem. Phys.*, 1998, **108**, 950.
6. M. Hankel, S. C. Smith, R. J. Allan, S. K. Gray and G. G. Balint-Kurti, State-to-state reactive differential cross sections for the H + H_2 → H_2 + H

reaction on five different potential energy surfaces employing a new quantum wavepacket computer code: DIFFREALWAVE, *J. Chem. Phys.*, 2006, **125**, 164303.

7. G. G. Balint-Kurti, R. N. Dixon and C. C. Marston, Time-dependent quantum dynamics of molecular photofragmentation processes, *J. Chem. Soc. Faraday Trans.*, 1990, **86**, 1741.
8. C. Cohen-Tannoudji, B. Diu and F. Laloë, *Quantum mechanics, Vol. 2, Appendix II*, Wiley, New York, 1977.
9. S. K. Gray, E. M. Goldfield, G. C. Schatz and G. G. Balint-Kurti, Helicity decoupled quantum dynamics and capture model cross sections and rate constants for $O(^1D) + H_2 \rightarrow OH + H$, *Phys. Chem. Chem. Phys.*, 1999, **1**, 1141.
10. S. C. Althorpe, Quantum wavepacket method for state-to-state reactive cross sections, *J. Chem. Phys.*, 2001, **114**, 1601.
11. I. S. Gradshteyn and I. M. Ryhzik, *Table of Integrals, Series and Products*, Academic Press, New York, 4th edn, 1965.

APPENDIX I

Technical Aspects of Time-Dependent Quantum Dynamics

In this appendix we discuss in various technical topics needed for the study of time-dependent quantum dynamics. These include the use of a grid representation for the wavepacket, the scaling of the Hamiltonian to permit the use of the Chebyshev expansion of the time propagation operator, the operation of the Hamiltonian on a wavepacket and a description of the Fourier Grid Hamiltonian method for the calculations of diatomic vibrational wavefunctions.

I.1 The Angular Discrete Variable Representation (DVR) Approach for $J = 0$

We now consider how to operate with the angle dependent operator $\hat{j}^2$ when the wavepacket is defined by its amplitude at a set of grid points. We will choose the so-called Gauss–Legendre grid points[1] for reasons which will become clear below. The Discrete Variable Representation approach which we introduce here was popularized in the quantum dynamics field by Lill, Parker and Light[2,3] but is based on earlier work.[4,5]

Let us consider the action of the operator $\hat{j}^2$ on a function of the angle γ as encountered in eqn (H-16) and (7.19). For convenience we will treat all the angular functions as a function of $x = \cos\gamma$. As γ takes values only from 0 to π there is a single valued, unique, relationship between the two variables:

$$F(x = \cos\gamma) = \hat{j}^2 f(x) \tag{I-1}$$

RSC Theoretical and Computational Chemistry Series No. 7
Theory of Molecular Collisions
By Gabriel G. Balint-Kurti and Alexander P. Palov

Published by the Royal Society of Chemistry, www.rsc.org

The natural way to execute the action of the $\hat{j}^2$ on the right hand side of this equation is to expand the function $f(x)$ in terms of the eigenfunctions of the operator. But when $J = 0$ these eigenfunctions are just the Legendre polynomials, $P_j(x)$, or, more generally, when $J \neq 0$ they are the associated Legendre Polynomials, $P_j^{\Omega}(x)$. As in Chapter 7 we will use a normalized version of these polynomials[6]

$$\Theta_{j,\Omega}(x) = (-1)^{-\Omega}\left[\frac{(2j+1)}{2}\frac{(j-\Omega)!}{(j+\Omega)!}\right]^{\frac{1}{2}} P_j^{\Omega}(x). \tag{I-2}$$

When $J = 0$, Ω must also be zero so we can write

$$\Theta_{j,0}(x) = \left[\frac{(2j+1)}{2}\right]^{\frac{1}{2}} P_j(x) \quad \text{when } J = 0. \tag{I-3}$$

In common with many other sets of orthogonal polynomials, the Legendre polynomials form a complete orthogonal set. The orthogonality property is

$$\int_{-1}^{1} \Theta_{j0}(x)\Theta_{j'0}(x)\,dx = \delta_{j,j'}. \tag{I-4}$$

The expansion of $f(x = \cos\gamma)$ in terms of the Legendre polynomials $\Theta_{j',0}(x)$ may be written in the form

$$f(x) = \sum_j f_j\,\Theta_{j,0}(x). \tag{I-5}$$

The expansion coefficients f_j are given by

$$f_j = \int_{-1}^{1} f(x)\,\Theta_{j,0}(x)\,\mathrm{d}x. \tag{I-6}$$

The theory of Gauss–Legendre quadrature[1] states that we may replace the integral in the above equation by the summation

$$f_j = \sum_{i=0}^{n} f(x_i)\,\Theta_{j,0}(x_i)\,w_i \tag{I-7}$$

where the w_i form a set of specially determined weight functions and that this representation of the integral will be exact if the function $f(x)$ can be expressed as a linear combination of Legendre polynomials of order up to $(2n+1)$.[1] The angular grid points, $x_i = \cos\gamma_i$, and the weights, w_i, are both specially chosen to ensure this accuracy. They are referred to as the Gauss–Legendre quadrature points and weights.[1]

In an important paper Dickinson and Certain[4] proved that the Gauss–Legendre quadrature points were the eigenvalues of a matrix whose elements are the matrix elements of "x" between the Legendre polynomials, *i.e.* in the equation below the secular determinant is equal to zero when $\lambda = x_i$ is any of the Gauss–Legendre grid points

$$\text{Det}\left[\left\langle \Theta_{j,0}(x)\right| x \left|\Theta_{\ell,0}(x)\right\rangle - \lambda\delta_{j,\ell}\right] = 0. \tag{I-8}$$

Furthermore, they showed that the corresponding eigenvectors were

$$T_{j,i} = \Theta_{j,0}(x_i)\sqrt{w_i}. \tag{I-9}$$

Because these are eigenvectors of a Hermitian matrix the array consisting of them constitutes an orthogonal matrix with the properties

$$\sum_{i=0}^{n} T_{j,i}T_{j',i} = \delta_{j,j'}, \tag{I-10}$$

$$\sum_{j=0}^{n} T_{j,i}T_{j,k} = \delta_{i,k}. \tag{I-11}$$

The orthogonal matrix **T** is a transformation matrix between two orthonormal basis sets. One is the set of $(n+1)$ normalized Legendre polynomials, $\Theta_{j,0}(x)$ with $j = 0, 1, \ldots, n$, and the other is a set of functions which we will denote as $\chi_i(x)$, which are the best approximation to Dirac delta functions localized at the Gauss–Legendre quadrature points (*i.e.* $\delta(x - x_i)$) which it is possible to construct from the set of $n+1$ normalized Legendre polynomials. We may write these relationships as

$$\chi_i(x) = \sum_{j=0}^{n} T_{j,i}\Theta_{j,0}(x), \tag{I-12}$$

$$\Theta_{j,0}(x) = \sum_{i=0}^{n} T_{j,i}\chi_i(x). \tag{I-13}$$

The Legendre polynomials are eigenfunctions of the angular momentum operator $\hat{\boldsymbol{j}}^2$ which satisfy the equation

$$\hat{\boldsymbol{j}}^2\Theta_{j,0}(x) = \hbar^2 j(j+1)\Theta_{j0}(x). \tag{I-14}$$

For use with a grid based wavepacket we require a grid representation of the angular momentum operator $\hat{\boldsymbol{j}}^2$ and we are now in a position to be able to find this representation

$$\begin{aligned}\left[\hat{\boldsymbol{j}}^2\right]_{i,k} &= \langle\chi_i(x)|\,\hat{\boldsymbol{j}}^2\,|\chi_k(x)\rangle \\ &= \left\langle\left\{\sum_{j'=0}^{n} T_{j',i}\Theta_{j',0}(x)\right\}\right|\hat{\boldsymbol{j}}^2\left|\left\{\sum_{j=0}^{n} T_{j,k}\Theta_{j,0}(x)\right\}\right\rangle\end{aligned}$$

$$= \sum_{j'=0}^{n} \sum_{j=0}^{n} \langle \{T_{j',i}\Theta_{j',0}(x)\} | \hbar^2 j(j+1) | \{T_{j,k}\Theta_{j,0}(x)\} \rangle$$

$$= \sum_{j'=0}^{n} \sum_{j=0}^{n} \hbar^2 j(j+1) T_{j',i} \langle \Theta_{j',0}(x) | \Theta_{j,0}(x) \rangle T_{j,k}$$

$$= \sum_{j'=0}^{n} \sum_{j=0}^{n} T_{j',i}\hbar^2 j(j+1) \delta_{j',j} T_{j,k}$$

$$= \sum_{j=0}^{n} T_{j,i}\hbar^2 j(j+1) T_{j,k} \tag{I-15}$$

where we have used the expansion of $\chi_i(x)$ in terms of $\Theta_{j,0}(x)$ from eqn (I-12).

So if we have an angular wavefunction $f(x_i)$ specified by its values at the Gauss–Legendre quadrature points x_i, then the action of the $\hat{\boldsymbol{j}}^2$ is given by a matrix multiplication

$$F(x_i) = \left[\hat{\boldsymbol{j}}^2 \boldsymbol{f}\right]_i = \sum_k \left\{ \sum_{j=0}^{n} T_{j,i}\hbar^2 j(j+1) T_{j,k} \right\} f(x_k). \tag{I-16}$$

I.2 The Angular Discrete Variable Representation (DVR) Approach for $J > 0$

When $J > 0$, Ω may also take on non-zero values. In this case the Legendre polynomials are no longer the correct eigenfunctions of the rotational angular momentum operator $\hat{\boldsymbol{j}}^2$ [7–9] but the correct eigenfunctions are now rather the associated Legendre polynomials $\Theta_{j,\Omega}(\cos\gamma)$ (see eqn (I-2)). Ω can take on values from $-J$ to $+J$, and for each value of Ω a different set of associated Legendre polynomials should be used. Each of these sets has their own distinct Gauss–Legendre grid points and weights. If we adopted the same strategy as for the $J = 0$ case above, we would end up with a different angular grid for each value of Ω. Some researchers have pursued this route, but Corey and Lemoine,[10] based on earlier work by Light, Hamilton and Lill[2] and by Leforestier,[11] have shown how it is possible to use the same angular grid points for all values of Ω. Below, without much justification, we will state how this is done and outline the evaluation of the action of the various terms in the Hamiltonian on the grid based wavepacket.

The same angular grid (quadrature) points and weights are used as for the Legendre polynomials, however the transformation matrix that transforms between the grid and the fixed basis set representations is different for each value of Ω (see eqn (I-9)) and now becomes

$$T^{\Omega}_{j,i} = \Theta_{j,\Omega}(x_i)\sqrt{w_i}. \tag{I-17}$$

Another subtle difference arises from the fact that for a non-zero value of Ω the j value of the associated Legendre polynomial cannot be less than $|\Omega|$. Therefore the summations over j start at $j = |\Omega|$.

I.2.1 The $\hat{j}^2$ Operator

The operator $\hat{j}^2$ does not couple different values of Ω. Its action on a function $\boldsymbol{f}_\Omega$ associated with a body-fixed Z component of the angular momentum of Ω is given by (see eqn (I-16)

$$F_\Omega(x_i) = \left[\hat{j}^2 \boldsymbol{f}_\Omega\right]_i = \sum_k \left\{ \sum_{j=|\Omega|}^{n} T^{\Omega}_{j,i} \hbar^2 \, j(j+1) \, T^{\Omega}_{j,k} \right\} f_\Omega(x_k). \tag{I-18}$$

I.2.2 The $\hat{j}_+$ and $\hat{j}_-$ Operators

We now also have to deal with the action of the $\hat{j}_+$ and $\hat{j}_-$ operators on a grid based wavefunction. These operators couple wavepackets associated with different Ω' values. Let us look at the action of the $\hat{j}_+$ operator. It acts on a wavepacket associated with quantum number $\Omega' - 1$ so as to make a contribution to the wavepacket with quantum number Ω'. The matrix elements of the operator in the grid representation are therefore (see comment below eqn (H-14) of Appendix H)

$$\begin{aligned}
\left[\hat{j}_+\right]_{i,k} &= \left\langle \chi_i^{\Omega'}(x) \middle| \hat{j}_+ \middle| \chi_k^{\Omega'-1}(x) \right\rangle \\
&= \left\langle \left\{ \sum_{j'=|\Omega'|}^{n} T^{\Omega'}_{j',i} \Theta_{j',\Omega'}(x) \right\} \middle| \hat{j}_+ \middle| \left\{ \sum_{j=|\Omega'-1|}^{n} T^{\Omega'-1}_{j,k} \Theta_{j,\Omega'-1}(x) \right\} \right\rangle \\
&= \left\langle \left\{ \sum_{j'=|\Omega'|}^{n} T^{\Omega'}_{j',i} \Theta_{j',\Omega'}(x) \right\} \middle| \left\{ \sum_{j=|\Omega'-1|}^{n} T^{\Omega'-1}_{j,k} \hat{j}_+ \Theta_{j,\Omega'-1}(x) \right\} \right\rangle \\
&= \left\langle \left\{ \sum_{j'=|\Omega'|}^{n} T^{\Omega'}_{j',i} \Theta_{j',\Omega'}(x) \right\} \middle| \left\{ \sum_{j=|\Omega'-1|}^{n} T^{\Omega'-1}_{j,k} \hbar \lambda_+(j, \Omega'-1) \Theta_{j,\Omega'}(x) \right\} \right\rangle \\
&= \hbar \sum_{j'=|\Omega'|}^{n} T^{\Omega'}_{j',i} \sum_{j=|\Omega'-1|}^{n} \lambda_-(j,\Omega') T^{\Omega'-1}_{j,k} \left\langle \Theta_{j',\Omega'}(x) \middle| \Theta_{j,\Omega'}(x) \right\rangle \\
&= \hbar \sum_{j'=|\Omega'|}^{n} T^{\Omega'}_{j',i} \sum_{j=|\Omega'-1|}^{n} \lambda_-(j,\Omega') T^{\Omega'-1}_{j,k} \, \delta_{j,j'} \\
&= \hbar \sum_{j=|\Omega'|}^{n} T^{\Omega'}_{j,i} \lambda_-(j,\Omega') T^{\Omega'-1}_{j,k}.
\end{aligned} \tag{I-19}$$

Therefore the action of the $\hat{j}_+$ operator on a function associated with helicity quantum number $(\Omega' - 1)$ in a grid representation takes the form

$$
\begin{aligned}
F_{\Omega'}(x_i) &= \left[\hat{j}_+ \boldsymbol{f}_{\Omega'-1}\right]_i \\
&= \sum_k \left[\hat{j}_+\right]_{i,k} \boldsymbol{f}_{\Omega'-1}(x_k) \\
&= \sum_k \left\{ \hbar \sum_{j=|\Omega'|}^{n} T_{j,i}^{\Omega'} \lambda_-(j,\Omega') T_{j,k}^{\Omega'-1} \right\} \boldsymbol{f}_{\Omega'-1}(x_k).
\end{aligned}
\quad \text{(I-20)}
$$

A similar formula applies to the action of the operator $\hat{j}_-$ with all the + and − signs interchanged.

I.3 Use of Discrete Fourier Transforms to Evaluate Action of Radial Differential Operators

We now need to discuss the action of the radial second order derivatives in the Hamiltonian on a wavepacket. From eqn (H-16) we see that the action of these operators is of the form

$$
F(R) = \frac{\partial^2}{\partial R^2} f(R). \quad \text{(I-21)}
$$

We will first discuss the evaluation of the second derivative while treating $f(r)$ as a continuous function of R. We will then address the problem of implementing the action of the operators within a grid based representation. In both cases Fourier transforms will form the key to the process. The use of Fourier transforms in this context (for time-dependent quantum scattering theory) was first advocated by Kosloff and Kosloff.[12] Fourier transforms come in pairs. They connect functions of two different variables and convert one to the other and *vice versa*. In the present case the two variables we consider are the distance R and the wavenumber k, where the momentum is given by $k\hbar$. The Fourier transform pair is[1]

$$
H(k) = \frac{1}{\sqrt{2\pi}} \int_{-\infty}^{\infty} \mathrm{e}^{ikR} f(R) \mathrm{d}R = \mathrm{FFT}\{f(R)\} \quad \text{(I-22)}
$$

and

$$
f(R) = \frac{1}{\sqrt{2\pi}} \int_{-\infty}^{\infty} \mathrm{e}^{-ikR} H(k) \mathrm{d}k = \mathrm{FFT}^{-1}\{H(k)\} \quad \text{(I-23)}
$$

where we have introduced the notation "FFT" standing for "Fast Fourier Transform".

Using eqn (I-22) and (I-23) in eqn (I-21) we obtain

$$
\begin{aligned}
F(R) &= \frac{\partial^2}{\partial R^2} f(R) \\
&= \frac{\partial^2}{\partial R^2} \left\{ \frac{1}{\sqrt{2\pi}} \int_{-\infty}^{\infty} \mathrm{e}^{-ikR} H(k)\, \mathrm{d}k \right\}
\end{aligned}
$$

$$
\begin{aligned}
&= \frac{1}{\sqrt{2\pi}} \int_{-\infty}^{\infty} \left[\frac{\partial^2}{\partial R^2} \mathrm{e}^{-ikR} \right] H(k)\,\mathrm{d}k \\
&= \frac{1}{\sqrt{2\pi}} \int_{-\infty}^{\infty} \left[-k^2 \mathrm{e}^{-ikR} \right] H(k)\,\mathrm{d}k \\
&= \frac{1}{\sqrt{2\pi}} \int_{-\infty}^{\infty} \mathrm{e}^{-ikR} \left[-k^2 \left\{ \frac{1}{\sqrt{2\pi}} \int_{-\infty}^{\infty} \mathrm{e}^{ikR} f(R)\mathrm{d}R \right\} \right] \mathrm{d}k. \qquad \text{(I-24)}
\end{aligned}
$$

We can describe the evaluation of the second derivative of $f(R)$ in words by the steps:

1. Fourier transform $f(R)$,
2. multiply by $-k^2$,
3. back Fourier transform.

Equation (I-24) is written for a function of a continuous variable $f(R)$. In practice the wavepackets are specified by their values on a set of evenly spaced grid points. The Fourier transform operations are then accomplished through use of discrete fast Fourier computer codes.[1,13]

I.3.1 Discrete Fourier Transforms

The grids needed for Fourier transforms are evenly spaced. If the grid spacing is Δx and the overall grid length is L, the number of grid points in the one-dimensional grid is $N = L/\Delta x$. Note that Fourier transform theory assumes that the function being analyzed is cyclic, so that the value of the function at the first and last grid point are identical. The final or grid point $(N + 1)$ is not counted because the function value at this grid point is the same as that at the first grid point.[1] The maximum value of the momentum which can be represented on the grid is related to the grid spacing. The maximum value of the linear momentum is $\hbar k_{\mathrm{max}} = \pi\hbar/\Delta x$. The maximum value of the kinetic energy is therefore related to the grid spacing by $E_{\mathrm{max}} = \hbar^2\pi^2/2\mu\Delta x^2$. It is important when using Fourier transforms to take account of the fact that the theory assumes that the function being transformed is cyclic. That is, the value of the function at the end of the grid is assumed to equal that at the beginning of the grid. It is therefore vital that the wavepacket should never reach the end of the grid. The problems arising if or when it does so are referred to as "aliasing". To avoid the possibility of the wavepacket reaching the edge of the grid, an absorbing potential must be introduced and a sufficient length at the end of the of the grid must be allocated to this purpose. Absorbing potentials are discussed in Chapter (7).

I.4 Scaling the Hamiltonian

The use of the Chebyshev expansion of the time evolution operator requires us to scale the Hamiltonian so that its eigenvalue spectrum lies between -1 and $+1$ using eqn (7.6). In order to do this we need to set a value for both

E_{min} and E_{max}. For $J = 0$ this is relatively simple. The maximum kinetic energy for the atom–diatom relative motion is $E^R_{max} = \hbar^2\pi^2/2\mu_R\Delta R^2$ where ΔR is the grid spacing for the evenly spaced grid in the R coordinate. Similarly, the maximum kinetic energy for the vibrational motion is $E^r_{max} = \hbar^2\pi^2/2\mu_r\Delta r^2$. The minimum value of the energy is just the minimum value of the potential, V_{min}. The potential can become extremely large in the classically forbidden regions when either R or r become very small. We define a cut-off value V_{cut} and all values of the potential which are larger than this are set equal to V_{cut}.

The centrifugal terms involving either $1/R^2$ or $1/r^2$ can also become extremely large (see eqn (H-13)). In order to limit the magnitudes of these terms we must also cap them at some maximum value. But care must be taken because the potential itself can have a deep well and it is only when the combination of the potential and the centrifugal term exceed V_{cut} that we should cap the value of the centrifugal term itself. For each fixed value of R, J and Ω we find the minimum value of the potential, which we designate as $V^{J\Omega}_{min}(R)$. We then examine the quantity $V^{J\Omega}_{min}(R) + \dfrac{\hbar^2\left(J(J+1) + j'(j'+1) - 2\Omega'^2\right)}{2\mu_R R^2}$. If this quantity exceeds V_{cut}, we cap the centrifugal element of this at V_{cut}. In this way we ensure that we only cap the centrifugal potential when the combined potential plus centrifugal potential exceeds the set value. Care must be taken to note the maximum value of the centrifugal potential determined in this way may be greater that V_{cut} and consequently an additional contribution must be added to the maximum energy in the spectrum of the Hamiltonian. Similar considerations apply to the centrifugal terms involving $1/r^2$ in eqn (H-13). E_{max} is then equal to the sum of all the terms mentioned above.

An important problem associated with capping the value of the potential relates to the centrifugal coupling terms connecting different values of Ω when $J > 0$. These terms can also become very large as they contain a factor of $1/R^2$ (see eqn (H-14)). Whenever we cap the diagonal centrifugal terms involving $1/R^2$ as described above we also set the centrifugal coupling terms which are linked to them to zero.

I.5 The Fourier Grid Hamiltonian Method

The Fourier Grid Hamiltonian method[14–17] provides a convenient way of finding the bound state solutions of a one-dimensional Schrödinger equation with the wavefunction represented by its values on an evenly spaced grid. The derivation of the method will not be repeated here, but the main equations will be given.

Let us consider the equation

$$\left[-\frac{\hbar^2}{2\mu}\frac{\partial^2}{\partial^2 r} + \frac{j(j+1)\hbar^2}{2\mu\, r^2} + V(r)\right]\chi_{vj}(r) = \varepsilon_{vj}\chi_{vj}(r). \tag{I-25}$$

If we combine the centrifugal potential with the potential to form an effective potential we can write

$$\left[-\frac{\hbar^2}{2\mu}\frac{\partial^2}{\partial^2 r} + V_{\text{eff}}(r)\right]\chi_{vj}(r) = \varepsilon_{vj}\chi_{vj}(r). \tag{I-26}$$

I.5.1 Odd Number of Grid Points

Let us suppose that the wavefunction is represented on a grid of N evenly spaced grid points, where $N = 2n + 1$. The Hamiltonian matrix is defined as

$$H_{i,j} = \frac{4}{N\mu}\sum_{\ell=1}^{n}\cos\left(\frac{\ell\, 2\pi(i-j)}{N}\right)\left(\frac{\hbar\pi\ell}{N\Delta r}\right)^2 + V_{\text{eff}}(r_i)\delta_{i,j} \tag{I-27}$$

where r_i are the grid points and Δr is the grid spacing.

The eigenvalues of this Hamiltonian matrix are the vibrational–rotational energy levels and the eigenvectors are the vibrational–rotational eigenfunctions evaluated at the grid points.

I.5.2 Even Number of Grid Points

If the number of grid points N is even, *i.e.* $N = 2n$, the Hamiltonian takes the form†

$$H_{i,j} = \begin{cases} \dfrac{h^2}{4\mu L^2}\left[\dfrac{(N+2)(N-2)}{6} + 1\right] + V_{\text{eff}}(r_i)\delta_{i,j} & \text{for } i = j \\ \dfrac{(-1)^{(i-j)}}{\mu}\left\{\dfrac{h}{2L\sin(\pi(i-j)/N)}\right\}^2 & \text{for } i \neq j. \end{cases} \tag{I-28}$$

References

1. W. A. Vetterling, W. H. Press, S. A. Teukolsky and B. P. Flannery, *Numerical Recipes in Fortran*, Cambridge University Press, Cambridge, 2nd edn, 1992.
2. J. C. Light, I. P. Hamilton and V. J. Lill, Generalized discrete variable approximation in quantum mechanics, *J. Chem. Phys.*, 1985, **82**, 1400.
3. J. V. Lill, G. A. Parker and J. C. Lightl, Discrete variable representations and sudden models in quantum scattering theory, *Chem. Phys. Lett.*, 1985, **89**, 483.
4. A. S. Dickinson and P. R. Certain, Calculation of matrix elements for one-dimensional quantum-mechanical problems, *J. Chem. Phys.*, 1968, **49**, 4209.

†Note that there is a misprint in eqn (52) of ref. 16. $(N + 1)$ should be replaced by $(N + 2)$ in the first line of the equation.

5. D. O. Harris, G. G. Engerholm and W. D. Gwinn, Calculation of matrix elements for one-dimensional quantum-mechanical problems and the application to anharmonic oscillators, *J. Chem. Phys.*, 1965, **43**, 1515.
6. R. N. Zare, *Angular Momentum*, John Wiley and Sons, New York, 1988. See also list of corrections at web address: http://web.stanford.edu/group/Zarelab/Errata.pdf.
7. L. Pauling and E. B. Wilson, *Introduction to Quantum Mechanics*, McGraw-Hill Book Co., 1935.
8. H. Eyring, J. Walter and G. E. Kimball, *Quantum Chemistry*, John Wiley and Sons Inc., 1st edn, 1944.
9. C. Leforestier, R. H. Bisseling, C. Cerjan, M. D. Feit, R. Friesner, A. Guldberg, A. Hammerich, G. Jolicard, W. Karrlein, H.-D. Meyer, N. Lipkin, O. Roncero and R. Kosloff, A comparison of different propagation schemes for the time dependent Schrödinger equation, *J. Comput. Phys.*, 1991, **94**, 59.
10. G. Corey and D. Lemoine, Pseudospectral method for solving the time-dependent Schrodinger equation in spherical coordinates, *J. Chem. Phys.*, 1992, **97**, 4115.
11. C. Leforestier, Grid representation of rotating triatomics, *J. Chem. Phys.*, 1991, **94**, 6388.
12. D. Kosloff and R. Kosloff, A fourier method solution for the time dependent schrodinger equation as a tool in molecular dynamics, *J. Comput. Phys.*, 1983, **52**, 35.
13. P. N. Swarztrauber. FFTPACK is a package of Fortran subprograms for performing fast Fourier transforms, web site: http://www.netlib.org/fftpack/.
14. C. C. Marston and G. G. Balint-Kurti, The Fourier grid Hamiltonian method for bound state eigenvalues and eigenfunctions, *J. Chem. Phys.*, 1989, **91**, 3571.
15. G. G. Balint-Kurti, C. L. Ward and C. C. Marston, Two computer programs for solving the Schrödinger equation for bound-state eigenvalues and eigenfunctions using the Fourier grid Hamiltonian method, *Comput. Phys. Commun.*, 1991, **67**, 285.
16. G. G. Balint-Kurti, R. N. Dixon and C. C. Marston, Grid methods for solving the Schrodinger equation and time dependent quantum dynamics of molecular photofragmentation and reactive scattering processes, *Int. Rev. Phys. Chem.*, 1992, **11**, 317.
17. F. Gögtas, G. G. Balint-Kurti and C. C. Marston, The Fourier grid Hamiltonian method based on even grid numbers, *Turk. J. Phys.*, 1997, **21**, 923.

APPENDIX J

Technical Aspects of the Real Wavepacket Method

J.1 Analysis of the Real Part of the Wavepacket to Determine the S Matrix

In Section H.2 of Appendix H we described one method of analyzing the complex time-dependent wavepacket to extract the **S** matrix. In this appendix we discuss how this analysis is changed when we consider the analysis of only the real part of the time-dependent wavepacket. The real part of the wavepacket is given by

$$Re\left[\Phi_{\Omega'}^{Jvj\Omega}(R_\beta, r_\beta, \gamma_\beta, t)\right] = \frac{1}{2}\left[\Phi_{\Omega'}^{Jvj\Omega}(R_\beta, r_\beta, \gamma_\beta, t) + \Phi_{\Omega'}^{Jvj\Omega\,*}(R_\beta, r_\beta, \gamma_\beta, t)\right]. \tag{J-1}$$

We now follow all the arguments of Section H.2 but replace Φ with its real part. The analysis proceeds exactly as before until eqn (H-22) which is replaced by

$$\begin{aligned}
&A^{J}_{vj\Omega\to v'j'\Omega'}(E)\\
&\quad= \frac{1}{2}\frac{1}{2\pi}\int_0^\infty e^{iEt/\hbar}\int \chi_{v'j'}(r_\beta)\,\Theta_{j'\Omega'}(\gamma_\beta)\int_0^\infty dR_\beta\,\delta(R_\beta - R_{\beta,\infty})\\
&\qquad\times e^{-i\hat{H}t/\hbar}\left[\Phi^{Jvj\Omega}(R_\alpha, r_\alpha, \gamma_\alpha, t=0)\right.\\
&\qquad\left.+\,\Phi^{Jvj\Omega\,*}(R_\alpha, r_\alpha, \gamma_\alpha, t=0)\right] dr_\beta \sin\gamma_\beta\, d\gamma_\beta\, dt\\
&\quad= \frac{1}{2}\frac{1}{2\pi}\int_0^\infty e^{iEt/\hbar}\int \chi_{v'j'}(r_\beta)\,\Theta_{j'\Omega'}(\gamma_\beta)\int_0^\infty dR_\beta\,\delta(R_\beta - R_{\beta,\infty})\\
&\qquad\times\left[e^{-i\hat{H}t/\hbar}\Phi^{Jvj\Omega}(R_\alpha, r_\alpha, \gamma_\alpha, t=0)\right.
\end{aligned}$$

RSC Theoretical and Computational Chemistry Series No. 7
Theory of Molecular Collisions
By Gabriel G. Balint-Kurti and Alexander P. Palov

Published by the Royal Society of Chemistry, www.rsc.org

$$+ \left\{ e^{i\hat{H}t/\hbar} \Phi^{J\nu j\Omega}(R_\alpha, r_\alpha, \gamma_\alpha, t=0) \right\}^* \Big]$$
$$\times \, dr_\beta \sin\gamma_\beta \, d\gamma_\beta \, dt \qquad \text{(J-2)}$$

where $\frac{1}{2}\left[\Phi^{J\nu j\Omega}(R_\alpha, r_\alpha, \gamma_\alpha, t=0) + \Phi^{J\nu j\Omega *}(R_\alpha, r_\alpha, \gamma_\alpha, t=0)\right]$ is the real part of the initial wavepacket (see eqn (7.15)) and note that $\chi_{\nu' j'}(r_\beta)$ and $\Theta_{j'\Omega'}(\gamma_\beta)$ are real.

The term $\left\{e^{i\hat{H}t/\hbar}\Phi^{J\nu j\Omega}(R_\beta, r_\beta, \gamma_\beta, t=0)\right\}$ in eqn (J-2) is the initial wavepacket propagated *backwards* to a time "$-t$". As the initial wavepacket has been given a momentum which brings the two collision partners together, propagation backwards in time will take the collision partners further apart, so that no collision will take place. As the analysis line is placed in the product channel the backwards propagated term will never reach the analysis line $R_\beta = R_{\beta,\infty}$ and because of the term $\delta(R_\beta - R_{\beta,\infty})$ in eqn (J-2) this term does not contribute. If we neglect this second term in the square brackets then the remaining expression is the same as that in eqn (H-22) except for an extra factor of $\frac{1}{2}$. Our conclusion is that if we analyze the real part of the wavepacket to extract the **S** matrix in exactly the same way that we would analyze the full complex wavepacket, then we must multiply by a factor of 2 to obtain the correct result.

List of Symbols

Physical Constants

h	Plank constant
c	Speed of light
k_B	Boltzmann constant
$\hbar$	Plank constant divided by 2π
a_0	Bohr radius
π	3.14159

Special Symbols

$\hat{f}$	f-Operator
$\boldsymbol{f}$	f-Vector or matrix (bold)
f^*	Conjugate complex number of f
$\hat{f}^\dagger$	An adjoint f-operator
$\mathrm{d}f$	Differential of a function/variable f
$\nabla^2 f$	The Laplacian operator acting on f
δf	A infinitesimal increment of an f-function
$\dot{f}$	Time derivative of a function f

Latin Symbols

A	A Hermitian matrix
$A(\rho, \theta)$	A "rotational constant" in APH coordinates corresponding to the smallest principal moment of inertia
$Ai(x)$	An Airy function (of the first kind)
$A^J_{\nu j\Omega \to \nu' j' \Omega'}(E)$	A Fourier transform of the **C** matrix

RSC Theoretical and Computational Chemistry Series No. 7
Theory of Molecular Collisions
By Gabriel G. Balint-Kurti and Alexander P. Palov

Published by the Royal Society of Chemistry, www.rsc.org

$\hat{\mathbf{A}}$	An absorbing operator
$\mathbf{A}_q^{(k)}$	A complex orientation/alignment parameter of rank k and component q
$A_J(t)$	A configuration coefficient in MCTDH theory
$B(\rho,\theta)$	A "rotational constant" in APH coordinates corresponding to the principal moment of inertia with the middle value
$\mathbf{B}$	A unitary transformation matrix relating angular basis functions in the space-fixed and body-fixed coordinate systems
b	An impact parameter
$Bi(x)$	An Airy function (of the second kind)
$C(\rho,\theta)$	A "rotational constant" in APH coordinates corresponding to the largest principal moment of inertia
$C_1(r)$ and $C_2(r)$	Wavefunction expansion coefficients used in variable phase method
$C^J_{vj\Omega\rightarrow v'j'\Omega'}(t)$	A cut through a wavepacket, along an analysis line, in the asymptotic region, projected onto particular product vibrational–rotational states
$C_f(t)$	The flux–flux autocorrelation function
$D^J_{\Omega,\Omega'}(\phi,\theta,\gamma)$	A Wigner **D** matrix element
$d^J_{m_j,\Omega}(\theta_R)$	A Wigner (small) **d** matrix element
d_α	The dimensionless scaling factor for mass scaled Jacobi coordinates
E	A total energy of a system
$\bar{\boldsymbol{F}}$	The symmetrized flux operator
FFT and FFT^{-1}	A Fourier and inverse Fourier transform
$f_{v'j'm'_j\leftarrow vjm_j}(\theta_R,\phi_R)$	The scattering amplitude
f_j	Eigenvalues of the thermal flux operator
f^j	Eigenvectors of the thermal flux operator
$\hat{\boldsymbol{F}}_{2T}$	Thermal flux operator
$f(\hat{\boldsymbol{H}}_s)$	A function of the scaled and shifted Hamiltonian
$G_{vjl}(R,R')$	The Green's function
$\hat{\boldsymbol{H}}$	A Hamiltonian operator
$\hat{\boldsymbol{H}}_s$	A scaled and shifted Hamiltonian operator

$\mathbf{I}$	A unit matrix
$\hat{\mathcal{I}}$	The inversion operator
i	Either an index or the imaginary quantity $\sqrt{-1}$
J	Total angular momentum quantum number or a combined index
j	A rotational quantum number of a diatomic or an index
$j_\ell(kr)$	A spherical Bessel function of the first kind
$\hat{j}_\ell(kr)$	A Riccati–Bessel function of the first kind
$J_q^{(k)}$	An angular momentum spherical-tensor operator
$\boldsymbol{k}$	Wave vector of a particle
$\hat{\boldsymbol{k}}$	A unit vector in the direction $\boldsymbol{k}$
$k(T)$	A reaction rate constant
$K(T)$	Equilibrium constant for a reaction
K	Rank of state multipole
L	Either the Lagrange's function or the length of the absorbing region
ℓ	An orbital angular momentum quantum number
M	Either the mass of a system or the quantum number for the projection of the total angular momentum onto the Z axis of the laboratory coordinate system
m	Either a mass of an atom or the quantum number for the projection of the rotational angular momentum of a diatomic onto the Z axis of the laboratory coordinate system
$N(E)$	The cumulative reaction probability
n	An index or a number of grid points
$n_\ell(kr)$	A spherical Bessel function of the second kind (Neumann)
$\hat{n}_\ell(kr)$	A Riccati–Bessel function of the second kind
$\hat{\boldsymbol{P}}$	A permutation operator
$\hat{\boldsymbol{P}}^{(k)}$	A projection operator onto the space spanned by the single particle functions
P_i	The probabilities of finding a molecule in its i-th quantum states
$P_J(j', j)$	An opacity function

$P_\ell(\cos\theta)$	The ℓ-th Legendre polynomials
p	The generalized momentum
$p(\tau)$	Imaginary part of a wavepacket
$Q_n(x)$	The Chebyshev polynomials of complex argument
$Q_\mathrm{R}(T)$	The total partition function per unit volume for the reactants
Q	The quantum number for the projection of the rank of a state multipole onto the direction of the Z-axis
q_F	The "internal" partition function for an F atom
$\boldsymbol{q}_k$	A generalized Jacobi coordinate
$q(\tau)$	Real part of a wavepacket
R	Distance between two scattering partners or a scattering coordinate
$\hat{\boldsymbol{R}}$	Unit vector in the direction of the interparticle separation
R_abs	A starting point of absorbing region
$\boldsymbol{R}_\mathrm{c}$	Vector position of center-of-mass of a system
$\tilde{\boldsymbol{R}}_\alpha$	A mass-scaled Jacobi scattering coordinate for arrangement α
R_∞	Radial coordinate defining analysis line for taking cut through wavepacket
r	Distance between two atoms in a diatomic
$\tilde{\boldsymbol{r}}_\alpha$	Mass-scaled Jacobi diatomic coordinate for arrangement α
$\mathbf{S}$	A scattering matrix
T	Either temperature or the kinetic energy of a system
$\mathbf{T}(\chi_{\gamma\alpha})$	A unitary transformation matrix between reactive channels α and γ in mass-scaled Jacobi coordinates
$\mathbf{T}$	Scattering T matrix or the transformation matrix between the grid and the fixed basis set representations
$T_n(x)$	The n-th Chebyshev polynomial
t	Time
$\boldsymbol{u}$	A particle velocity vector in the center-of-mass frame or an eigenvector

u^1_{vjl}, u^2_{vjl} and $u(r)$	Solutions of the homogeneous second order radial differential equation
V	Potential energy of a system
v	The absolute value of a velocity
$\mathbf{v}$	A particle velocity vector
ν	A vibrational quantum number
W or $\mathcal{W}$	A Wronskian
$\mathbf{W}(R)$	"Potential" matrix
X	X coordinate of center-of-mass, also a direction in a Cartesian coordinate system
x	X coordinate relative to the center-of-mass
Y	Y coordinate of center-of-mass, also a direction in a Cartesian coordinate system
y	Y coordinate of a particle in the center-of-mass coordinate system
$\mathbf{Y}$	log-derivative matrix
$Y_{\ell m}(\theta, \phi)$	The ℓ, m-th spherical harmonics
$\mathcal{Y}^{JM}_{j\ell}(\hat{\boldsymbol{R}}, \hat{\boldsymbol{r}})$	Angular function which is an eigenfunction of the total angular momentum and its space-fixed Z component. Obtained by coupling rotational and orbital angular momenta.
Z	Z coordinate of a center-of-mass, also a direction in a Cartesian coordinate system
z	Z coordinate relative to the center-of-mass

Greek Symbols

α	Either an angle between the laboratory product velocity and the center-of-mass velocity or an arrangement channel index in reactive scattering
α_{Q}	A hyperangle of hyperspherical coordinate
β	An arrangement channel index in reactive scattering or an anisotropy parameter of a single photon photodissociation process
γ	Either an arrangement channel index in reactive scattering, an Euler angle or a Jacobi angle
$\boldsymbol{\Gamma}^k_1(R^k_{\mathrm{R}}, R^k_{\mathrm{L}})$	Diagonal matrices to propagate a wavefunction

Δ_Q	The APH angular momentum operator
δ_{lk}	The Kronecker symbol
$\delta(x - x')$	The Dirac delta-function
ε	Well depth of the LJ potential
ε_i	Energy of quantum state i
η_ℓ	The ℓ-th phase shift
$\Theta_{j,\Omega}(x)$	A normalized version of the associated Legendre Polynomials
$\Theta^{JMp}_{j\lambda}(\hat{\boldsymbol{r}}, \hat{\boldsymbol{R}})$	Parity adapted angular basis functions
θ	Polar scattering angle
ϑ	An Euler angle
λ	An absolute value of helicity
λ_{DB}	The de Broglie wave length
μ	Reduced mass of a molecule
ν	A photon frequency
$\Xi^{JM}_{j\Omega}(\hat{\boldsymbol{R}}, \hat{\boldsymbol{r}}')$	The body-fixed angular basis
ξ	An Euler angle
ρ	Either a hyperradius or a density matrix
$\rho_{j,m_j;j,m'_j}$	A density-matrix element describing the polarized ensemble of molecules
$\rho^{(k)}_q(J)$	The state-multipole moments
σ	A collision cross section
τ	A time step
$\Phi^{J\nu j\Omega}_{\nu' j' \Omega'}(R, E)$	A body-fixed radial wavefunction
$\Phi_J(x_1, x_2, \cdots x_f, t)$	A configuration wave function
ϕ	Azimuthal angle or scattering angle
$\phi^{(2)}_{j_2}(x_2, t)$	A single particle wave function
φ	An Euler angle
χ	A continuous kinematic transformation angle of adiabatically adjusting, principal axes hyperspherical (APH) coordinates
$\chi(E, b)$	A deflection function
$\chi(r)$	A radial wave function

$\chi_{\gamma\alpha}$	The obtuse kinematic angle
Ψ or ψ	Wave function of a system
Ω	Either a solid angle or a helicity quantum number
ω_i	An eigenvalue

Glossary

Adiabatically adjusting, principal axes hyperspherical coordinates. A universal coordinate system where the direction of the axes is independent of any particular choice of reactive channel and is defined by the principal axes of inertia of the system.

Alkali age. The early era of crossed molecular beam experiments based on the use of surface ionization detectors, which could detect only alkali atoms and could distinguish between them and the product alkali halides.

Body-fixed coordinates. A coordinate system defined in relation to the relative positions of the atoms and not fixed in space.

Born approximation. An approximation based on using the integral equations of scattering and substituting the initial unperturbed scattering wavefunction for the exact wavefunction. The method is most valid for weak interaction potentials or high collision energies.

Centrifugal potential. An effective potential arising from the rotational angular momentum of a system that takes the form of a repulsion tending to force any two bodies further apart.

Chebyshev expansion of the time evolution operator. Expansion of the time evolution operator as a sum of Chebyshev polynomials of the Hamiltonian operator.

Classical turning point. The point on a classical trajectory at which the collision partners are closest to each other.

Close coupling equations. The set of second order radial Schrödinger equations coupled either through the matrix elements of the potential or through terms arising from the kinetic energy operator.

Conical intersection. These can occur in molecules composed of three or more atoms. They correspond to positions of the atoms at which two potential energy surfaces of the same symmetry posses the same energy.

RSC Theoretical and Computational Chemistry Series No. 7
Theory of Molecular Collisions
By Gabriel G. Balint-Kurti and Alexander P. Palov

Published by the Royal Society of Chemistry, www.rsc.org

For a system of $N (N \geq 3)$ atoms a conical intersection will consist of a $(3N - 8)$-dimensional surface in the $(3N - 6)$-dimensional coordinate space describing the relative positions of the atoms.

Cumulative reaction probability. The sum of the squares of all **S** matrix elements weighted by a $(2J + 1)$ degeneracy factor. When averaged over a Boltzmann energy distribution it yields the rate constant.

Deflection function. The angle of deflection as a function of the impact parameter for a fixed relative kinetic energy of two atoms.

Detailed balance. The relation between the detailed rate constants of direct and inverse reactions based on symmetry of the scattering **S** matrix.

Detailed rate constant. The rate of the reaction of reactants in certain initial quantum states to produce products in definite quantum states.

Differential cross section. The number of atoms scattered into an increment of solid angle per unit time divided by the incident flux of the collision partners.

Discrete variable representation. Representation of a function by its values at a specially chosen set of coordinate points such that there exists a one-to-one relationship between these values and the expansion coefficients of the function in terms of the chosen basis set of functions.

Distorted wave approximation. An approximation based on the use of the integral equations of scattering theory in which the exact scattering wavefunction is replaced by the solution of the scattering problem in which the terms coupling different channels have been ignored.

Doppler spectroscopy. Use of high resolution spectral methods, often laser induced fluorescence or resonance enhanced multi-photon ionization, to utilize the change in the frequency of a spectral line due to the motion of the observed molecule or atom. Can be used to study orientation and alignment of a molecular ensemble.

Effective potential. The sum of a potential energy and centrifugal potential of a colliding system.

Elastic scattering. A scattering process where the relative kinetic energy of the collision partners remains unchanged but their direction of motion may change.

Euler angles. The three angles standardly used to determine the rotations required to rotate one axis system or frame of reference into another one.

Feschbach resonance. Sharp peaks in collision cross sections that have their origin in an exchange of energy from relative translational motion to the internal excitation of the molecule resulting in the system temporarily not possessing sufficient energy of relative motion to break up.

Flux conservation. The conservation law that reflects the equality of incoming and outgoing particle flux to preserve the number of particles or collision partners.

Flux–flux autocorrelation method. A method for computing the rate constant as a function of temperature without the need to compute all the **S** matrix elements.

Flux–flux autocorrelation function. A function that is related to the "cumulative reaction probability" and that may be used to compute the rate constant without calculating the state-to-state specific **S** matrix elements. Normally used for large systems.

Glory scattering. The singularity of the differential scattering cross section in the forward and backward directions. Oscillations in the integral cross section as a function of energy are related to this process.

Hard sphere collision. A model collision process in which the two collision partners are treated as impenetrable hard spheres, like billiard balls. Normally used only in the context of classical mechanics.

Helicity quantum number. The quantum number for the component of the total angular momentum along the Z direction in a body-fixed coordinate system.

Hyperspherical coordinates. A system of internal coordinates with only a single radial coordinate which determines the overall size of the N-body entity, and $(3N - 7)$ angular coordinates.

Impact parameter. The closest distance of approach of two scattering partners in the hypothetical situation that there were no interaction potential acting between them.

Incident flux of atoms. The number of atoms crossing a unit area perpendicular to their direction of motion per unit time.

Inelastic scattering. A scattering process in which the internal quantum state of the collision partners is altered and their relative kinetic energy also changes.

Integral cross-section. This is the integral over all solid angles of the differential cross section. It can be associated with quantum numbers of the initial and final states of the scattering partners.

Jacobi coordinates. A coordinate system for the relative positions of the atoms in a system in which both the classical and quantum expressions for the kinetic energy posses no cross terms coupling different coordinates. The coordinate system is built up by taking a pair of atoms at a time and then linking their center-of-mass to another atom.

JWKB approximation. See "Semiclassical approximation"

K matrix. The scattering matrix that gives the relative amplitude of $\sin(kR - \ell\pi/2)$ to $\cos(kR - \ell\pi/2)$ in the asymptotic form of the scattering wavefunction.

Kinematic rotation. A unitary transformation of mass-scaled Jacobi coordinates to switch between different reaction channels.

Laser-induced fluorescence (LIF). A highly selective spectroscopic technique used to detect atoms or molecules in a specific quantum state. It is based on measuring fluorescence occurring following an initial state-selective excitation.

Lennard-Jones potential. A simple model potential for the interaction potential between two atoms. In its standard form the repulsive and attractive parts depend on the internuclear distance as a power of -12 and -6 respectively.

Lifetime matrix. A Hermitian matrix whose eigenvalues give the lifetimes of the states of thc collision complex involved.

Mass-scaled Jacobi coordinates. Modified Jacobi coordinates that lead to a simple coordinate transformation, called a "kinematic rotation", between the coordinates for different reactive channels.

MCTDH. Multi-Configuration Time Dependent Hartree method.

Mean free path. The distance, on average, that an atom or molecule can travel in a gas before a collision.

Newton diagram. A diagram that relates the velocities of the collision partners in the laboratory and the center-of-mass reference frames.

Orbital angular momentum. The angular momentum resulting from the realtive motion of two collision partners.

Parity. The behavior of the wavefunction under the symmetry operation of inversion through a point in space. Under this operation the wavefunction must either remain unchanged or change sign.

Partial cross-section. Contribution to a cross section arising from a particular value of the orbital or total angular momentum.

Partial wave expansion. The expansion or representation of the scattering wavefunction as a sum over Legendre polynomials of the scattering angle. The coefficients in the expansion are called partial waves or amplitudes.

Partition function. The sum over all states of the Boltzmann factors. It is a function of temperature and is a measure of the number of quantum states accessible to a system at a given temperature.

Phase shift. The change or shift of the phase of a partial wave arising in a scattering process as compared to the phase it would have in the absence of an interaction potential. Normally this term is used in atom–atom scattering only.

Photoloc. The abbreviation of PHOTO initiated reaction analyzed *via* the Law Of Cosines.

Polarization Dependent Differential Cross Sections. Differential cross section that depends on an initial and/or final orientation of the internal angular momentum of the collision partners.

Potential energy surface. The set of electronic energies calculated at fixed nuclear geometries that govern the motion of the nuclei. The concept is only defined within the Born–Oppenheimer approximation.

Quasi-bound states. The states of a system temporarily trapped in a cooperative motion in such a way that no one atom can escape from the scattering complex.

Rate constant. A function of a temperature that, when multiplied by the concentrations of the reactants, yields the rate of a chemical reaction or of some other process (such as an inelastic transition).

Rainbow scattering. The singularity of the differential cross sections which occurs, in classical mechanics, at scattering angles such that the deflection function has a minimum. Associated maxima and scattering patterns also arise in the quantum treatment.

Reactive scattering. A scattering process in which one chemical bond is broken and another is formed.

Real wavepacket method. This is a method for solving the time-dependent Schrödinger equation using only real, as opposed to complex, algebra. It is based on a recurrence relationship which propagates the wavepacket forward in time.

REMPI. Acronym for Resonantly-Enhanced Multiphoton Ionization. A laser based method that involves the use of one or more photons to excite a molecule (or atom) to a high lying quantum level followed by absorption of a further photon which results in ionization.

REMPI-TOF. The REMPI Time-Of-Flight technique is a combination of the REMPI method, which is highly state-selective, and Time-Of-Flight mass spectrometry. After ionization and extraction from the ionization region, the ions fly through a field free region and their time of arrival at the detector is noted. This permits the calculation of their velocity in the direction of the detector.

Rotating bond approximation. An approximation used in the theory of reactive scattering for systems containing four or more atoms. It reduces the number of variables in the calculation by removing the two angles governing the rotation of one or more bonds.

S matrix. The ratio of the radial outgoing wave to that of the radial incoming one in the asymptotic form of the scattering wavefunction.

Scattering amplitude. The amplitude of the radially outgoing wave associated with a particular set of quantum states of the collision partners as a function of the polar scattering angle.

Scattering length. A parameter having the physical significance of a measure of the range of the potential (no inner structure) and related to the low energy limit of the elastic scattering cross section.

Semiclassical approximation. An approximation designed particularly for collisions of heavy partners and high energies when the de Broglie wavelength of the relative motion is small compared to the dimensions or range of the interaction potential. Commonly known as the JWKB (Jeffreys–Wentzel–Kramers–Brillouin) approximation.

Shape resonance. This is a resonance feature, *i.e.* a sharp peak in a cross section, that is caused by the shape of the potential rather than by the transfer of kinetic energy of relative motion into fragment excitation as in the case of a Feschbach resonance.

Space-fixed reference frame. A reference frame or coordinate system fixed relative to the laboratory rather than relative to the positions of the atoms.

Space-fixed to body-fixed transformation. The transformation of the coordinates and the wavefunction from a space-fixed to the body-fixed coordinate system. It involves a rotation of the angular basis functions.

T matrix. The ratio of the radial outgoing wave to that of the radial sine wave that is present in the asymptotic form of the scattering wavefunction when no potential is present.

Time evolution operator. The quantum mechanical operator which when it acts on a wavefunction creates a wavefunction corresponding to a later time. *i.e.* it changes $\psi(t)$ to $\psi(t + \delta t)$.

Total angular momentum. The total angular momentum of a system is a conserved quantity. It is associated with two "good" quantum numbers, J and M, related to its magnitude and its Z component. In the case of an atom–diatom system, the total angular momentum is the vector sum of the rotational angular momentum of the diatom and the angular momentum of the relative motion of the two partners.

Total cross section. The effective cross sectional area that can lead to a collision. It is the sum over all product quantum states of the integral cross section.

Universal detectors. Detectors used in crossed molecular beam experiments that are based on mass spectrometers with electron impact ionization and are not limited to detecting alkali atoms only.

Variable phase approach. A method of solving the Schrödinger equation for atom–atom scattering by directly solving for the phase shift of the wavefunction.

Vector correlation. The correlations between the various vector quantities which characterize the reactants and products in a chemical reaction, such as the relative velocities of the reactants and products and their "internal" angular momenta, consisting of their rotational, spin and any other angular momentum components.

Wavepacket analysis. The extraction of the **S** matrix from the evolution of a wavepacket. This may involve watching the wavepacket traverse a line drawn across the exit channel in the asymptotic region so that the outgoing flux can be calculated.

Wigner D matrix. A matrix that is used to transform angular momentum eigenfunctions to a new coordinate system. It is a function of the three angles which are needed to rotate one coordinate system into the other.

Subject Index

References to tables and charts are in **bold** type.